T0297655

More
Math Into LaTeX

George Grätzer

More
Math Into LaTeX

Fifth Edition

George Grätzer
Toronto, ON, Canada

ISBN 978-3-319-23795-4 ISBN 978-3-319-23796-1 (eBook)
DOI 10.1007/978-3-319-23796-1

Library of Congress Control Number: 2015953672

Springer Cham Heidelberg New York Dordrecht London

Printed on acid-free paper

Springer International Publishing AG Switzerland is part of Springer Science+Business Media (www.springer.com)

To the **Volunteers**

without whose dedication,
over 25 years,
this book could not have been done

and to the young ones

Emma (10),
Kate (8),
Jay (3)

Short Contents

Contents

Foreword

It was the autumn of 1989—a few weeks before the Berlin wall came down, President George H. W. Bush was president, and the American Mathematical Society decided to outsource TEX programming to Frank Mittelbach and me.

Why did the AMS outsource TEX programming to us? This was, after all, a decade before the words "outsourcing" and "off-shore" entered the lexicon. There were many American TEX experts. Why turn elsewhere?

For a number of years, the AMS tried to port the mathematical typesetting features of AMS-TEX to LATEX, but they made little progress with the AMSFonts. Frank and I had just published the New Font Selection Scheme for LATEX, which went a long way to satisfy what they wanted to accomplish. So it was logical that the AMS turned to us to add AMSFonts to LATEX. Being young and enthusiastic, we convinced the AMS that the AMS-TEX commands should be changed to conform to the LATEX standards. Michael Downes was assigned as our AMS contact; his insight was a tremendous help.

We already had LATEX-NFSS, which could be run in two modes: compatible with the old LATEX or enabled with the new font features. We added the reworked AMS-TEX code to LATEX-NFSS, thus giving birth to AMS-LATEX , released by the AMS at the August 1990 meeting of the International Mathematical Union in Kyoto.

AMS-LATEX was another variant of LATEX. Many installations had several LATEX variants to satisfy the needs of their users: with old and new font changing commands, with and without AMS-LATEX , a single and a multi-language version. We decided to develop a Standard LATEX that would reconcile all the variants. Out of a group of interested people grew what was later called the LATEX3 team—and the LATEX3 project got underway. The team's first major accomplishment was the release of LATEXe in June 1994. This standard LATEX incorporates all the improvements we wanted back in 1989. It is now very stable and it is uniformly used.

Under the direction of Michael Downes, our AMS-LATEX code was turned into AMS packages that run under LATEX just like other packages. Of course, the LATEX3 team recognizes that these are special; we call them "required packages" because they are part and parcel of a mathematician's standard toolbox.

Since then a lot has been achieved to make an author's task easier. A tremendous number of additional packages are available today. The LaTeX *Companion,* 2nd edition, describes many of my favorite packages.

George Grätzer got involved with these developments in 1990, when he got his copy of AMS-LaTeX in Kyoto. The documentation he received explained that AMS-LaTeX is a LaTeX variant—read Lamport's LaTeX book to get the proper background. AMS-LaTeX is not AMS-TeX either—read Spivak's AMS-TeX book to get the proper background. The rest of the document explained in what way AMS-LaTeX differs from LaTeX and AMS-TeX. Talk about a steep learning curve ...

Luckily, George's frustration working through this nightmare was eased by his lengthy e-mail correspondence with Frank and lots of telephone calls to Michael. Three years of labor turned into his first book on LaTeX, providing a "simple introduction to AMS-LaTeX ". This edition is more mature, but preserves what made his first book such a success. Just as in the first book, Part I, *Mission Impossible,* is a short introduction for the beginner. Chapter 1, *Short Course*, dramatically reducing the steep learning curve of a few weeks to a few hours in only 30 pages. Chapter 2, *And a few more things...* adds a few more advanced topics useful already at this early stage.

The rest of the book is a detailed presentation of everything you may need to know. George "teaches by example". You find in this book many illustrations of even the simplest concepts. For articles, he presents the LaTeX source file and the typeset result. For formulas, he discusses the building blocks with examples, presents a *Formula Gallery,* and a *Visual Guide* for multiline formulas.

Going forth and creating "masterpieces of the typesetting art"—as Donald Knuth put it at the end of the TeXbook—requires a fair bit of initiation. This is the book for the LaTeX beginner as well as for the advanced user. You just start at a different point.

The topics covered include everything you need for mathematical publishing.

- Instructions on creating articles, from the simple to the complex
- Converting an article to a presentation
- Customize LaTeX to your own needs
- The secrets of writing a book
- Where to turn to get more information

The many examples are complemented by a number of easily recognizable features:

Rules which you must follow

Tips on what to be careful about and how to achieve some specific results

Experiments to show what happens when you make mistakes—sometimes, it can be difficult to understand what went wrong when all you see is an obscure LaTeX message

This book teaches you how to convert your mathematical masterpieces into typographical ones, giving you a lot of useful advice on the way. How to avoid the traps for the unwary and how to make your editor happy. And hopefully, you'll experience the fascination of doing it right. Using good typography to better express your ideas.

If you want to learn LaTeX, buy this book and start with the *Short Course*. If you can have only one book on LaTeX next to your computer, this is the one to have. And if you want to learn about the world of LaTeX packages as of 2004, also buy a second book, the LaTeX *Companion,* 2nd edition.

Rainer Schöpf

Rainer Schöpf
LaTeX3 team

What's in the book?

Part I is *Mission Impossible;* it helps you to get started quickly with LaTeX, to type your first articles, to make your first presentations, and it prepares you to tackle LaTeX in more depth in the subsequent parts.

Chapter 1 is the *Short Course*. You start writing your *first article*—as typeset on page 4—and prepare your *first presentation*—see some of the slides typeset on page 27. This chapter introduces how LaTeX uses the *keyboard* and how to *type text*. You do not need to learn much to understand the basics. Text markup is quite easy. You also learn math markup, which is not so straightforward. Several sections in this chapter ease you into *mathematical typesetting*. There is a section on the basic building blocks of math formulas. Another one discusses equations. Finally, we present the two simplest multiline formulas, which should cover most of your everyday needs. We also cover the elements of presentations with a simple example.

In **Chapter 2**, we explain how things work, the structure of LaTeX, the auxiliary files, the logical and visual design of an article, LaTeX error messages. Finally, we present a long list of dos and don't to help you write good LaTeX.

Part II introduces the two most basic skills for writing with LaTeX in depth, *typing text* and *typing math.*

Chapters 3 and **4** introduce *text* and *displayed text.* Chapter 3 is especially important because, when you type a LaTeX document, most of your time is spent typing text. The topics covered include special characters and accents, hyphenation, fonts, and spacing. Chapter 4 covers displayed text, including *lists* and *tables,* and for the mathematician, *proclamations* (theorem-like structures) and *proofs*.

Typing math is the heart of any mathematical typesetting system. **Chapter 5** discusses inline formulas in detail, including basic constructs, delimiters, operators, math accents, and horizontally stretchable lines. The chapter concludes with the *Formula Gallery*.

Math symbols are covered in three sections in **Chapter 6**. How to space them, how to build new ones; we introduce the new set of some 2,000 STIX math symbols. We also look at the closely related subjects of math alphabets and fonts. Then we discuss tagging and grouping equations.

LaTeX knows a lot about typesetting an inline formula, but not much about how to display a multiline formula. **Chapter 7** presents the numerous tools LaTeX offers to help you do that. We start with a *Visual Guide* to help you get oriented.

Part III discusses the parts of a LaTeX document. In **Chapter 8**, you learn about the *structure* of a LaTeX document. The most important topics are *sectioning* and *cross-referencing*. In **Chapter 9**, we discuss the `amsart` *document class* for articles. In particular, I present the title page information. Chapter 9 also features `secondarticle.tex`, a sample article for `amsart`, somewhat more advanced than `firstarticle.tex` typeset on page 4. You can learn a lot about LaTeX just by reading the source file one paragraph at a time and seeing how that paragraph is typeset. We conclude this chapter with a brief description of the AMS distribution, the packages and document classes, of which amsart is a part.

In **Chapter 10** the most commonly used *legacy document classes* are presented, article, report, and letter (the book class is discussed in Chapter 17), along with a description of the standard LaTeX distribution. Although article is not as sophisticated as amsart, it is commonly used for articles not meant for publication.

In **Part IV**, we start with **Chapter 11**, discussing PDF *files, hyperlinks,* and the hyperref package. This prepares you for *presentations,* which are PDF files with hyperlinks. In **Chapter 12** we utilize the beamer *package* for making LaTeX presentations and **Chapter 13** introduces its sister package Ti*k*Z for illustrations.

Part V (**Chapter 14**) introduces techniques to *customize* LaTeX: custom commands and environments created by users, and command files. We present a sample command file, newlattice.sty, and a version of the second sample article utilizing this command file. You learn how parameters that affect LaTeX's behavior are stored in counters and length commands, how to change them, and how to design your own custom lists. A final section discusses the pitfalls of customization.

In **Part VI** (**Chapters 15** and **16**), we discuss the special needs of longer documents. Two applications, contained in the standard LaTeX distribution, BibTeX and *MakeIndex*, make compiling *large bibliographies* and *indexes* much easier.

LaTeX provides the book and the amsbook document classes to serve as foundations for well-designed books. We discuss these in **Chapter 17**. Better quality books have to use document classes designed by professionals.

You will probably find yourself referring to **Appendices A** and **B** time and again. They contain the *math and text symbol tables.* You can also find them in the samples file.

Appendix C relates some historical background material on LaTeX. It gives you some insight into how LaTeX developed and how it works. **Appendix D** discusses the many ways we can find LaTeX material on the *Internet.* **Appendix E** is a short introduction to the use of *PostScript fonts* in a LaTeX document. **Appendix F** briefly describes the use of LaTeX for languages other than American English.

LaTeX on an iPad is introduced in **Appendix G**.

Finally, **Appendix H** discusses what we left out, points you towards some areas for further reading, and mentions some recent developments.

Lots of sample files help you save typing examples, see Section 1.1.2. You also get PDF files of Mission Impossible, the symbol tables, and the STIX symbols.

Mission statement

This book is a guide for typesetting mathematical documents within the constraints imposed by LaTeX, an elaborate system with hundreds of rules. LaTeX allows you to perform almost any mathematical typesetting task through the appropriate application of its rules. You can customize LaTeX by introducing custom commands and environments and by changing LaTeX parameters. You can also extend LaTeX by invoking packages that accomplish special tasks.

It is *not my goal*

- to survey the hundreds of LaTeX packages you can utilize to enhance LaTeX
- to teach how to write TeX code to create your own packages
- to discuss how to design beautiful documents by writing document classes

The definitive book on the first topic, as of 2004, is Frank Mittelbach and Michel Goosens's *The LaTeX Companion,* 2nd edition [56] (in collaboration with Johannes Braams, David Carlisle, and Chris Rowley). The second and third topics still await authoritative treatment.

Conventions

To make this book easy to read, I use some simple conventions:

- Explanatory text is set in this typeface: Times.
- `Computer Modern typewriter is used to show what you should type, as well as messages from LaTeX. All the characters in this typeface have the same width, making it easy to recognize.`
- I also use Computer Modern typewriter to indicate
  - Commands (`\newpage`)
  - Environments (`\align`)
  - Documents (`firstarticle.tex`)
  - Document classes (`amsart`)
  - Document class options (`draft`)
  - Folders or directories (`work`)
  - The names of *packages*—extensions of LaTeX (`verbatim`)
- When I show you how something looks when typeset, I use Computer Modern, TeX's standard typeface:

I think you find this typeface sufficiently different from the other typefaces I have used. The strokes are much lighter so that you should not have much difficulty recognizing typeset LaTeX material. When the typeset material is a separate paragraph or paragraphs, corner brackets in the margin set it off from the rest of the text—unless it is a displayed formula.

- For explanations in the text, such as

  Compare iff with iff, typed as `iff` and `if{f}`, respectively.

  the same typefaces are used. Because they are not set off spatially, it may be a little more difficult to see that iff is set in Computer Modern roman (in Times, it looks like this: iff), whereas `iff` is set in the Computer Modern typewriter typeface. Compare: iff, iff, `iff`, and a larger version: iff, iff, `iff`.

- I usually introduce commands with examples, such as

  `\\[22pt]`

  However, it is sometimes necessary to define the syntax of a command more formally. For instance,

  `\\[`*`length`*`]`

  where *`length`*, typeset in Computer Modern typewriter italic font, represents the value you have to supply.

Good luck and have fun.

George Gratzer

E-mail:
`gratzer@me.com`
Home page:
`http://server.maths.umanitoba.ca/homepages/gratzer/`

PART I

Mission Impossible

CHAPTER

1

Short course

It happens to most of us. We live a happy life without LaTeX and then, all of a sudden, we have to do something urgent that requires it.

If you are a student, maybe your professor turned to you and said "I need the solutions to these exercises typed up and distributed to the class by tomorrow" and the solutions are chock-full of formulas, difficult to do in Word.

Or you are a researcher whose documents have always been typed up by a secretary. You have to attend a conference and give a presentation. Your secretary is on vacation.

In my case, it was a letter (this was before e-mail) from the American Mathematical Society, in which they informed me that my paper, written in Word, was accepted for publication. The AMS will publish the paper in nine months. However, a LaTeX version would be published in three months! So I had to learn LaTeX in a hurry.

> The mission, should you choose to accept it, is to get started really fast in LaTeX. Our goal is to produce in LaTeX the little article printed on the next page.

Relax, this chapter will not self-destruct in five seconds.

G. Grätzer, *More Math Into* LaTeX, DOI 10.1007/978-3-319-23796-1_1

A TECHNICAL RESULT FOR CONGRUENCES OF FINITE LATTICES

G. GRÄTZER

ABSTRACT. We present a technical result for congruences on finite lattices.

1. INTRODUCTION

In some recent research, G. Czédli and I, see [1] and [2], spent quite an effort in proving that some equivalence relations on a planar semimodular lattice are congruences. The number of cases we had to consider was dramatically cut by the following result.

Theorem 1. *Let L be a finite lattice. Let δ be an equivalence relation on L with intervals as equivalence classes. Then δ is a congruence relation iff the following condition and its dual hold:*

(C_+) *If x is covered by $y, z \in L$ and $x \equiv y \pmod{\delta}$, then $z \equiv y + z \pmod{\delta}$.*

2. THE PROOF

We prove the join-substitution property: if $x \leq y$ and $x \equiv y \pmod{\delta}$, then

$$x + z \equiv y + z \pmod{\delta}. \tag{1}$$

Let $U = [x, y + z]$. We induct on $\operatorname{length} U$, the length of U.

Let $I = [y_1, y + z]$ and $J = [z_1, y + z]$. Then $\operatorname{length} I$ and $\operatorname{length} J < \operatorname{length} U$. Hence, the induction hypothesis applies to I and $\delta \rceil I$, and we obtain that $w \equiv y + w \pmod{\delta}$. By the transitivity of δ, we conclude that

$$z_1 \equiv y + w \pmod{\delta}. \tag{2}$$

Therefore, applying the induction hypothesis to J and $\delta \rceil J$, we conclude (1).

REFERENCES

[1] G. Czédli, *Patch extensions and trajectory colorings of slim rectangular lattices.* Algebra Universalis **88** (2013), 255–280.
[2] G. Grätzer, *Congruences of fork extensions of lattices.* Acta Sci. Math. (Szeged), **57** (2014), 417–434.

DEPARTMENT OF MATHEMATICS, UNIVERSITY OF MANITOBA, WINNIPEG, MB R3T 2N2, CANADA
E-mail address, G. Grätzer: gratzer@me.com
URL, G. Grätzer: http://tinyurl.com/gratzerhomepage

Date: March 21, 2015.
2010 *Mathematics Subject Classification.* Primary: 06B10.
Key words and phrases. finite lattice, congruence.

1.1 Getting started

1.1.1 Your LaTeX

Are you sitting in front of your computer that has a LaTeX implementation? If you use a UNIX computer, you surely are. If you are in front of a Windows computer or a Mac, point your Internet browser at `tug.org`. Choose to download MikTeX for a Windows computer and MacTeX for a Mac. Follow the easy instructions (and be patient, these are big downloads) and you are done.

Even better, find a friend who can help.

1.1.2 Sample files

We work with a few sample documents. Download them from the Springer page for this book:

`http://www.springer.com/us/book/9783319237954`

I suggest you create a folder, `samples`, on your computer to store the downloaded sample files, and another folder called `work`, where you will keep your working files. Copy the documents from the `samples` to the `work` folder as needed. *In this book, the* `samples` *and* `work` *folders refer to the folders you created.*

One of the sample files is `sample.cls`. Make sure it is in the `work` folder when you typeset a sample document.

1.1.3 Editing cycle

Watch a friend type a document in LaTeX and learn the basic steps.

1. *A text editor is used to create a LaTeX source file*. A source file might look like this:

   ```
   \documentclass{amsart}
   \begin{document}
   Then $\delta$ is a congruence relation. I can type formulas!
   \end{document}
   ```

   Note that the source file is different from a typical word processor file. All characters are displayed in the same font and size.

2. *Your friend "typesets" the source file* (*tells the application to produce a typeset version*) *and views the result on the monitor*:

   Then δ is a congruence relation. I can type formulas!

3. *The editing cycle continues.* Your friend goes back and forth between the source file and the typeset version, making changes and observing the results of these changes.

4. *The file is viewed/printed.* View the typeset version as a pdf file or print it to get a paper version.

If LaTeX finds a mistake when typesetting the source file, it records this in the `log` *file*. The `log` *window* (some call it *console*) displays a shorter version.

Various LaTeX implementations have different names for the source file, the text editor, the typeset file, the typeset window, the `log` file, and the `log` window. Become familiar with these names, so you can follow along with our discussions.

1.1.4 Typing the source file

A source file is made up of *text, formulas,* and *instructions* (*commands*) *to* LaTeX.

For instance, consider the following variant of the first sentence of this paragraph:

```
A source file is made up of text, formulas (e.g.,
$\sqrt{5}$), and \emph{instructions to} \la.
```

This typesets as

A source file is made up of text, formulas (e.g., $\sqrt{5}$), and *instructions to* LaTeX.

In this sentence, the first part

```
A source file is made up of text, formulas (e.g.,
```

is text. Then

```
$\sqrt{5}$
```

is a formula

```
), and
```

is text again. Finally,

```
\emph{instructions to} \la.
```

The instruction `\emph` is a *command with an argument,* while the instruction `\LaTeX` is a *command without an argument.* Commands, as a rule, start with a backslash (\ ) and tell LaTeX to do something special. In this case, the command `\emph` emphasizes its *argument* (the text between the braces). Another kind of instruction to LaTeX is called an *environment.* For instance, the commands

```
\begin{center}
\end{center}
```

enclose a `center` environment; the *contents* (the text typed between these two commands) are centered when typeset.

In practice, text, formulas, and instructions (commands) are mixed. For example,

```
My first integral: $\int \zeta^{2}(x) \, dx$.
```

is a mixture of all three; it typesets as

My first integral: $\int \zeta^2(x)\, dx$.

Creating a document in LaTeX requires that we type in the source file. So we start with the keyboard, proceed to type a short note, and learn some simple rules for typing text in LaTeX.

1.2 The keyboard

The following keys are used to type the source file:

```
a-z   A-Z   0-9
+ = * / ( ) [ ]
```

You can also use the following punctuation marks:

```
, ; . ? ! : ` ' -
```

and the space bar, the Tab key, and the Return (or Enter) key.

Finally, there are thirteen special keys that are mostly used in LaTeX commands:

```
# $ % & ~ _ ^ \ { } @ " |
```

If you need to have these characters typeset in your document, there are commands to produce them. For instance, the dollar sign, $ is typed as `\$`, the underscore, _, is typed as `\_`, and the percent sign, %, is typed as `\%`. Only @ requires no special command, type `@` to print @; see Sections 3.1.2 and B.4.

There are also commands to produce composite characters, such as accented characters, for example ä, which is typed as `\"{a}`. LaTeX prohibits the use of other keys on your keyboard unless you have special support for it. See the text accent table in Sections 3.4.7 and B.2. If you want to use accented characters in your source file, then you must use the `inputenc` package.

Tip The text accent table looks formidable. Don't even dream of memorizing it. You will need very few. When you need a text accent, look it up. I know only one: `\"a` (LOL). If you use a name with accented characters, figure out once how to type it, and then any time you need it you can just copy and paste (chances are that the name is in your list of references).

1.3 Your first text notes

We start our discussion on how to type a note in LaTeX with a simple example. Suppose you want to use LaTeX to produce the following:

It is of some concern to me that the terminology used in multi-section math courses is not uniform.

In several sections of the course on matrix theory, the term "hamiltonian-reduced" is used. I, personally, would rather call these "hyper-simple". I invite others to comment on this problem.

To produce this typeset document, create a new file in your `work` folder with the name `textnote1.tex`. Type the following, including the spacing and linebreaks shown, but not the line numbers:

```
1   % Sample file: textnote1.tex
2   \documentclass{sample}
3
4   \begin{document}
5   It is of some concern to me   that
6   the terminology used in  multi-section
7    math courses is not uniform.
8
9   In several sections of the course on
10  matrix theory, the  term
11   ``hamiltonian-reduced'' is used.
12   I, personally, would rather call these
13  ``hyper-simple''. I invite others
14   to comment on this  problem.
15  \end{document}
```

Alternatively, copy the `textnote1.tex` file from the `samples` folder (see page 5).

The first line of `textnote1.tex` starts with `%`. Such lines are called *comments* and are ignored by LaTeX. Commenting is very useful. For example, if you want to add some notes to your source file and you do not want those notes to appear in the typeset version of your document, begin those lines with a `%`. You can also comment out part of a line:

```
simply put, we believe % actually, it's not so simple
```

Everything on the line after the `%` character is ignored by LaTeX.

Line 2 specifies the *document class*, `sample` (the special class we provided for the sample documents), which controls how the document is formatted.

The text of the note is typed within the `document` environment, that is, between `\begin{document}` and `\end{document}`.

Now typeset `textnote1.tex`. You should get the typeset document as shown. As you can see from this example, LaTeX is different from a word processor. It disregards the way you input and position the text, and follows only the formatting instructions given by the document class and the markup commands. LaTeX notices when you put a blank space in the text, but it ignores *how many blank spaces* have been typed. LaTeX does not distinguish between a blank space (hitting the space bar), a tab (hitting the Tab key), and a *single* carriage return (hitting Return once). However, hitting Return twice gives a blank line; *one or more* blank lines mark the end of a paragraph. There is also a command for a *new paragraph*: `\par`.

LaTeX, by default, fully justifies text by placing a flexible amount of space between words—the *interword space*—and a somewhat larger space between sentences—the *intersentence space*. If you have to force an interword space, you can use the `\␣` command (in LaTeX books, we use the symbol ␣ to mean a blank space). The ~ (tilde) command also forces an interword space, but with a difference: it keeps the words on the same line. This command produces a *tie* or *nonbreakable space*.

Note that on lines 11 and 13, the left double quotes is typed as two left single quotes and the right double quote is typed as two right single quotes, apostrophes.

We numbered the lines of the source file for easy reference. Sometimes you may want the same for the typeset file. This is really easy. Just add the two lines

```
\usepackage{lineno}
\linenumbers
```

after the `\documentclass` line and you get:

1 It is of some concern to me that the terminology used in multi-section math
2 courses is not uniform.
3 In several sections of the course on matrix theory, the term “hamiltonian-
4 reduced” is used. I, personally, would rather call these “hyper-simple”. I invite
5 others to comment on this problem.

Next, we produce the following note:

January 5, 2015

From the desk of George Grätzer

February 7–21 *please* use my temporary e-mail address:

George_Gratzer@yahoo.com

Type the source file, without the line numbers. Save it in your `work` folder as `textnote2.tex` (`textnote2.tex` can also be found in the `samples` folder):

```
1   % Sample file: textnote2.tex
2   \documentclass{sample}
3
4   \begin{document}
5   \begin{flushright}
6      \today
7   \end{flushright}
8   \textbf{From the desk of George Gr\"{a}tzer}
9
10  February 7--21 \emph{please} use my
11  temporary e-mail address:
12  \begin{center}
13     \texttt{George\_Gratzer@yahoo.com}
14  \end{center}
15  \end{document}
```

This note introduces several additional text features of LaTeX.

- The `\today` command (in line 6) to display the date on which the document is typeset, so you will see a date different from the date shown above in your own typeset document (see also Section 3.4.8).
- The environments to *right justify* (lines 5–7) and *center* (lines 12–14) text.
- The commands to change the text style, including the `\emph` command (line 10) to *emphasize* text, the `\textbf` command (line 8) for **bold** text (text bold font), and the `\texttt` command (line 13) to produce `typewriter style` text. These are *commands with arguments.*
- The form of the LaTeX commands. As we have noted already, almost all LaTeX *commands* start with a backslash (`\`) followed by the *command name*. For instance, `\textbf` is a command and `textbf` is the command name. The command name is terminated by the first *non-alphabetic character*, that is, by any character other than a–z or A–Z.

Tip `textnote2.tex` is a file name but `textbf1` is not a command name. `\textbf1` typesets as **1**. Let's look at this a bit more closely. `\textbf` is a valid command. If a command needs an argument and it is not followed by braces, then it takes the next character as its argument. So `\textbf1` is the command `\textbf` with the argument 1; it typesets as **1**.

- The multiple role of hyphens: Double hyphens are used for number ranges. For example, `7--21` (in line 10) typesets as 7–21. The punctuation mark – is called an *en dash*. Use triple hyphens for the *em dash* punctuation mark—such as the one in this sentence.
- Special rules for special characters (see Section 1.2), for *accented characters*, and for some *European characters*. For instance, the accented character ä is typed as `\"{a}`. (But I confess, I always type my name as `Gr\"atzer` without the braces.)

See Section 3.4 for more detail. In Appendix B, all the text symbols are organized into tables. We also have the `SymbolTables.pdf` in the `samples` folder.

Tip Keep `SymbolTables.pdf` handy on your computer!

1.4 Lines too wide

LaTeX reads the text in the source file one line at a time and typesets the entire paragraph when the end of a paragraph is reached. Occasionally, LaTeX gets into trouble when trying to split the paragraph into typeset lines. To illustrate this situation, modify `textnote1.tex`. In the second sentence, replace `term` by `strange term`. Now save this modified file in your `work` folder using the name `textnote1bad.tex` (or copy the file from the `samples` folder).

Typesetting `textnote1bad.tex`, you obtain the following:

1 It is of some concern to me that the terminology used in multi-section math
2 courses is not uniform.
3 In several sections of the course on matrix theory, the term "hamiltonian-
4 reduced" is used. I, personally, would rather call these "hyper-simple". I invite
5 others to comment on this problem.

The first line of paragraph two is too wide. In the `log` window, LaTeX displays the following messages:

```
Overfull \hbox (15.38948pt
too wide) in paragraph at lines 9--15 []\OT1/cmr/m/n/10 In sev-eral
sec-tions of the course on ma-trix the-ory, the strange term
``hamiltonian-
```

It informs you that the typeset version of this paragraph has a line that is 15.38948 points too wide. LaTeX uses *points* (pt) to measure distances; there are about 72 points in 1 inch. Then it identifies the source of the problem: LaTeX did not properly hyphenate the word `hamiltonian-reduced` because it (automatically) hyphenates a hyphenated word *only at the hyphen.*

What to do, when a line is too long?

Tip Your first line of defense: reword the offending line. Write

```
The strange term ``hamiltonian-reduced'' is used
in several sections of the course on matrix theory.
```

and the problem goes away.

Your second line of defense: insert one or more *optional hyphen commands* (`\-`), which tell LaTeX where it can hyphenate the word. Write:

```
hamil\-tonian-reduced
```

1.5 A note with formulas

In addition to the regular text keys and the 13 special keys discussed in Section 1.2, two more keys are used to type formulas: < and >. The formula $2 < |x| > y$ (typed as `$2 < |x| >y$`) uses both. Note that such a formula, called *inline*, is enclosed by a pair of `$` symbols.

We begin typesetting formulas with the following note:

In first-year calculus, we define intervals such as (u, v) and (u, ∞). Such an interval is a *neighborhood* of a if a is in the interval. Students should realize that ∞ is only a symbol, not a number. This is important since we soon introduce concepts such as $\lim_{x \to \infty} f(x)$.

When we introduce the derivative

$$\lim_{x \to a} \frac{f(x) - f(a)}{x - a},$$

we assume that the function is defined and continuous in a neighborhood of a.

To create the source file for this mixed text and formula note, create a new document with your text editor. Name it `formulanote.tex`, place it in the `work` folder, and type the following, without the line numbers (or simply copy `formulanote.tex` from the `samples` folder):

```
1   % Sample file: formulanote.tex
2   \documentclass{sample}
3
4   \begin{document}
5   In first-year calculus, we define intervals such
6   as  $(u, v)$ and $(u, \infty)$. Such an interval
7   is a \emph{neighborhood} of  $a$
8   if  $a$ is in the interval. Students should
9   realize that  $\infty$ is only a
10  symbol, not a number. This is important since
11  we soon introduce concepts
```

```
12     such as $\lim_{x \to \infty} f(x)$.
13
14    When we introduce the derivative
15    \[
16       \lim_{x \to a} \frac{f(x) - f(a)}{x - a},
17    \]
18    we assume that the function is defined and
19    continuous in a neighborhood of  $a$.
20    \end{document}
```

This note introduces several basic concepts of formulas in LaTeX.

- There are two kinds of math formulas and environments in `formulanote.tex`:
  - *Inline* formulas; they open and close with `$` or open with `\(` and close with `\)`.
  - *Displayed* math environments; they open with `\[` and close with `\]`. (We will introduce many other displayed math environments in Section 1.7 and Chapter 7.)
- LaTeX uses its own spacing rules within math environments, and completely ignores the white spaces you type, with two exceptions:
  - Spaces that terminate commands. So in `$\infty a$` the space is not ignored; `$\inftya$` produces an error.
  - Spaces in the arguments of commands that temporarily revert to regular text. `\text` is such a command; see Sections 1.6 and 5.4.6.

  The white space that you add when typing formulas is important only for the readability of the source file.
- A math symbol is invoked by a command. For example, the command for ∞ is `\infty` and the command for $\to$ is `\to`. The math symbols are organized into tables in Appendix A; see also `SymbolTables.pdf` in the `samples` folder.
- Some commands, such as `\sqrt`, need *arguments* enclosed by `{` and `}`. To typeset $\sqrt{5}$, type `$\sqrt{5}$`, where `\sqrt` is the command and `5` is the argument.

  Some commands need more than one argument. To get

$$\frac{3+x}{5}$$

  type

```
\[
   \frac{3+x}{5}
\]
```

  where `\frac` is the command, `3+x` and `5` are the arguments.
- There is no blank line before a displayed formula!

Tip Keep in mind that many spaces equal one space in text, whereas your spacing is ignored in formulas, unless the space terminates a command.

1.6 The building blocks of a formula

A formula (inline or displayed) is built from components. We group them as follows:

Arithmetic
Binomial coefficients
Congruences
Delimiters
Ellipses
Integrals
Math accents
Matrices
Operators
Roots
Text

In this section, I describe each of these groups, and provide examples illustrating their use. Read carefully the groups you need!

Arithmetic We type the arithmetic operations $a+b$, $a-b$, $-a$, a/b, and ab in the natural way: `$a + b$`, `$a - b$`, `$-a$`, `$a / b$`, and `$a b$` (the spaces are typed only for readability).

If you wish to use $\cdot$ or $\times$ for multiplication, as in $a \cdot b$ or $a \times b$, use `\cdot` or `\times`, respectively. The formulas $a \cdot b$ and $a \times b$ are typed as `$a \cdot b$` and `$a \times b$`.

Displayed fractions, such as

$$\frac{1 + 2x}{x + y + xy}$$

are typed with `\frac`:

```
\[
   \frac{1 + 2x}{x + y + xy}
\]
```

Subscripts and superscripts Subscripts are typed with `_` and superscripts with `^` (caret). Subscripts and superscripts should be enclosed in braces, that is, typed between `{` and `}`. To get a_1, type `$a_{1}$`. Omitting the braces in this example causes no harm, but to get a_{10}, you *must* type `$a_{10}$`. Indeed, `$a_10$` is typeset as $a_1 0$.

There is one symbol, the prime (`'`), that is automatically superscripted in a formula. To get $f'(x)$, just type `$f'(x)$`. (On many keyboards, the symbol on the key looks like this: `)

See Section 5.4.1 for more detail.

Binomial coefficients Binomial coefficients are typeset with the `\binom` command. `\binom{a}{b + c}` is here inline: $\binom{a}{b+c}$, whereas

$$\binom{a}{b + c}$$

is the displayed version.

See Section 5.4.2 for more detail.

Congruences The two most important forms are

$a \equiv v \pmod{\theta}$	typed as	`$a \equiv v \pmod{\theta}$`
$a \equiv v \pod{\theta}$	typed as	`$a \equiv v \pod{\theta}$`

See Section 5.6.2 for more detail.

Delimiters Parentheses and square brackets are examples of delimiters. They are used to delimit some subformulas, as in `$[(a*b)+(c*d)]^{2}$`, which typesets as $[(a * b) + (c * d)]^2$. LaTeX can be instructed to expand them vertically to enclose a formula such as

$$\left(\frac{1 + x}{2 + y^{2}} \right)^{2}$$

which is typed as

```
\[
   \left( \frac{1 + x}{2 + y^{2}} \right)^{2}
\]
```

The `\left(` and `\right)` commands tell LaTeX to size the parentheses correctly, relative to the size of the formula inside the parentheses; sometimes the result is pleasing, sometimes not.

We dedicate Section 5.5 to this topic.

Ellipses In a formula, the ellipsis is printed either as *low* (or *on-the-line*) *dots*:

$F(x_1, \dots, x_n)$ is typed as `$F(x_{1}, \dots, x_{n})$`

or as *centered dots*:

$x_1 + \cdots + x_n$ is typed as

```
$x_{1} + \dots + x_{n}$
```

Use `\cdots` and `\ldots` if `\dots` does not work as expected.

See Section 5.4.3 for more detail.

Integrals The command for an integral is `\int`. The lower limit is specified as a subscript and the upper limit is specified as a superscript. For example, the formula $\int_{0}^{\pi} \sin x \, dx = 2$ is typed as

```
$\int_{0}^{\pi} \sin x \, dx = 2$
```

where `\,` is a spacing command.

The formula looks bad without the spacing command: $\int_{0}^{\pi} \sin x dx = 2$.

See Section 5.4.4 for more complicated integrals.

Math accents The four most frequently used math accents are:

$\bar{a}$	typed as	`$\bar{a}$`	$\hat{a}$	typed as	`$\hat{a}$`
$\tilde{a}$	typed as	`$\tilde{a}$`	$\vec{a}$	typed as	`$\vec{a}$`

See Section 5.7 for more detail. See Sections 5.7 and A.8 for complete lists.

Matrices You type the matrix

$$\begin{matrix} a + b + c & uv & x - y & 27\\ a + b & u + v & z & 134 \end{matrix}$$

with the `\matrix` command

```
\[
   \begin{matrix}
      a + b + c & uv    & x - y & 27\\
      a + b     & u + v & z     & 134
   \end{matrix}
\]
```

The `matrix` environment separates adjacent matrix elements within a row with ampersands. Rows are *separated* by new line commands, `\\`.

Tip Do not end the last row with a new line command.

The `matrix` environment has to appear within a formula, as a rule, in a displayed formula. It can be used in the `align` environment discussed in Sections 1.7.3 and 7.5.

The `matrix` environment does not provide delimiters. Several variants do, including `pmatrix` and `vmatrix`. For example,

$$\mathbf{A} = \begin{pmatrix} a + b + c & uv\\ a + b & u + v \end{pmatrix} \begin{vmatrix} 30 & 7\\ 3 & 17 \end{vmatrix}$$

is typed as follows:

```
\[
  \mathbf{A} =
  \begin{pmatrix}
    a + b + c & uv\\
    a + b & u + v
  \end{pmatrix}
  \begin{vmatrix}
    30 & 7\\
    3 & 17
  \end{vmatrix}
\]
```

As you can see, `pmatrix` typesets as a `matrix` between a pair of `\left(` and `\right)` commands, while `vmatrix` typesets as a `matrix` between a pair of `\left|` and `\right|` commands. There is also `bmatrix` for square brackets.

See Section 7.7.1 for a listing of all the matrix variants and Sections 5.5 and A.6 for lists of delimiters.

Operators To typeset the sine function, $\sin x$, type `$\sin x$`. Note that `$sin x$` would be typeset as $sinx$—-how awful. LaTeX calls `\sin` an *operator*. Sections 5.6 and A.7 list a number of operators. Some are just like `\sin`. Others produce a more complex display, for example,

$$\lim_{x \to 0} f(x) = 0$$

is typed as

```
\[
  \lim_{x \to 0} f(x) = 0
\]
```

See Section 5.6 for more detail.

Large operators The command for *sum* is `\sum` and for *product* is `\prod`. The following two examples:

$$\sum_{i=1}^{n} x_{i}^{2}\ \ \prod_{i=1}^{n} x_{i}^{2}$$

are typed as

```
\[
  \sum_{i=1}^{n} x_{i}^{2}\ \ \prod_{i=1}^{n} x_{i}^{2}
\]
```

Sum and product are examples of *large operators*. They are typeset larger in displayed math than in an inline formula. They are listed in Sections 5.6.3 and A.7.1. See Section 5.6.3 for more detail.

Roots \sqrt produces a square root. $\sqrt{a + 2b}$ typesets as $\sqrt{a+2b}$. The n-th root, $\sqrt[n]{5}$, requires the use of an *optional argument,* which is specified in brackets: $\sqrt[n]{5}$. See Section 5.4.5.

Text You can include text in a formula with a \text command. For instance,

$$a = b, \quad \text{by assumption},$$

is typed as

```
\[
   a = b, \text{\quad by assumption},
\]
```

where \quad is a spacing command.

See Section 5.4.6 for more detail.

1.7 Displayed formulas

1.7.1 Equations

The equation environment creates a displayed formula and automatically generates an equation number. The equation

$$\int_{0}^{\pi} \sin x \, dx = 2 \tag{1}$$

is typed as

```
\begin{equation}\label{E:firstIntegral}
   \int_{0}^{\pi} \sin x \, dx = 2
\end{equation}
```

The equation number, which is automatically generated, depends on how many numbered displayed formulas occur before the given equation. You can choose to have equations numbered within each section—(1.1), (1.2), ..., in Section 1; (2.1), (2.2), ..., in Section 2; and so on—by including, in the preamble (see Sections 1.8 and 5.3), the command

```
\numberwithin{equation}{section}
```

You can choose to have the equation numbers on the right; see the reqno option of the amsart document class in Section 10.1.2.

The equation* environment is the same as the displayed formula opened with \[and closed with \] we discussed in Section 1.5. Sometimes you may want to use equation* for the ease of deleting the *-s if you wish.

1.7.2 Symbolic referencing

To reference a formula without having to remember a number—which can change when you edit your document—give the equation a symbolic label by using the `\label` command and refer to the equation in your document by using the symbolic label, the argument of the `\label` command. In this example, I have called the first equation `firstIntegral`, and used the convention that the label of an equation starts with `E:`, so that the complete `\label` command is `\label{E:firstIntegral}`.

The number of this formula is referenced with the `\ref` command. Its page is referenced using the `\pageref` command. For example, to get

see (1) on page 18.

type (see Sections 1.3 and Section 3.4.3 for ~)

```
see~(\ref{E:firstIntegral}) on page~\pageref{E:firstIntegral}.
```

The `\eqref` command provides the reference number in parentheses. So the last example could be typed

```
see~\eqref{E:firstIntegral} on page~\pageref{E:firstIntegral}.
```

The `\eqref` command is smart. Even if the equation number is referenced in emphasized or italicized text, the reference typesets upright (in roman type).

The main advantage of this cross-referencing system is that when you add, delete, or rearrange equations, LaTeX automatically renumbers the equations and adjusts the references that appear in your typeset document. For bibliographic references, LaTeX uses the `\bibitem` command to define a bibliographic item and the `\cite` command to cite it.

Tip For renumbering to work, you have to typeset **twice**.

Tip It is a good idea to check the LaTeX warnings periodically in the `log` file. If you forget to typeset the source file twice when necessary, LaTeX issues a warning.

What happens if you misspell a reference, e.g., typing `\ref{E:FirstIntegral}` instead of `\ref{E:firstIntegral}`? LaTeX typesets **??**. There are two warnings in the `log` file:

```
LaTeX Warning: Reference ‘E:FirstIntegral’ on page 39
        undefined on input line 475.
```

for the typeset page and the other one close to the end:

```
LaTeX Warning: There were undefined references.
```

If the argument of \cite is misspelled, you get **[?]** and similar warnings.

Check the **Tip** on page 69.

Absolute referencing

Equations can also be *tagged* by attaching a name to the formula with the \tag command. The tag replaces the equation number.

For example,

$$\int_{0}^{\pi} \sin x \, dx = 2 \tag{Int}$$

is typed as

```
\begin{equation}
  \int_{0}^{\pi} \sin x \, dx = 2 \tag{Int}
\end{equation}
```

Tags are *absolute.* This equation is *always* referred to as (Int). Equation numbers, on the other hand, are *relative,* they may change when the file is edited.

1.7.3 Aligned formulas

LaTeX has many ways to typeset multiline formulas. We discuss three constructs in this section: *simple alignment, annotated alignment,* and *cases*. For more constructs, see Chapter 7.

For simple and annotated alignment we use the align environment. Each line in the align environment is a separate equation, which LaTeX automatically numbers.

Simple alignment

Simple alignment is used to align two or more formulas. To obtain the formulas

$$\begin{align} r^{2} &= s^{2} + t^{2}, \tag{2}\\ 2u + 1 &= v + w^{\alpha}. \tag{3} \end{align}$$

type the following, using \\ as the *line separator* and & as the *alignment point*:

```
\begin{align}
  r^{2}  &= s^{2} + t^{2},        \label{E:Pyth}\\%\eqref{E:Pyth}
  2u + 1 &= v + w^{\alpha}.       \label{E:alpha}%\eqref{E:alpha}
\end{align}
```

Figure 1.1 may help visualize the placements of the ampersands.

Tip In this displayed formula, \\ is a *line separator*, not a new line command. Do not place a \\ to terminate the last line!

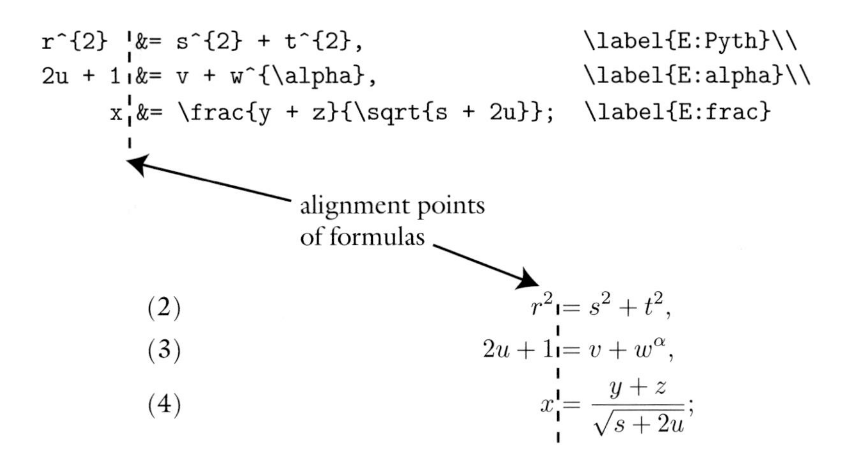

Figure 1.1: Simple alignment: source and typeset.

These formulas are numbered (2) and (3) because they are preceded by one numbered equation earlier in this section.

The `align` environment can also be used to break a long formula into two or more parts. Since numbering both lines in such a case would be undesirable, you can prevent the numbering of the second line by using the `\notag` command in the second part of the formula. For example,

$$h(x) = \int \left(\frac{f(x) + g(x)}{1 + f^{2}(x)} + \frac{1 + f(x)g(x)}{\sqrt{1 - \sin x}} \right) dx \tag{4}$$
$$= \int \frac{1 + f(x)}{1 + g(x)}\, dx - 2\tan^{-1}(x - 2)$$

is typed as follows:

```
\begin{align}
  h(x) &= \int \left( \frac{f(x) + g(x)}{1+ f^{2}(x)}
            + \frac{1+ f(x)g(x)}{\sqrt{1 - \sin x}}
            \right) \, dx\label{E:longInt}\\%\eqref{E:longInt}
         &= \int \frac{1 + f(x)}{1 + g(x) } \, dx
           - 2 \tan^{-1}(x-2)\notag
\end{align}
```

The rules for simple alignment are easy to remember.

Rule ■ Simple alignments

- Use the `align` environment.
- *Separate* the lines with `\\`.

- In each line, indicate the alignment point with `&`, one `&` per line. If the alignment point is adjacent to an =, +, and so on, place the `&` *before* to ensure proper spacing.
- Place a `\notag` command in each line that you do not wish numbered.
- If no line should be numbered, use the `align*` environment.
- Place a `\label` command in each numbered line you can want to reference with `\ref`, `\eqref`, or `\pageref`.

Annotated alignment

Annotated alignment allows you to align formulas and their annotations, that is, explanatory text, separately:

$$\begin{aligned} x &= x \wedge (y \vee z) && \text{(by distributivity)} \qquad (5)\\ &= (x \wedge y) \vee (x \wedge z) && \text{(by condition (M))}\\ &= y \vee z \end{aligned}$$

This is typed as

```
\begin{align}
   x &= x \wedge (y \vee z)
    &&\text{(by distributivity)}\label{E:Align}%\eqref{E:Align}\\
   &= (x \wedge y) \vee (x \wedge z)
    &&\text{(by condition (M))} \notag\\
   &= y \vee z \notag
\end{align}
```

Figure 1.2 may help visualize the placements of the ampersands.

Rule ■ **Annotated alignment**

The rules for annotated alignment are similar to the rules of simple alignment. In each line, in addition to the alignment point marked by `&`, there is also a mark for the start of the annotation: `&&`.

1.7.4 Cases

The `cases` construct is a specialized matrix. It has to appear within a math environment such as the `equation` environment or the `align` environment. Here is a typical example:

$$f(x) = \begin{cases} -x^2, & \text{if } x < 0;\\ \alpha + x, & \text{if } 0 \leq x \leq 1;\\ x^2, & \text{otherwise.} \end{cases}$$

It is typed as follows:

```
\[
  f(x)=
  \begin{cases}
    -x^{2},          &\text{if $x < 0$;}\\
    \alpha + x,      &\text{if $0 \leq x \leq 1$;}\\
    x^{2},           &\text{otherwise.}
  \end{cases}
\]
```

The rules for using the `cases` environment are the same as for matrices. Separate the lines with `\\` and indicate the annotation with `&`.

1.8 The anatomy of a document

To begin, we use the sample document `firstarticle.tex` (in the `samples` folder) to examine the anatomy of an document.

Every LaTeX document has two parts, the preamble and the body. The *preamble* of a document is everything from the first line of the source file down to the line

```
\begin{document}
```

The *body* is the contents of the `document` environment. For a schematic view of a document, see Figure 1.3.

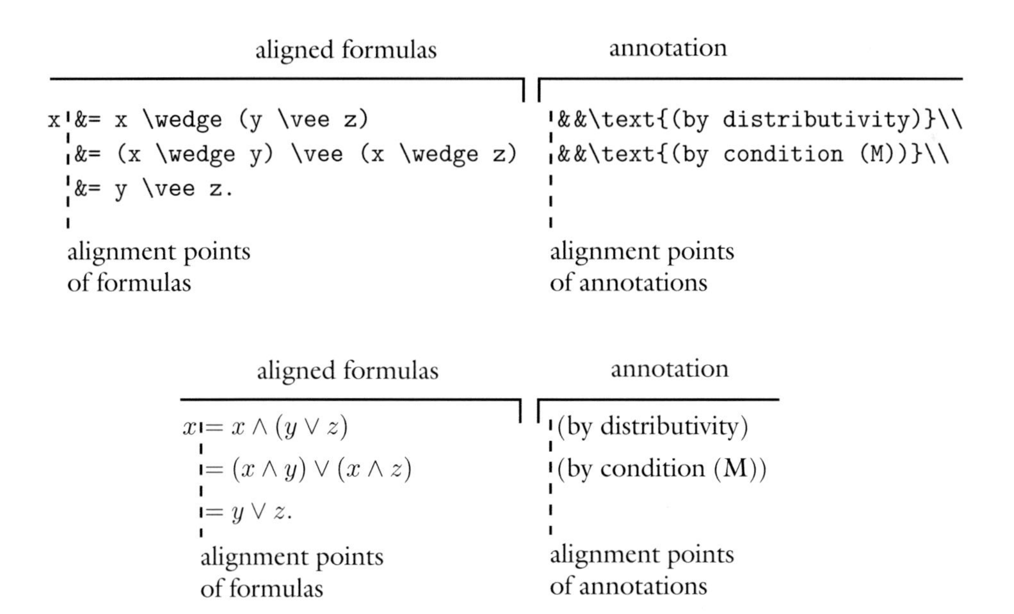

Figure 1.2: Annotated alignment: source and typeset.

The preamble contains instructions affecting the entire document. The *only* required command in the preamble is the \documentclass command. There are other commands (such as the \usepackage commands, see Section 8.2) that must be placed in the preamble if they are used, but such commands do not have to be present in every document.

Here is the preamble and top matter of firstarticle:

```
%First document, firstarticle.tex
\documentclass{amsart}
\usepackage{amssymb,latexsym}

\newtheorem{theorem}{Theorem}

\begin{document}
\title{A technical result\\ for congruences of finite lattices}
\author{G. Gr\"atzer}
\address{Department of Mathematics\\
  University of Manitoba\\
  Winnipeg, MB R3T 2N2\\
  Canada}
```

```
\documentclass{...}
\usepackage{...}
...                                   preamble

\begin{document}

             \title{...}
             \author{...}
             \address{...}
top matter   \date{...}

             \begin{abstract}
             ...                 abstract
             \end{abstract}
             \maketitle                              body

             \section{...}
main matter
             \section{...}

             \begin{thebibliography}{9}
back matter  ...
             \end{thebibliography}

\end{document}
```

Figure 1.3: A schematic view of a document.

```
\email[G. Gr\"atzer]{gratzer@me.com}
\urladdr[G. Gr\"atzer]{http://tinyurl.com/gratzerhomepage}
\date{March 21, 2015}
\subjclass[2010]{Primary: 06B10.}
\keywords{finite lattice, congruence.}
\maketitle

\begin{abstract}
We present a technical result for congruences on finite lattices.
\end{abstract}
```

You find the source file, `firstarticle.tex`, in the `samples` folder and the typeset document on page 4.

To simplify the discussion in Part I, we discuss only one document class for articles: `amsart`. You may come across its predecessor, `article`, which handles a limited set of commands for the preamble and the top matter and displays them differently. We shall discuss in detail the `amsart` document class in Chapter 9. For the `article` document class, see Section 10.1.

1.9 Your own commands

Over time, LaTeX can be adjusted to fit your needs. You add packages to enable LaTeX to do new things (such as the `graphicx` package, see Sections 1.10 and 8.4.3) and introduce your own commands to facilitate typing and make the source file more readable.

We can add two new commands to the sample article `firstarticle.tex`:

```
\newcommand{\pdelta}{\pmod{\delta}}
\DeclareMathOperator{\length}{length}
```

So instead of

```
$x \equiv y \pmod{\delta}$+
```

we can type

```
$x \equiv y \pdelta$
```

and instead of `length\,$U$`, we can type `$\length U$` (see Section 14.1.6). Notice how the spacing is now done by LaTeX!

We'll dedicate Chapter 14 to customizing LaTeX.

1.10 Adding an illustration

"And what is the use of a book," thought Alice, "without pictures or conversations?" I am not sure what to suggest about conversations, but illustrations we can tackle with ease. Let us add an illustration, `covers.pdf` to `firstarticle`. First, add

```
\usepackage{graphicx}
```

as the fourth line of the document, to the preamble. This will enable LaTeX to tackle illustrations. Secondly, add the following lines to `firstarticle.tex`, say, as the second paragraph of the introduction:

```
\begin{figure}[hbt]
{\centering\includegraphics{covers}}
\caption{Theorem~\ref{T:technical} illustrated}\label{F:Theorem}
\end{figure}
```

We place the illustration `covers.pdf` in the same folder as `firstarticle.tex`. That's it. You find `covers.pdf` and `firstarticleill.tex` in the `samples` folder.

Tip Make sure that the `\label` command follows the `\caption` command! You may have hard to explain troubles otherwise.

See Section 8.4.3 for more information.

Most people in my field used the vector graphics application Adobe Illustrator to produce the PDF files for illustrations. Quite recently, it became prohibitively expensive. Luckily, many reasonably priced alternatives are available. In Chapter 13, we discuss an alternative, Ti*k*Z, built for LaTeX. Inkspace is an alternative, available for all platforms.

1.11 The anatomy of a presentation

Chances are, one of your first exposures to LaTeX was watching a *presentation*. The presenter used a pdf document produced by LaTeX and opened it with Adobe Reader. He went from "slide" to "slide" by pressing the space bar. Figure 1.4 and Figure 1.5 show four slides of a presentation.

In LaTeX, you use a presentation package—really, a document class—to prepare the PDF file. We use Till Tantau's BEAMER.

Here are the first few lines—the preamble and the Title slide—of the source file of our sample presentation, `firstpresentation.tex` (see `firstpresentation.tex` in the `samples` folder, along with `Louisville.tex`, the full presentation):

```
\documentclass[leqno]{beamer}
\usetheme{Warsaw}

\DeclareMathOperator{\Princ}{Princ}

\begin{document}
\title{The order of principal congruences}
\author{G. Gr\"atzer}
\date{}
\maketitle
```

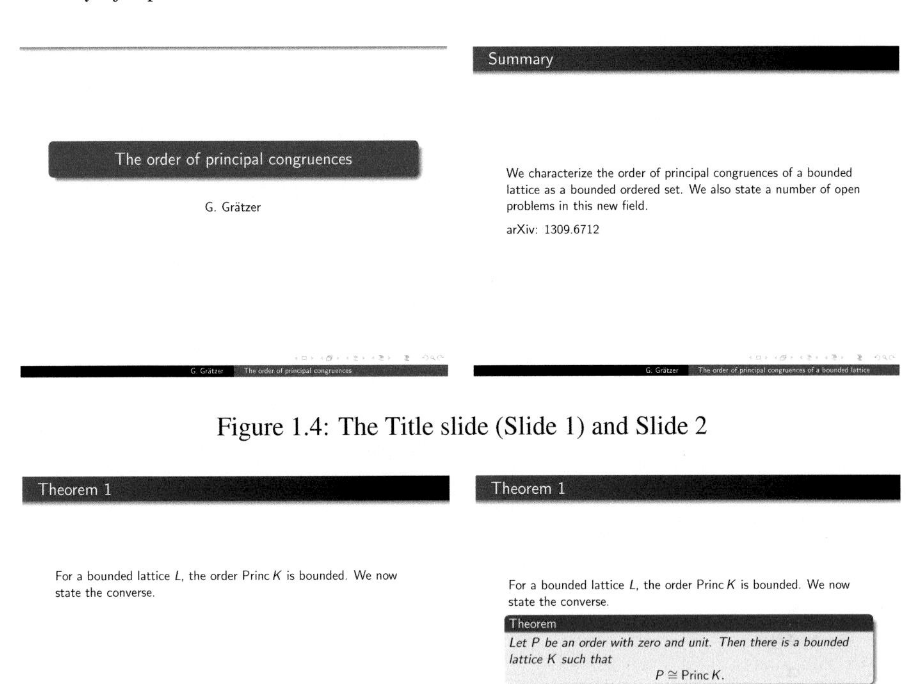

Figure 1.4: The Title slide (Slide 1) and Slide 2

Figure 1.5: Slides 3 and 4

`\usetheme{Warsaw}` provides a flavor. It is followed by the Title slide, providing the title and the author.

The `\title` command may be longer, it may contain all the additional information you may want to display. Here is the `\title` command of `Louisville.tex`:

```
\title[The order of principal congruences of a bounded lattice]
{The order of principal congruences\\
of a bounded lattice.\\
AMS Fall Southeastern Sectional Meeting\\
University of Louisville, Louisville, KY\\
October 5-6, 2013}
```

Note that the `\title` has two parts. The first, in `[]`, is the short title, repeated in the bottom line on every slide. The second, in `{}`, is the title for the front page.

The rest of the presentation source file is divided into two *frames* with the structure:

```
\begin{frame}
\frametitle{}
\end{frame}
```

Each frame produces a "slide" (or more). Here is the first frame:

```
\begin{frame}
\frametitle{Summary}
We characterize the order of principal congruences
of a bounded lattice
as a bounded ordered set.
We~also state a number of open problems in this new field.
\medskip

arXiv: 1309.6712
\end{frame}
```

The command `\frametitle` gives the slide its title: Summary, see Slide 2 in Figure 1.4. In the body of the `frame`, you type regular LaTeX.

To produce Slides 3 and 4, it would be natural to try

```
\begin{frame}
\frametitle{Theorem 1}
For a bounded lattice $L$, the order $\Princ K$ is bounded.
We now state the converse.
\end{frame}

\begin{frame}
\frametitle{Theorem 1}
For a bounded lattice $L$, the order $\Princ K$ is bounded.
We now state the converse.
\begin{theorem}
Let $P$ be an order with zero and unit.
Then there is a bounded lattice~$K$ such that
\[
   P \cong \Princ  K.
\]
If $P$ is finite, we can construct $K$ as a finite lattice.
\end{theorem}
\end{frame}
```

which produces the two frames of Figure 1.6.

This is really jarring to watch. The two lines of the new Slide 3 jump up more than two lines as they transition to Slide 4.

Here is how we produce Slides 3 and 4 of Figure 1.5:

```
\begin{frame}
\frametitle{Theorem 1}
For a bounded lattice $L$, the order $\Princ K$ is bounded.
We now state the converse.
\pause
\begin{theorem}
```

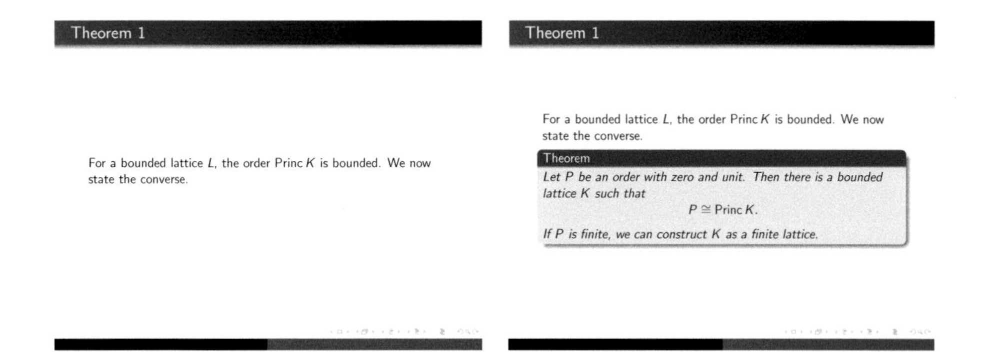

Figure 1.6: Slides 3 and 4, first try

```
Let $P$ be an order with zero and unit.
Then there is a bounded lattice~$K$ such that
\[
   P \cong \Princ  K.
\]
If $P$ is finite, we can construct $K$ as a finite lattice.
\end{theorem}
\end{frame}
\end{document}
```

There is only one new command to learn: `\pause`; it produces from this frame **two** slides.

The `\pause` in this frame splits the contents of the frame into two parts. The first slide is typeset from the first part as if the second part was also present. The second slide is typeset from both parts. So the transition from the first slide to the second is smooth, see Figure 1.5.

You can have more than one `\pause` in a frame. Use `\pause` also to display a list one item at a time.

Chapter 12 discusses BEAMER in more detail.

CHAPTER

2

And a few more things...

If life was perfect, we would not need this chapter. You would write perfect LaTeX, based on Chapter 1, no need to study how LaTeX works, what error messages mean... But life is not perfect, you will make mistakes, LaTeX will send messages, plain and mysterious.

In this chapter, we briefly explain how things work, the structure of LaTeX, the auxiliary files, the logical and visual design of an article, LaTeX error messages. See Appendix C for more detail. Finally, we present a long list of dos and don't to help you write good LaTeX.

2.1 Structure

LaTeX's core is a programming language called TeX, created by Donald E. Knuth, which provides low-level typesetting instructions. TeX comes with a set of fonts called *Computer Modern* (CM). The CM fonts and the TeX programming language form the foundation of a typical TeX system. TeX is extensible—new commands can be defined in terms of more basic ones. LaTeX is one of the best known extensions of TeX.

G. Grätzer, *More Math Into* LaTeX, DOI 10.1007/978-3-319-23796-1_2

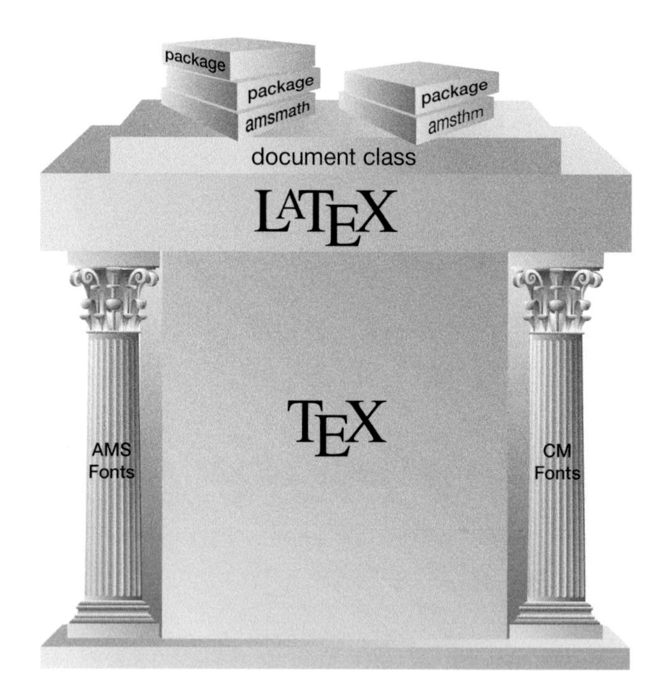

Figure 2.1: The structure of LaTeX.

The visual layout of a LaTeX document is primarily determined by the *document class*, such as `amsart`, `article` for articles, `amsbook`, `book` for books. Many journals, publishers, and schools have their own document classes for formatting articles, books, and theses.

Extensions of LaTeX are called *packages*. They provide additional functionality by adding new commands and environments, or by changing the way previously defined commands and environments work. It is essential that you find the packages that make your work easier. *The* LaTeX *Companion,* 2nd edition [56] discusses a large number of the most useful packages as of 2004.

The structure of LaTeX is illustrated in Figure 2.1. This figure suggests that in order to work with a LaTeX document, you first have to install TeX and the CM fonts, then LaTeX, and finally specify the document class and the necessary packages. The packages must include `amsmath`, `amsthm`, and so on. Of course, your LaTeX installation already includes all of these.

2.2 Auxiliary files

Figure 2.2 illustrates the steps in the production of a typeset document.

You start by opening an existing LaTeX source file or creating a new one with a text editor. For this discussion, the source file is called `myart.tex`. Once the source file is ready, you typeset it. Depending on the document class options you choose and the packages the document loads, you end up with at least three additional files:

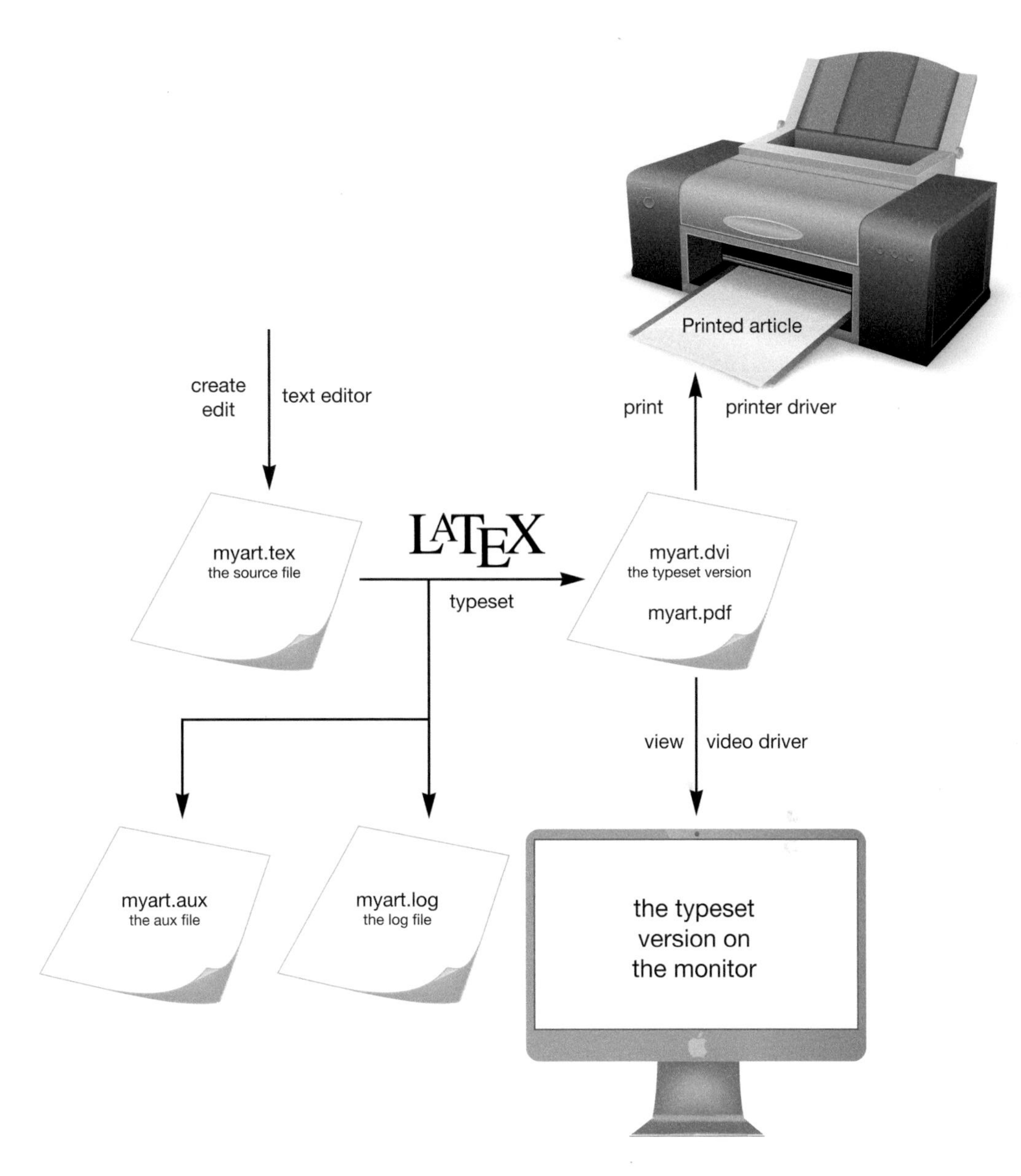

Figure 2.2: Using LaTeX.

1. `myart.pdf` The typeset article in PDF format.

2. `myart.aux` The auxiliary file, used by LaTeX for internal bookkeeping, including cross-references and bibliographic citations.

3. `myart.log` The `log` file. LaTeX records the typesetting session in the `log` file, including any warnings and messages that appear on your monitor in the `log` window.

Your computer uses a *video driver* to display the typeset article on your monitor and a *printer driver* to print the typeset article on a printer. The video and printer drivers are computer and LaTeX implementation dependent.

It should be emphasized that of the three applications used, only one is the same for all computers and all implementations.

LaTeX always uses the aux file from the last typesetting. Here is an example. Your article has Theorems 1 (with `\label{T:first}`) and 2 (with `\label{T:main}`). The aux file has the two lines:

```
\newlabel{T:first}{{1}{1}}
\newlabel{T:main}{{2}{1}}
```

`\newlabel{T:first}{{1}{1}}` means that the label `T:first` is assigned the value 1 and appears on page 1. `\newlabel{T:main}{{2}{1}}` means that the label `T:main` is assigned the value 2 and appears on page 1. So the reference

```
see Theorems \ref{T:first} and \ref{T:first}.
```

is typeset as

see Theorems 1 and 2.

Now add a new theorem between Theorems 1 and 2. Typeset the article. In the typeset article, the three theorems are properly numbered, but it still contains the same typeset line:

see Theorems 1 and 2.

The aux file has the lines:

```
\newlabel{T:first}{{1}{1}}
\newlabel{T:main}{{3}{1}}
```

So at the next typesetting, the reference is displayed as

see Theorems 1 and 3.

2.3 Logical and visual design

The typeset version of `firstarticle.tex` looks impressive on p. 4. To produce such articles, you need to understand that there are two aspects of article design: *visual* and *logical*.

As an example, let us look at a theorem from `firstarticle.tex` (see the typeset form of the theorem on page 4). You tell LaTeX that you want to state a theorem by using a `theorem` environment:

```
\begin{theorem}\label{T:technical}
Let $L$ be a finite lattice.


...
\end{theorem}
```

The logical part of the design is choosing to define a theorem by placing material inside a `theorem` environment. For the visual design, LaTeX makes hundreds of decisions. Could you have specified all of the spacing, font size changes, centering, numbering, and so on? Maybe, but would you *want* to? And would you want to repeat that process for every theorem in your document?

Even if you did, you would have spent a great deal of time and energy on the *visual design* of the theorem rather than on the *logical design* of your article. The idea behind LaTeX is that you should concentrate on what you have to say and let LaTeX take care of the visual design.

This approach allows you to easily alter the visual design by changing the document class (or its options, see Sections 9.5, 10.1.2, and 17.1). Section 9.1 provides some examples. If you code the visual design into the article—hard coding it, as a programmer would say—such changes are much harder to accomplish, for you and for the journal publishing the article.

For more on this topic, see Section C.4.

2.4 General error messages

Now that you are ready to type your first document, we give you some pointers on using LaTeX.

You will probably make a number of mistakes in your first document. These mistakes fall into the following categories:

1. Typographical errors, which LaTeX blindly typesets.

2. Errors in formulas or in the formatting of the text.

3. Errors in your instructions to LaTeX, that is, in commands and environments.

Typographical errors can be corrected by viewing and spell checking the source file, finding the errors, and then editing the typeset file. Mistakes in the second and third categories may trigger errors during the typesetting process, such as lines too wide of Section 1.4.

We now look at some examples of the third class of errors by deliberately introducing a number of mistakes into `firstarticle.tex` and examining the messages.

Experiment 1. In `firstarticle.tex`, go to line 19 (use the `Go to Line` command of your editor) and remove the closing brace so that it reads `\begin{abstract`

When you typeset `firstarticle.tex`, LaTeX reports a problem:

```
{abstract We present a technical result for congruences on\ETC.
./firstarticle.tex:23:
Paragraph ended before \begin was complete.
<to be read again>
                   \par
l.23
```

Line 23 of the file is the line after `\maketitle`. The message informs you that the environment name was not completed.

`Runaway argument?` is a message that comes up often. It means that the argument of a command is either longer than expected or it contains material the argument cannot accept. Most often a closing brace solves the problem, as in this experiment.

Experiment 2. Now restore line 19, then go to line 21 and change `\end{abstract}` to `\end{abstrac}` and typeset again. LaTeX informs you of another error:

```
./firstarticle.tex:21: LaTeX Error: \begin{abstract}
on input line 19 ended by \end{abstrac}.

See the LaTeX manual or LaTeX Companion for explanation.
Type  H <return>  for immediate help.
 ...
l.21 \end{abstrac}
```

This is perfect. LaTeX correctly analyzes the problem and tells you where to make the change.

Experiment 3. Correct the error in line 21, and introduce a new error in line 61. This line reads

```
z_1 \equiv y+ w \pmod{\delta}.
```

Change `\delta` to `\deta`. Now, when you typeset the document, LaTeX reports

```
./firstarticle.tex:61: Undefined control sequence.
<argument> {\operator@font mod}\mkern 6mu\deta

l.61 z_1 \equiv y+ w \pmod{\deta}
```

This mistake is easy to identify: `\deta` is a misspelling of `\delta`.

Experiment 4. In line 38, delete the closing brace of the `\label` command. This results in a message:

```
Runaway definition?
->E:cover\text {If $x$ is covered by $y,z \in L$ and\ETC.
! File ended while scanning definition of \df@label.
<inserted text>
                  }
<*> firstarticle.tex
```

Undo the change to line 38.

Experiment 5. Add a blank line following line 61:

```
    x+ z = z + z_1 \equiv z + (y+ w) = y+ z \pmod{\delta},
```

This change results in the message

```
./firstarticle.tex:62: Missing $ inserted.
<inserted text>
                  $
l.62
```

There can be no blank lines within a displayed math environment. LaTeX catches the mistake, but the message itself is misleading.

Experiment 6. Add a $ before `\pmod` in line 61 (such an error often occurs when cutting and pasting a formula). You get the message:

```
./firstarticle.tex:61: Display math should end with $$.
<to be read again>
                      \penalty
l.61 z_1 \equiv y+ w $\pmod{\delta}
```

Maybe this could be more to the point?

Tip LaTeX's messages are not very useful with displayed formulas. Comment out some of the lines to try to localize the problem.

Tip Typeset often.

Typesetting my book *First Steps into* LaTeX with the closing brace of the first `\caption` command on line 480 of the source file missing, I get the error message

```
! Text line contains an invalid character.
l.1227 ...pletely irreducible^^?
```

where the reference is to line 1227, about 700 lines removed from the actual error. However, if the only thing I did before typesetting was to insert that figure with its incorrect caption command, at least I would know where to look for errors. If you make a dozen changes and then typeset, you may not know where to start.

2.5 *Errors in math*

Even in such a simple note there are opportunities for errors. To help familiarize yourself with some of the most commonly seen LaTeX errors in formulas, we introduce mistakes into `formulanote.tex`.

Experiment 1 In line 6 of `formulanote.tex`, delete the third `$` symbol; save the file under the name `formulanotebad1.tex` in the `work` folder.

Typeset `formulanotebad1.tex`. LaTeX generates the following message:

```
! Missing $ inserted.
<inserted text>
                $
l.6 as $(u, v)$ and  (u, \infty
                              )$. Such an interval
```

LaTeX reads `(u, \infty)` as text; but the `\infty` command instructs LaTeX to typeset a math symbol, which can only be done in a formula. So LaTeX offers to put a `$` in front of `\infty` while typesetting the source file—it does not put the `$` in the source file itself. LaTeX attempts a cure, but in this example it comes too late, because the formula *should* start just before `(u`.

Experiment 2 In line 16 of `formulanote.tex`, delete the second `}` symbol and save it under the name `formulanotebad2.tex` in the `work` folder. This introduces an error: the closing brace of the subscript (see page 14) is missing. Now typeset the note. You get the message

```
Missing } inserted.
<inserted text>
                }
l.12 such as $\lim_{x \to \infty f(x)$
```

LaTeX reports that a closing brace (`}`) is missing, but it is not sure where the brace should be. LaTeX noticed that a subscript started with `{`, but it reached the end of the formula before finding a closing brace `}`. To remedy this, you must look in the formula for an opening brace `{` that is not balanced, and insert the missing closing brace `}`. Make the necessary change and typeset again to view the difference.

Experiment 3 In `mathnote.tex`, delete the two `$` signs in line 19, that is, replace `$a$` by a. Typeset the file. It typesets with no errors. Here is the last line of the typeset file you get:

we assume that the function is defined and continuous in a neighborhood of a.

instead of

we assume that the function is defined and continuous in a neighborhood of a.

This is probably the error most often made by beginners. There is no message by LaTeX and the typeset version looks good. Notice the difference in the shape of the letter a in the two cases. You need sharp eyes to catch such an error.

Tip After an error is corrected, LaTeX can refuse to typeset your document. If your document is `document.tex`, look in the same folder for the *auxiliary file* `document.aux` that was created by LaTeX. Delete `document.aux` and **typeset twice**. See Section 2.2.

2.6 Your errors: Davey's Dos and Don'ts

Based on his many years of experience correcting LaTeX articles for the journal *Algebra Universalis*, Brian Davey collected the LaTeX mistakes most often made by authors. Here are some items from his list, divided into three categories.

Commands

1. Place ALL custom commands and environments in the preamble!
   If you have trouble with custom commands, then you know where to find them.

2. Don't use `\def`; rather use `\newcommand` or `\renewcommand`.
   `\def` is a TeX command. It is like `\newcommand` (see Sections 1.9 and 14.1), but it can redefine an existing command. Redefining your own commands is bad enough, redefining a TeX command can be a disaster.

3. Do not simply type the name of an operator into a formula. Declare the appropriate operator; see Sections 1.9 and 14.1.
   For instance, do not type `$length I$`; it typesets as $length I$. It should be length I, typed as `$\length I$`. Of course, you have to add

   ```
   \DeclareMathOperator{\length}{length}
   ```

   to the preamble (see Section 1.8).

4. When you send a document to a coauthor or submit an article to a journal, remove all the custom commands not used.
   This is a real time saver for your coauthor and editor.

Text

1. Do not produce a list with horizontal and vertical spacing commands. Use a list environment; see Sections 3.8 and 4.2.

2. Do not type numbers for citations and internal references. Use `\cite{...}` for citations and `\ref{...}` for references. For references to equations, use `\eqref`; see Sections 1.7.1 and 5.3.

3. Do not number proclamations (see Section 4.4). Use the standard `amsart` environments for theorems, and so on, and let LaTeX number them.

4. When writing a document for a journal requiring a document class file, **do not**

   (a) change any of the size parameters: for instance, do not use options like `12pt` to change the font size or the `\setlength` command to change any parameter of the page size;

   (b) insert vertical white space via `\bigskip`, `\smallskip`, `\vskip`, `\vspace`, etc, nor via your own custom commands. Do not adjust horizontal space without a very good reason.

   So if you want to display some text:

   Please, display this text.

   don't do this:

   ```
   \medskip
   \hspace*{6pt} Please, display this text.
   \medskip
   ```

   but rather

   ```
   \begin{itemize}
   \item[] Please, display this text.
   \end{itemize}
   ```

   or

   ```
   \begin{quote}
   Please, display this text.
   \end{quote}
   ```

5. Do not leave a blank line before `\end{proof}` or before a text environment (see Section 4.1).

6. Do not use the `geometry` package.

Formulas

1. Do not leave a blank line before a displayed formula.

2. Don't use the symbol `|` in a set description, use the binary relation `\mid`; see Section 5.5.4.
   For instance, `\{ x | x^2 < 2 \}` typesets as $\{x|x^2 < 2\}$. The correct form is $\{x \mid x^2 < 2\}$, typed as `$\{x \mid x^2 < 2\}$`.

3. Don't put punctuation marks inside an inline math environment.
   For instance, $\sin x.$ typed as `$\sin x.$`; use `$\sin x$`. This typesets as $\sin x$. Notice the smaller space between "$\sin x$." and "typed" and the wider space between "`$\sin x$.`" and "This"; see Sections 1.3 and 3.2.2.

4. Don't use two or more displayed formulas one after another. Use an appropriate environment such as `\align`, `\alignat`, `\gather`, and so on (see Section 7.1.1).

5. Don't use `\left\{`, `\right\}`, `\left(`, `\right)`, and so on, by default (see page 15 and Section 5.5.1 for the commands `\left` and `\right`). Even when `\left` and `\right` do not change the size of the symbol, they add extra space after the closing delimiter.

6. Use `\colon` for functions. For instance, `$f(x) \colon x \to x^2$` typesets as $f(x)\colon x \to x^2$. If you type `$f(x) : x \to x^2$`, you get $f(x) : x \to x^2$; the spacing is bad.

7. Use `\[` and `\]` (or `equation*`) to type a displayed math environment (see Section 1.7) rather than the old TeX `$$` matched by `$$`. While display math produced via the latter does work properly most of the time, there are some LaTeX commands that do not; for example, `\qedhere`.

8. Do not use the `center` environment to display formulas.

9. Use `\dots` first and let LaTeX make the decision whether to use `\dots` or `\cdots`; see page 15 and Section 5.4.3. If LaTeX gets it wrong, then use `\cdots` or `\ldots`.

10. If you can, avoid constructs (for instance, $\stackrel{\text{up}}{\to}$) in inline formulas that disrupt the regular line spacing. Although LaTeX automatically leaves room for it, it does not look good, as a rule.

PART II

Text and Math

CHAPTER

3

Typing text

In Chapter 1, we briefly discussed how to type text in a document. Now we take up this topic more fully.

This chapter starts with a discussion of the keyboard in Section 3.1 and continues with the rules for spaces in Section 3.2. We cover a very important topic that must precede any in-depth discussion of LaTeX, how to control LaTeX with commands and environments, in Section 3.3.

A document may contain symbols that cannot be found on your keyboard. In Section 3.4, we show how to get these symbols in our typeset documents by using commands.

Some other characters are defined by LaTeX as command characters. For example, the `%` character plays a special role in the source document. In Section 3.5.1, you will see how `%` is used to comment out lines. In Section 3.5.2, we introduce the command for footnotes.

In Section 3.6, we discuss the commands (and environments) for changing fonts, their shapes and sizes. In Section 3.7, you learn about lines, paragraphs, and pages. The judicious use of horizontal and vertical spacing is an important part of document

G. Grätzer, *More Math Into* LaTeX, DOI 10.1007/978-3-319-23796-1_3

formatting, and also the topic of Section 3.8. In Section 3.9, you learn how to typeset text in a "box", which behaves as if it were a single large character.

To help the discussion along, we shall use the terms *text mode* and *math mode* to distinguish between typesetting text and math.

3.1 The keyboard

Most of the keys on your computer's keyboard produce characters, while others are function or modifier keys.

3.1.1 Basic keys

The basic keys are grouped as follows:

Letters The 52 letter keys:

```
a b c ... z    A B C ... Z
```

Digits The ten digits:

```
1 2 ... 9 0
```

Old-style digits are available with the `\oldstylenums` command. The next line shows the default digits followed by the old style digits:
1234567890 1234567890
typed as

```
1234567890 \quad \oldstylenums{1234567890}
```

Punctuation There are nine punctuation marks:

```
, ; . ? ! : ' ' -
```

The first six are the usual punctuation marks. The ' is the *left single quote*—also known as the *grave accent*—while ' doubles as the *right single quote* and *apostrophe* (see Section 3.4.1). The - key is the *dash* or *hyphen* (see Sections 3.4.2 and 3.4.9).

Parentheses There are four:

```
( ) [ ]
```

(and) are *parentheses;* [and] are called (*square*) *brackets.*

Math symbols Seven math symbols correspond to keys. The math symbols are:

$* \quad + \quad = \quad - \quad < \quad > \quad /$

The last four characters have a role also in text mode:

- The minus sign – corresponds to the hyphen key, - (see Section 3.4.9).
- The math symbols < and > correspond to the keys < and >; use them only in math mode.

Note that there is also a version of colon (:) for math formulas (see Sections 6.1 and A.2).

Space keys Pressing the space bar gives the *space character.* Pressing the tab key gives the *tab character.* When typesetting the source file, LaTeX does not distinguish between these two. Pressing the *Return* key gives the *end-of-line character.* These keys produce *invisible characters* that are normally not displayed on your monitor by the text editor. Different computer systems have different end-of-line characters, which may cause some problems when transferring files from one system to another. A good text editor translates end-of-line characters automatically or on demand. Section 3.2.1 explains how LaTeX handles the invisible characters.

When explaining some rules of LaTeX, sometimes it is important to show if a space is required. In such cases, I use the symbol verb* ␣ to indicate a space, for instance, `\in␣ut` and `\␣`.

The tilde ~ produces a *nonbreakable space* or *tie* (see Section 3.4.3).

3.1.2 *Special keys*

There are 13 special keys on the keyboard:

```
# $ % & ~ _ ^ \ { } @ " |
```

They are mostly used to give instructions to LaTeX and some are used in math mode (see Chapter 5), and some in BibTeX (see Chapter 15). See Section 3.4.4 on how to print these characters in text. Only @ requires no special command, type `@` to print @.

3.1.3 *Prohibited keys*

Keys other than those discussed in Sections 3.1.1 and 3.1.2 are prohibited! Specifically, do not use the computer's modifier keys—Control, Alt, Escape, and others—to produce special characters, such as accented characters. LaTeX will either reject or misunderstand them. Prohibited characters may not cause problems in some newer LaTeX implementations. They may just print ï£¡ if your source file has ï£¡, and ignore the invisible invalid characters. However, for portability reasons, you should avoid using prohibited characters. The `babel` package provides support for using some modifier keys (see Appendix F).

Tip If there is a prohibited character in your document, you may receive a message such as

```
! Text line contains an invalid character.
l.222 completely irreducible^^?
                                    ^^?
```

Delete and retype the offending word or line until the error goes away.

3.2 Words, sentences, and paragraphs

Text consists of words, sentences, and paragraphs. In text, *words* are separated by one or more spaces, which may include a single end-of-line character (see the rule, **Spacing in text**), or by parentheses and punctuation marks. A group of words terminated by a period, exclamation point, or question mark forms a *sentence* (not all periods terminate a sentence, see the discussion in Section 3.2.2). A group of sentences terminated by one or more blank lines constitutes a *paragraph*.

3.2.1 Spacing rules

Here are the most important LaTeX rules about spaces in text in the source file.

Rule ■ Spacing in text

1. Two or more spaces in text are the same as one.
2. A tab or end-of-line character is the same as a space.
3. A blank line, that is, two end-of-line characters separated only by spaces and tabs, indicates the end of a paragraph. The `\par` command is equivalent.
4. Spaces at the beginning of a line are ignored.

Rules 1 and 2 make cutting and pasting text less error-prone. In your source file, you do not have to worry about the line length or the number of spaces separating words or sentences, as long as there is at least one space or end-of-line character separating any two words. Thus

```
 You   do not have to    worry
 about the number of   spaces
separating words, as long as there
is   at least one space or end-of-line character
separating  any two words.
```

produces the same typeset text as

```
You do not have to worry about the number of spaces
separating words, as long as there is at least one space
or end-of-line character separating any two words.
```

However,

```
the number of    spaces separating words,
as long
```

and

```
the number of    spaces separating words,
as long
```

produce different results:

the number of spaces separating words, as long
the number of spaces separating words , as long

Notice the space between "words" and the comma in the second line. That space was produced by the end-of-line character in accordance with Rule 2.

It is very important to maintain the readability of your source file. LaTeX may not care about the number of spaces or line length, but you, your coauthor, and your editor might.

Rule 3 contradicts Rules 1 and 2, consider it an exception. Sometimes—especially when defining commands and environments (see Sections 14.1 and 14.2)—it is more convenient to indicate the end of a paragraph with `\par`.

3.2.2 Periods

LaTeX places a certain size space between words—the *interword space*—and a somewhat larger space between sentences—the *intersentence space*. To know which space to use, LaTeX must decide whether or not a period indicates the end of a sentence.

Rule 1 ■ Period

To LaTeX, a period after a capital letter, for instance, A. or caT., signifies an abbreviation or an initial. Generally, every other period signifies the end of a sentence.

This rule works most of the time. When it fails—for instance, twice with e.g.—you need to specify the type of space you want, using the following two rules.

Rule 2 ■ Period

If an abbreviation does not end with a capital letter, for instance, `etc.`, and it is not the last word in the sentence, then follow the period by an interword space (`\␣`) or a tie (`~`), if appropriate (see Section 3.4.3).

Recall that \␣ provides an interword space.

```
The result was first published, in a first approximation,
in the Combin.\ Journal. The result was first published,
in a first approximation, in the Combin. Journal.
```

prints as

The result was first published, in a first approximation, in the Combin. Journal.
The result was first published, in a first approximation, in the Combin. Journal.

Notice that Combin. in the first line is followed by a regular interword space. The intersentence space following Combin. in the second line is a little wider; it is an error.

A tie (or nonbreakable space)—see also Section 3.4.3—is more appropriate than \␣ in phrases such as Prof. Smith, typed as `Prof.~Smith`, and pp. 271–292, typed as `pp.~271--292`.

Tip The `thebibliography` environment handles periods properly. You do not have to mark periods for abbreviations (in the form `.\␣`) in the name of a journal, so

```
Acta Math. Acad. Sci. Hungar.
```

is correct.

Rule 3 ■ Period

If a capital letter is followed by a period and is at the end of a sentence, precede the period with `\@`.

For example,

```
(1) follows from condition~H\@. We can proceed\\
(1) follows from condition~H. We can proceed
```

prints:

(1) follows from condition H. We can proceed
(1) follows from condition H. We can proceed

Notice that there is not enough space after H. in the second line.

Most typographers agree on the following rule (see, for instance, *The Elements of Typographic Style* by Robert Bringhurst [8], p. 30):

Rule 4 ■ Period

Add no space or a thin space (`\,`) within strings of initials and be consistent.

So W.H. Lampstone with no space or W. H. Lampstone with thin space is preferred over W. H. Lampstone. My personal choice is W. H. Lampstone with thin space.

To make all intersentence spaces equal to the interword space—as required in French typography—you can use the command

```
\frenchspacing
```

To switch back to using spaces of different sizes, give the command

```
\nonfrenchspacing
```

3.3 Commanding LaTeX

How do you command LaTeX to do something special for you, such as starting a new line, changing emphasis, or displaying the next theorem? You use *commands* and special pairs of commands called *environments*, both briefly introduced in Section 1.1.4.

Most, but not all, commands have *arguments,* which are usually fairly brief. Environments have *contents*, the text between the `\begin` and `\end` commands. The contents of an environment can be several paragraphs long.

3.3.1 Commands and environments

The `\emph{`*text*`}` *command* instructs LaTeX to emphasize its argument, *text*. The `\&` command has no argument; it instructs LaTeX to typeset & (see Section 3.4.4).

The `flushright` *environment* instructs LaTeX to right justify the content, the text between the two commands

```
\begin{flushright}
\end{flushright}
```

The content of the `document` environment is the body of the article (see Section 1.8) and the content of the `abstract` environment is the abstract.

Rule ■ Environments

An environment starts with the command

```
\begin{name}
```

and ends with

```
\end{name}
```

Between these two commands is the *content* of the environment, affected by the definition of the environment.

Rule ■ Commands

A LaTeX command starts with a backslash, \, and is followed by the *command name*. The *name* of a command is either a *single non-alphabetic character* other than a tab or end-of-line character or a *string of letters*, that is, one or more letters.

So # and ' are valid command names. The corresponding commands \# and \' are used in Sections 3.4.4 and 3.4.7, respectively. `input` and `date` are also valid command names. However, `input3`, `in#ut`, and `in␣ut` are not valid names because 3, #, and ␣ should not occur in a multicharacter command name. Note that ␣ is a command name, the command \␣ produces a blank.

LaTeX has a few commands, for instance, $ (see Section 5.1) that do not follow this naming scheme, that is, they are not of the form *name*. See also Section 14.1.9 for special commands with special termination rules.

Rule ■ Command termination

LaTeX finds the end of a command name as follows:

- If the first character of the name is not a letter, the name is the first character.
- If the first character of the name is a letter, the command name is terminated by the first nonletter.

If the command name is a string of letters, and is terminated by a space, then LaTeX discards all spaces following the command name.

While `input3` is an invalid name, `\input3` is not an incorrect command. It is the `\input` command followed by the character 3, which is either part of the text following the command or the argument of the command.

LaTeX also allows some command names to be modified with *. Such commands are referred to as **-ed commands*. Many commands have *-ed variants. `\hspace*` is an often-used *-ed command (see Section 3.8.1).

Rule ■ Command and environment names

Command and environment names are *case sensitive*. `\ShowLabels` is not the same as `\showlabels`.

Rule ■ Arguments

Arguments are enclosed in braces, { }.
Optional arguments are enclosed in brackets, [ ].

Commands may have *arguments,* typed in braces immediately after the command. The argument(s) are used in processing the command.

Accents provide very simple examples. For instance, `\'{o}`—which produces ó—consists of the command `\'` and the argument `o` (see Section 3.4.7). In the command

```
\bibliography{article1}
```

the command is `\bibliography` and the argument is `article1` (see Section 15.2.2).

Sometimes, if the argument is a single character, the braces can be dropped: `\'o` also typesets as ó.

Some environments also have arguments. For example, the `alignat` environment (see Section 7.5.4) is delimited by the commands

```
\begin{alignat}{2}
```

and

```
\end{alignat}
```

The argument, 2, is the number of columns—it could be any number 1, 2, ... A command or environment may have more than one argument. The `\frac` command (see Section 5.4.1) has two, `$\frac{1}{2}$` typesets as $\frac{1}{2}$. The custom command `\con` has three (see Section 14.1.2).

Some commands and environments have one or more *optional arguments,* that is, arguments that may or may not be present. The `\sqrt` command (see Section 5.4.5) has an optional argument for specifying roots other than the square root. To get $\sqrt[3]{25}$, type `\sqrt[3]{25}`. The `\documentclass` command has an argument, the name of a document class, and an optional argument, a list of options (see Section 8.2), for instance,

```
\documentclass[12pt,draft,leqno]{amsart}
```

Tip If you get an error when using a command, check that:

1. The command is spelled correctly, including the use of uppercase and lowercase letters.
2. You have specified all required arguments in braces.
3. Any optional argument is in brackets, not braces or parentheses.
4. The command is properly terminated.
5. The package providing the command is loaded with the `\usepackage` command.

Most errors in the use of commands are caused by breaking the termination rule. We can illustrate some of these errors with the `\today` command, which produces today's date. You have already seen this command in Section 1.3 (see also Section 3.4.8). The correct usage is

```
\today\ is the day
```

or

```
\today{} is the day
```

which both typeset in the following form

July 19, 2015 is the day

In the first case, `\today` was terminated by `\␣`, the command that produces an interword space. In the second case, it was terminated by the *empty group* `{}`.

If there is no space after the `\today` command, as in

```
\todayis␣the␣day
```

you get the message

```
! Undefined control sequence.
1.3 \todayis
             the day
```

LaTeX thinks that `\todayis` is the command, and, of course, does not recognize it.

If you type one or more spaces after `\today`:

```
\today␣␣is␣the␣day
```

LaTeX interprets the two spaces as a single space by the first space rule (see page 48), and uses that one space to delimit `\today` from the text that follows it. So LaTeX produces

July 19, 2015is the day

Section 14.1.9 discusses how best to avoid such errors.

Tip If a command—or environment—can have an optional argument and

- none is given, and
- the text following the command starts with `[`,

then type this as `{[}`.

This may happen, for instance, with the command `\item` (see page 103). To get an example for an environment, see Section 7.6.1 for subsidiary math environments and page 211 for the alignment options. See what happens if no option is given but the math starts, say, with `[x]`.

3.3.2 *Scope*

A command issued inside a pair of braces `{ }` has no effect beyond the right brace, except for the rare *global* commands (see Section 3.3.3). You can have any number of pairs of braces:

```
{ ... { ... { ... } ... } ... }
```

The innermost pair containing a command is the *scope* of that command. The command has no effect outside its scope. We can illustrate this concept using the `\bfseries` command that switches the font to boldface:

```
{some text \bfseries bold text} no more bold
```

typesets as

some text **bold text** no more bold

The commands `\begin{name}` and `\end{name}` bracketing an environment act also as a pair of braces. In particular, `$`, `\[`, and `\]` are special braces.

Rule ■ Braces

1. Braces must be balanced: An opening brace has to be closed, and a closing brace must have a matching opening brace.
2. Pairs of braces cannot overlap.

Violating the first brace rule generates warnings and error messages. If there is one more opening brace than closing brace, the document typesets, but you get a warning:

```
(\end occurred inside a group at level 1)
```

For two or more unmatched opening braces, you are warned that \end occurred inside a group at level 2, and so on. There is a tendency to disregard such warnings since your article is already typeset and the unmatched opening brace may be difficult to find. However, such errors may have strange consequences. At one point in the writing of my second LaTeX book, there were two extra opening braces in Chapter 2.4. As a result, the title of Chapter 7 was placed on a page by itself! So it is best not to disregard such warnings.

If you have one unmatched closing brace, you get a message such as

```
! Too many }'s
```

If special braces, say, \begin{name} and \end{name}, do not balance, you get an error message:

```
! LaTeX Error: \begin{name} on input line 21
ended by \end{document}.
```

or

```
! LaTeX Error: \begin{document} ended by \end{name}.
```

To illustrate the second rule, here are two simple examples of overlapping braces.

Example 1

```
{\bfseries some text
\begin{lemma}
   more text} final text
\end{lemma}
```

Example 2

```
{some \bfseries text, then math: $\sqrt{2} }, \sqrt{3}$
```

In Example 1, the scope of \bfseries overlaps the braces \begin{lemma}, \end{lemma} In Example 2, the scope of \bfseries overlaps the special braces $ and $. Example 1 is easy to correct:

```
{\bfseries some text}
\begin{lemma}
   {\bfseries more text}
    final text
\end{lemma}
```

Example 2 may be corrected as follows:

```
{some \bfseries text, then math:} $\sqrt{2}, \sqrt{3}$
```

Actually, $\sqrt{2}$ does not even have a bold version (see Section 6.4.3).

If the braces do overlap and they are of the same kind, LaTeX simply misunderstands the instructions. The closing brace of the first pair is regarded as the closing brace of the second pair, an error that may be difficult to detect. LaTeX can help if special braces overlap. Typesetting Example 1 gives the message

```
! Extra }, or forgotten \endgroup.
l.7 more text }
               final text
```

3.3.3 Types of commands

It may be useful at this point to note that commands can be of various types.

Some commands have arguments, and some do not. Some commands effect change only in their arguments, while some commands declare a change.

For instance, `\textbf{This is bold}` typesets the phrase `This is bold` in bold type: **This is bold** and has no effect on the text following the argument of the command. On the other hand, the command `\bfseries` declares that the text that follows should be bold. This command has no argument. I call a command that declares change a *command declaration.* So `\bfseries` is a command declaration, while `\textbf` is not. As a rule, command declarations are commands without arguments.

Commands with arguments are called *long* if their argument(s) can contain a blank line or a `\par` command; otherwise they are *short.* For example, `\textbf` is a short command. So are all the top matter commands discussed in Section 9.2. The `\parbox` command, discussed in Section 3.9.4, is long.

Finally, as discussed in Section 3.3.2, the effect of a command remains within its scope. This is true only of *local* commands. There are also some *global* commands, such as the `\setcounter` command described in Section 14.5.1.

Fragile commands

As a rule, LaTeX reads a paragraph of the source file, typesets it, and then goes on to the next paragraph (see Section C.4). Some information from the source file, however, is separately stored for later use. Examples include the title of an article, which is reused as a running head (Section 9.2.1); titles of parts, sections, subsections, and other sectioning commands, which are used in the table of contents (Sections 17.2 and 8.4.1); footnotes (Section 3.5.2); table and figure captions (Section 8.4.3), which are used in lists of tables and figures (Section 8.4.3); and index entries (Chapter 16).

These are *movable arguments,* and certain commands embedded in them must be protected from damage while being moved. LaTeX commands that need such protection are called *fragile.* The inline math delimiter commands `\(` and `\)` are fragile, while $ is not.

In a movable argument, fragile commands must be protected with a `\protect` command. Thus

```
The function \( f(x^{2}) \)
```

is not an appropriate section title, but

```
The function \protect \( f(x^{2}) \protect \)
```

is. So is

```
The function $f(x^{2})$
```

To be on the safe side, you should protect every command that might cause problems in a movable argument. Section 17.2 shows an example of what happens if a fragile command is not protected. Alternatively, use commands declared with

```
\DeclareRobustCommand
```

This command works the same way as `\newcommand` but the command defined is *robust*, that is, not fragile.

3.4 Symbols not on the keyboard

A typeset document may contain symbols that cannot be typed. Some of these symbols may even be available on the keyboard but you are prohibited from using them (see Section 3.1.3). In this section, we discuss the commands that typeset some of these symbols in text.

3.4.1 Quotation marks

To produce single and double quotes, as in

‘subdirectly irreducible’ and “subdirectly irreducible”

type

```
`subdirectly irreducible' and ``subdirectly irreducible''
```

Here, ` is the left single quote and ' is the right single quote. Note that the double quote is obtained by pressing the single quote key twice, and *not* by using the double quote key. If you need single and double quotes together, as in “She replied, ‘No.’ ”, separate them with `\,` (which provides a thin horizontal space, see Section 3.8.1):

```
``She replied, `No.'\,''
```

3.4.2 Dashes

Dashes come in three lengths. The shortest dash, called a *hyphen*, is used to connect words:

Mean-Value Theorem

This phrase is typed with a single dash:

```
Mean-Value Theorem
```

A medium-sized dash, called an *en dash*, is typed as -- and is used

- For number ranges; for instance, the phrase see pages 23–45, is typed as

  ```
  see pages~23--45
  ```

  Note: ~ is a nonbreakable space or tie (see Section 3.4.3).

- In place of a hyphen in a compound adjective when one of the elements of the adjective is an open compound (such as New York) or hyphenated (such as non-English). For instance, the phrase Schmidt–Freid adjoint, is typed as

  ```
  Schmidt--Freid adjoint
  ```

A long dash—called an *em dash*—is used to mark a change in thought or to add emphasis to a parenthetical clause, as in this sentence. The two em dashes in the last sentence are typed as follows:

```
A long dash---called an \emph{em dash}---is used
```

In math mode, a single dash is typeset as the minus sign $-$ (a binary operation) with some spacing on both sides, as in $15 - 3$ or the "negative" as in -3; this is discussed in Sections 3.1.1 and 5.4.1.

Note that there is no space before or after an en dash or em dash.

3.4.3 Ties or nonbreakable spaces

A *tie* or *nonbreakable space* or *blue space* is an interword space that cannot be broken across lines. For instance, when referencing P. Neukomm in an article, you do not want the initial P. at the end of a line and the surname Neukomm at the beginning of the next line. To prevent such an occurrence, you should type `P.~Neukomm`.

If your keyboard does not have ~, use the `\nobreakspace` command instead, and type `P.\nobreakspace Neukomm`. The following examples show some typical uses:

```
Theorem~\ref{T:main} in Section~\ref{S:intro}

Donald~E. Knuth

assume that $f(x)$ is (a)~continuous, (b)~bounded

the lattice~$L$

Sections~\ref{S:modular} and~\ref{S:distributive}

In~$L$, we find
```

Of course, if you add too many ties, as in

```
Peter~G.~Neukomm% Incorrect!
```

LaTeX may send you a line too wide message (see Sections 1.4 and 3.7.1).

The tie (`~`) absorbs spaces, so typing `P.␣~␣Neukomm` works just as well. This feature is convenient when you add a tie during editing.

3.4.4 Special characters

The characters corresponding to nine of the 13 special keys (see Sections 1.2 and 3.1.2) are produced by typing a backslash (\) and then the key, as shown in Table 3.1.

If for some reason you want to typeset a backslash in your document, type the command `\textbackslash`, which typesets as \. You might think that you could get a typewriter style backslash by utilizing the `\texttt` command introduced in Section 1.3

```
\texttt{\textbackslash}
```

but this is not the case, `\textbackslash` and `\texttt{\textbackslash}` produce the same symbol, \, which is different from the typewriter style backslash: `\`. Look at them side-by-side: \ `\`. For a typewriter style backslash you can use the `\bsl` command introduced in Section 14.1.1 or the `\texttt{\symbol{92}}` command introduced later in this section.

The | key is seldom used in text. If you need to typeset the math symbol | in text, type `\textbar`.

Name	Type	Typeset
Ampersand	`\&`	&
Caret	`\^{}`	^
Dollar Sign	`\$`	$
Left Brace	`\{`	{
Right Brace	`\}`	}
Underscore (or Lowline)	`\_`	_
Octothorp	`\#`	#
Percent	`\%`	%
Tilde	`\~{}`	~

Table 3.1: Nine special characters.

Note that in text, `*` typesets as *, whereas in a formula it typesets centered as $*$. To typeset a centered star in text, use the command `\textasteriskcentered`. And `@` typesets as @.

Finally, the `"` key should never be used in text. See Section 3.4.1 for the proper way to typeset double quotes (see also Section 1.3). Nevertheless, sometimes `"` may be used to typeset `"`, as in the computer code segment `print("Hello!")`. In BIBTEX and *MakeIndex*, `"` has special meanings (see Chapters 15 and 16).

Tip Be careful when typing `\{` and `\}` to typeset the braces { }. Typing a brace without its backslash results in unbalanced braces, in violation of the first brace rule in Section 3.3.2.

We illustrated in Section 3.3.2 some consequences of unbalanced braces. You may avoid some of these problems by introducing custom commands, as introduced in Section 14.3.

You can also produce special characters with the `\symbol` command:

`\symbol{94}` typesets as ^

`\symbol{126}` typesets as ~

`\texttt{\symbol{92}}` typesets as \

The argument of the `\symbol` command is a number matching the slot of the symbol in the layout (encoding) of the font. The layout for the Computer Modern typewriter style font is shown in Table 3.2.

Alternatively, instead of `\texttt{\symbol{92}}`, we can use

```
\texttt{\char`\\}
```

	0	1	2	3	4	5	6	7	8	9
x	`	´	ˆ	˜	¨	˝	˚	ˇ	˘	¯
1x	˙	¸	˛	‚	‹	›	“	”	„	«
2x	»	–	—		₀	ı	ȷ	ff	fi	fl
3x	ffi	ffl	␣	!	"	#	$	%	&	’
4x	(	)	*	+	,	-	.	/	0	1
5x	2	3	4	5	6	7	8	9	:	;
6x	<	=	>	?	@	A	B	C	D	E
7x	F	G	H	I	J	K	L	M	N	O
8x	P	Q	R	S	T	U	V	W	X	Y
9x	Z	[	\	]	^	_	‘	a	b	c
10x	d	e	f	g	h	i	j	k	l	m
11x	n	o	p	q	r	s	t	u	v	w
12x	x	y	z	{	|	}	~	-		

Table 3.2: Font table for the Computer Modern typewriter style font.

Any character x in the font can be accessed by typing the character itself as `\x. This way you don't have to look up the position of the symbol.

You can obtain similar tables for any font in your LaTeX implementation by using the `fonttbl.tex` file in your `samples` folder. The table format in this file is used in Section 4.6 as an example of the `tabular` environment.

For more about font tables, see the `nfssfont.tex` file, part of the standard LaTeX distribution (see Section 10.3) and see also Section 7.5.7 of *The LaTeX Companion,* 2nd edition [56].

3.4.5 Ellipses

The text ellipsis, …, is produced using the `\dots` command. Typing three periods produces ... (notice that the spacing is wrong).`\dots` is one of several commands that can be used to create ellipses in formulas (see Section 5.4.3).

3.4.6 Ligatures

Certain groups of characters, when typeset, are joined together—such compound characters are called *ligatures*. There are five ligatures that LaTeX typesets automatically (if you use the Computer Modern fonts): ff, fi, fl, ffi, and ffl.

If you want to prevent LaTeX from forming a ligature, separate the characters with the command`\textcompwordmark`. Compare iff with iff, typed as `iff` and

```
if\textcompwordmark f
```

Enclosing the second character in braces (`{}`) is a crude method of preventing the ligature, as used in Formula 4 of the *Formula Gallery* (see Section 5.10). This method, in some instances, may interfere with LaTeX's hyphenation algorithm.

3.4.7 Accents and symbols in text

LaTeX provides 15 European accents. Type the command for the accent (\ and a character), followed by the letter (in braces) on which you want the accent placed (see Table 3.3). For example, to get Grätzer György, type

```
Gr\"{a}tzer Gy\"{o}rgy
```

and to get Ö type `\"{O}`.

To place an accent on top of an i or a j, you must use the *dotless* version of i and j. These are obtained by the commands `\i` and `\j`: `\'{\i}` typesets as í and `\v{\j}` typesets as ǰ. Tables 3.4 and 3.5 list some additional text symbols and European characters available in LaTeX when typing text. Using localized versions of LaTeX, you get more accented and combined characters such as the Catalan geminated ell (see Appendix F).

Note that the `\textcircled` command (in Table 3.5) takes an argument. It seems to work best with a single lowercase character, like ⓐ or ⓐ. Capitals such as Ⓐ are not very satisfactory. Section 3.9.6 explains how to create the symbol Ⓐ.

3.4.8 Logos and dates

`\TeX` produces TeX, `\LaTeX` produces LaTeX, and `\LaTeXe` produces LaTeX2e (the original name of the current version of LaTeX). The `\AmS` command produces the logo AMS.

Remember to type `\TeX\␣` or `\TeX{}` if you need a space after TeX (similarly for the others). A better way to handle this problem is discussed in Section 14.1.1.

LaTeX also stores some useful numbers:

Name	Type	Typeset	Name	Type	Typeset
acute	`\'{o}`	ó	macron	`\={o}`	ō
breve	`\u{o}`	ŏ	overdot	`\.{g}`	ġ
caron/haček	`\v{o}`	ǒ	ring	`\r{u}`	ů
cedilla	`\c{c}`	ç	tie	`\t{oo}`	o͡o
circumflex	`\^{o}`	ô	tilde	`\~{n}`	ñ
dieresis/umlaut	`\"{u}`	ü	underdot	`\d{m}`	ṃ
double acute	`\H{o}`	ő	underbar	`\b{o}`	o̲
grave	`` \`{o} ``	ò			
dotless i	`\i`	ı	dotless j	`\j`	ȷ
	`\'{\i}`	í		`\v{\j}`	ǰ

Table 3.3: European accents.

Name	Type	Typeset	Type	Typeset
a-ring	`\aa`	å	`\AA`	Å
aesc	`\ae`	æ	`\AE`	Æ
ethel	`\oe`	œ	`\OE`	Œ
eszett	`\ss`	ß	`\SS`	SS
inverted question mark	`` ?` ``	¿		
inverted exclamation mark	`` !` ``	¡		
slashed L	`\l`	ł	`\L`	Ł
slashed O	`\o`	ø	`\O`	Ø

Table 3.4: European characters.

- `\time` is the time of day in minutes since midnight
- `\day` is the day of the month
- `\month` is the month of the year
- `\year` is the current year

You can include these numbers in your document by using the `\the` command:

```
Year: \the\year; month: \the\month; day: \the\day
```

produces a result such as

Year: 2015; month: 7; day: 11

Name	Type	Typeset
ampersand	`\&`	&
asterisk bullet	`\textasteriskcentered`	∗
backslash	`\textbackslash`	\
bar (caesura)	`\textbar`	|
brace left	`\{`	{
brace right	`\}`	}
bullet	`\textbullet`	•
circled a	`\textcircled{a}`	ⓐ
circumflex	`\textasciicircum`	^
copyright	`\copyright`	©
dagger	`\dag`	†
double dagger (diesis)	`\ddag`	‡
dollar	`\$`	$
double quotation left	`\textquotedblleft` or ``` `` ```	“
double quotation right	`\textquotedblright` or `''`	”
em dash	`\textemdash` or `---`	—
en dash	`\textendash` or `--`	–
exclamation down	`\textexclamdown` or ``!` ``	¡
greater than	`\textgreater`	>
less than	`\textless`	<
lowline	`\_`	_
midpoint	`\textperiodcentered`	·
octothorp	`\#`	#
percent	`\%`	%
pilcrow (paragraph)	`\P`	¶
question down	`\textquestiondown` or ``?` ``	¿
registered trademark	`\textregistered`	®
section	`\S`	§

Table 3.5: Extra text symbols.

Of more interest is the `\today` command, which produces today's date in the form: July 11, 2015. It is often used as the argument of the `\date` command (see Section 9.2.1).

Remember the termination rule (Rule 3 in Section 3.3.1).

```
today's date in the form: \today (you may want
```

produces

today's date in the form: July 11, 2006(you may want

To get the desired effect, type `\␣` or `{}` after the `\today` command:

```
today's date in the form: \today\ (you may want
```

3.4.9 Hyphenation

LaTeX reads the source file one line at a time until it reaches the end of the current paragraph and then tries to balance the lines (see Section C.2.2). To achieve this goal, LaTeX hyphenates long words using a built-in hyphenation algorithm, a database stored in the `hyphen.tex` file, and a long `\hyphenation` list in the AMS document classes. If you use a document class not containing such a list, copy the hyphenation list from `amsart.cls` to your document.

Rule ■ Optional hyphen

If you find that LaTeX cannot properly hyphenate a word, put *optional hyphens* in the word. An optional hyphen is typed as `\-`, and *allows* LaTeX to hyphenate the word where the optional hyphen is placed—and only at such points—if the need arises.

Examples: `data\-base`,`an\-ti\-thet\-ic`, `set\-up`

Note that:

- Optional hyphens prevent hyphenation at any other point in the word.
- Placing an optional hyphen in a particular occurrence of a word does not affect the hyphenation of any other occurrences of that word.

Rule ■ Hyphenation specifications

List the words that often need help in a command:

```
\hyphenation{set-up as-so-ciate}
```

All occurrences of the listed words following this command in your document are hyphenated as specified.

Note that in the `\hyphenation` command the hyphens are designated by `-` and not by `\-`, and that the words are separated by spaces not by commas.

You must use optional hyphens for words with accented characters, as in

```
Gr\"{a}t\-zer
```

Such words cannot be included in a `\hyphenation` list (unless you use the T1 font encoding—see Appendix F).

Rule ■ Preventing hyphenation

To *prevent* hyphenation of a word, put it in the argument of a `\text` command or place it unhyphenated in a `\hyphenation` command.

For example, type

```
\text{database}
```

if you do not want this instance of `database` hyphenated, or type

```
\hyphenation{database}
```

if you do not want LaTeX to hyphenate any occurrence of the word after this command in your document. Of course, typing `data\-base` overrides the general prohibition for this one instance.

You can have any number of `\hyphenation` commands in your document.

Tip LaTeX does not break a hyphenated word except at the hyphen, nor does it break a word followed by or preceding an em dash or en dash (see Section 3.4.2). LaTeX often needs help with such words.

Sometimes a hyphen in a phrase should not be broken. For instance, the phrase $\mathfrak{m}$-complete lattice should not be broken after $\mathfrak{m}$; so type it as

```
\text{$\mathfrak{m}$-com}\-plete lattice
```

(see Section 6.4.2 for `\mathfrak`).

Use the `\nobreakdash` command (placed before the hyphen)

```
\nobreakdash-   \nobreakdash--   \nobreakdash---
```

to prevent such breaks. For example,

```
pages~24\nobreakdash--47
```

Since LaTeX does not hyphenate a hyphenated word except at the hyphen,

```
\nobreakdash-
```

prevents the hyphenation of the whole word as though it were enclosed in a `\text` command. The form

```
\nobreakdash-\hspace{0pt}
```

allows the normal hyphenation of the word that follows the hyphen. For example,

```
$\mathfrak{m}$\nobreakdash-\hspace{0pt}complete lattice
```

allows the word `complete` to be hyphenated.

This coding of the phrase $\mathfrak{m}$-complete lattice is a natural candidate for a custom command (see Section 14.1.1).

Tip If you want to know how LaTeX would hyphenate a list of words, place it in the argument of a `\showhyphens` command.

For instance,

```
\showhyphens{summation reducible latticoid}
```

The result,

```
sum-ma-tion re-ducible lat-ti-coid
```

is shown in the `log` file.

Tip Some text editors wrap lines in a source file by breaking them at a hyphen, introducing errors in your typeset document.

For instance,

```
It follows from Theorem~\ref{T:M} that complete-
simple lattices are very large.
```

is typeset by LaTeX as follows:

It follows from Theorem 2 that complete- simple lattices are very large.

As you can see, there is a space between the hyphen and the word simple. The text editor inserted an end-of-line character after the hyphen (by the second space rule, see Section 3.2.1). This end-of-line character was interpreted by LaTeX as a space. To correct the error, make sure that there is no such line break, or comment out (see Section 3.5.1) the end-of-line character:

```
It follows from Theorem~\ref{T:M} that complete-%
simple lattices are very large.
```

Better yet, rearrange the two lines:

```
It follows from Theorem~\ref{T:M} that
complete-simple lattices are very large.
```

Of course, LaTeX does not know everything about the complicated hyphenation rules of the English language. Consult *The Chicago Manual of Style,* 16th edition [11] and Lyn Dupré's *BUGS in Writing: A Guide to Debugging Your Prose,* 2nd edition [13] for additional guidance.

3.5 Comments and footnotes

Various parts of your source file do not get typeset like most of the rest. The two primary examples are comments that do not get typeset at all and footnotes that get typeset at the bottom of the page.

3.5.1 Comments

The % symbol tells LaTeX to ignore the rest of the line. A common use might be a comment to yourself in the source file:

```
therefore, a reference to Theorem~1 % check this!
```

The % symbol has many uses. For instance, a document class command (see Section 9.5),

```
\documentclass[twocolumn,twoside,legalpaper]{amsart}
```

may be typed with explanations, as

```
\documentclass[%
twocolumn,%  option for two-column pages
twoside,%    format for two-sided printing
legalpaper%  print on legal-size paper
]{amsart}
```

so you can easily comment out some at a later time, as in

```
\documentclass[%
twocolumn,%  option for two-column pages
twoside,%    format for two-sided printing
legalpaper%  print on legal-size paper
]{amsart}
```

Notice that the first line is terminated with a % to comment out the end-of-line character.

Tip Some command arguments do not allow any spaces. If you want to break a line within an argument list, you can terminate the line with a %, as shown in the previous example.

See also the example at the end of Section 3.4.9.

It is often useful to start a document with a comment line giving the file name and identifying the earliest version of LaTeX that must be used to typeset it.

```
%This is article.tex
\NeedsTeXFormat{LaTeX2e}[1994/12/01]
```

The second line specifies the December 1, 1994 (or later) release of LaTeX. You may need to use such a declaration if your document uses a feature that was not available in earlier releases. Since LaTeX changes very little these days, this command is of limited use. (See, however, the discussion on page 293.)

Other uses of % include marking parts of the article for your own reference. For instance, you may include comments to explain command definitions (as in

Section 14.3). If something goes wrong inside a multiline math display (see Chapter 7), LaTeX does not tell you precisely where the error is. You can try commenting out all but one of the lines, until each line works separately.

Note that % does not comment out lines in a BibTeX database document (see Section 15.2.5).

Tip **Symbolic referencing** With every `\label` command I add the commented out form of the symbolic reference, see Section 1.7.2.

So if I start a new theorem, I type `\the` and my text expander inserts the following in the article:

```
\begin{theorem}\label{T:xx}
%Theorem~\ref{T:xx}

\end{theorem}
```

And do remember:

Tip **The 25% rule**
If you want a % sign in text, make sure you type it as `\%`. Otherwise, % comments out the rest of the line. LaTeX does not produce a warning.

Using % to comment out large blocks of text can be tedious even with block comment. The `verbatim` package includes the `comment` environment:

```
\begin{comment}
   ...the commented out text...
\end{comment}
```

Rule ■ `comment` **environments**

1. `\end{comment}` must be on a line by itself.

2. There can be no `comment` within a `comment`.

In other words,

```
\begin{comment}
   commented out text...
   \begin{comment}
      some more commented out text...
   \end{comment}
   and some more commented out text...
\end{comment}
```

is not allowed. LaTeX may give one of several messages, depending on the circumstances. For instance,

```
! LaTeX Error: \begin{document} ended by \end{comment}.

l.175 \end{comment}
```

Tip Locating errors

The comment environment can be very useful in locating errors.

Suppose you have unbalanced braces in your source file (see Section 3.3.2). Working with a *copy* of your source file, comment out the first half at a safe point (not within an environment!) and typeset. If you still get the same message, the error is in the second half. If there is no error message, the error is in the first half. Comment out the half that has no error.

Now comment out half of the remaining text and typeset again. Check to see whether the error appears in the first half of the remaining text or the second. Continue applying this method until you narrow down the error to a paragraph that you can inspect visually.

Since the comment environment requires the verbatim package, you must include the line

```
\usepackage{verbatim}
```

in the preamble of the source file (see Sections 1.8, 10.3.1, and D.1).

3.5.2 Footnotes

The text of a footnote is typed as the argument of a \footnote command. To illustrate the use of footnotes, I have placed one here.[1] This footnote is typed as

```
\footnote{Footnotes are easy to place.}
```

If you want to use symbols to designate the footnotes, instead of numbers, type the command

```
\renewcommand{\thefootnote}
        {\ensuremath{\fnsymbol{footnote}}}
```

before the first footnote; this provides up to nine symbols. In Section 14.1.1, we discuss the \ensuremath command. Section 3.2 of *The LaTeX Companion,* 2nd edition [56] describes how to further customize footnotes.

In addition, there are title-page footnotes, such as the \thanks and \date commands in the top matter. See page 4 for a typeset example of \date. See also Section 9.2 and the typeset title page footnotes on page 272.

[1]Footnotes are easy to place.

You can add a footnote marked by * to the title of an article. For instance, type the title

```
\title[Complete congruence lattices]%
  {Complete congruence lattices$^*$}
```

and add the lines

```
{\renewcommand{\thefootnote}{\fnsymbol{footnote}}
\setcounter{footnote}{1}
\footnotetext{Lecture delivered at the \AMS
              annual meeting in Brandon.}
\setcounter{footnote}{0}
}
```

The footnote will appear as the first footnote on page 1 marked by *. All the other footnotes are unmarked.

3.6 Changing font characteristics

Although a document class and its options determine how LaTeX typesets characters, there are occasions when you want control over the shape or size of the font used.

3.6.1 Basic font characteristics

You do not have to be a typesetting expert to recognize the following basic font attributes:

Shape Normal text is typeset:

upright (or *roman*)	as this text
slanted	*as this text*
italic	*as this text*
small caps	AS THIS TEXT

Monospaced and proportional Typewriters use *monospaced* fonts, that is, fonts all of whose characters are of the same width. Most text editors display text using a monospaced font. LaTeX calls monospaced fonts *typewriter style*. In this book, such a font is used to represent user input and LaTeX's response, such as "`typewriter style text`". Whereas, normal text is typeset in a *proportional* font, such as "proportional text with ii and mm", in which ii is narrow and mm is wide:

Monospaced and proportional Typewriters use *monospaced* fonts,

`mmmmmm`
`iiiiii` } monospaced

mmmmmm
iiiiii } proportional

Serifs A *serif* is a small horizontal (sometimes vertical) stroke used to finish off a vertical stroke of a letter, as on the top and bottom of the letter M. LaTeX's standard serif font is Computer Modern roman, such as "serif text". Fonts without serifs are called *sans serif*, such as "sans serif text". Sans serif fonts are often used for titles or for special emphasis.

Series: weight and width The *series* is the combination of weight and width. A font's *weight* is the thickness of the strokes and the *width* is how wide the characters are.

Light, *medium* (or *normal*), and *bold* often describe weight.

Narrow (or *condensed*), *medium* (or *normal*), and *extended* often describe width.

The Computer Modern family includes **bold fonts**. Traditionally, when the user asks for bold CM fonts, LaTeX actually provides *bold extended* (a somewhat wider version).

Size Most LaTeX articles are typeset with 10 point text unless otherwise instructed. Larger sizes are used for titles, section titles, and so on. Abstracts and footnotes are often set in 8-point type.

Font family The collections of all sizes of a font is called a *font family*.

3.6.2 Document font families

In a document class, the style designer designates three document font families:

1. *Roman* (upright and serifed) document font family
2. *Sans serif* document font family
3. *Typewriter style* document font family

and picks one of these (for articles, as a rule, the roman document font family) as the *document font family* or *normal family*. In all the examples in this book, the document font family is the roman document font family except for presentations which use sans serif (see Section 1.11 and Chapter 12). When you use Computer Modern fonts in LaTeX, which is the default, the three document font families are Computer Modern roman, Computer Modern sans serif, and Computer Modern typewriter. The document font family is Computer Modern roman.

In this book, the roman document font family is Times, the sans serif document font family is Helvetica, and the typewriter style document font family is Computer Modern typewriter. The document font family is the roman document font family Times.

The document font family (normal family) is the default font. You can always switch back to it with

```
\textnormal{...}
```

or

```
{\normalfont ...}
```

Table 3.6 shows these two commands and three additional pairs of commands to help you switch among the three basic document font families. It also shows the command pairs for the basic font shapes.

Command with Argument	Command Declaration	Switches to the font family
`\textnormal{...}`	`{\normalfont ...}`	document
`\emph{...}`	`{\em ...}`	*emphasis*
`\textrm{...}`	`{\rmfamily ...}`	roman
`\textsf{...}`	`{\sffamily ...}`	sans serif
`\texttt{...}`	`{\ttfamily ...}`	`typewriter style`
`\textup{...}`	`{\upshape ...}`	upright shape
`\textit{...}`	`{\itshape ...}`	*italic shape*
`\textsl{...}`	`{\slshape ...}`	*slanted shape*
`\textsc{...}`	`{\scshape ...}`	SMALL CAPITALS
`\textbf{...}`	`{\bfseries ...}`	**bold**
`\textmd{...}`	`{\mdseries ...}`	normal weight and width

Table 3.6: Font family switching commands.

The font-changing commands of Table 3.6 come in two forms:

- A command with an argument, such as `\textrm{...}`, changes its argument. These are short commands, i.e., they cannot contain a blank line or a `\par` command.
- A command declaration, such as `\rmfamily`, carries out the font change following the command and within its scope (see Section 3.3.2).

You should always use commands with arguments for small changes within a paragraph. They have two advantages:

- You are less likely to forget to change back to the normal font.
- You do not have to worry about italic corrections (see Section 3.6.4).

Note that *MakeIndex* requires you to use commands with arguments to change the font in which page numbers are typeset (see Section 16.1).

For font changes involving more than one paragraph, use command declarations. These commands are preferred if you want to create custom commands and environments (see Chapter 14).

3.6.3 *Shape commands*

There are five pairs of commands to change the font shape:

- `\textup{...}` or `{\upshape ...}` switch to the upright shape.
- `\textit{...}` or `{\itshape ...}` switch to the *italic shape*.

- `\textsl{...}` or `{\slshape ...}` switch to the *slanted shape*.
- `\textsc{...}` or `{\scshape ...}` switch to SMALL CAPITALS.
- `\emph{...}` or `{\em ...}` switch to *emphasis*.

The document class specifies how emphasis is typeset. As a rule, it is italic or slanted unless the surrounding text is italic or slanted, in which case it is upright. For instance,

```
\emph{Rubin space}
```

in the statement of a theorem is typeset as

the space satisfies all three conditions, a so-called Rubin space *that ...*

The emphasis changed the style of Rubin space from italic to upright.

Tip Be careful not to interchange the command pairs. For instance, if by mistake you type `{\textit serif}`, the result is *s*erif. Only the *s* is italicized since `\textit` takes `s` as its argument.

Rule ■ Abbreviations and acronyms
For abbreviations and acronyms use small caps, except for two-letter geographical acronyms.

So Submitted to TUG should be typed as

```
Submitted to \textsc{tug}
```

Note that only the lowercase characters in the argument of the `\textsc` command are printed as small caps.

3.6.4 Italic corrections

The phrase

when using a *serif* font

may be typed as follows:

```
when using a {\itshape serif\/} font
```

The `\/` command before the closing brace is called an *italic correction.* Notice that `{\itshape M}M` typesets as *M*M, where the *M* is leaning into the M.

Type `{\itshape M\/}M` to get the correct spacing MM. Compare the typeset phrase from the previous example with and without an italic correction:

when using a *serif* font
when using a *serif* font

The latter is not as pleasing to the eye.

Rule 1 ■ Italic correction
If the emphasized text is followed by a period or comma, you should not type the italic correction.

For example,

Do not forget. My party is on Monday.

should be typed as

```
{\itshape Do not forget.} My party is on Monday.
```

Rule 2 ■ Italic correction
The shape commands with arguments do not require italic correction. The corrections are provided automatically where needed.

Thus you can type the phrase when using a *serif* font the easy way:

```
when using a \textit{serif} font
```

Whenever possible, let LaTeX take care of the italic correction. However, if LaTeX is adding an italic correction where you feel it is not needed, you can override the correction with the `\nocorr` command. LaTeX does not add an italic correction before a period or a comma. These two punctuation marks are stored in the `\nocorrlist` command. By redefining this command, you can modify LaTeX's behavior.

Rule 3 ■ Italic correction
The italic correction is required with the commands `\itshape`, `\slshape`, `\em`.

3.6.5 *Series*

These attributes play a very limited role with the Computer Modern fonts. There is only one important pair of commands,

```
\textbf{...}    {\bfseries ...}
```

to change the font to bold (actually, bold extended). The commands

```
\textmd{...}    {\mdseries ...}
```

which set both the weight and width to medium (normal) are seldom needed.

3.6.6 *Size changes*

Standard LaTeX documents are typeset in 10 point type. The 11 point and 12 point type are often used for better readability and some journals require 12 point—if this is the case, use the `12pt` document class option (see Sections 9.5, 10.1.2, and 17.1.3). The `8pt` and `9pt` document class options are rarely used. The sizes of titles, subscripts, and superscripts are automatically set by the document class, in accordance with the font size option.

If you must change the font size for some text—it is seldom necessary to do so in an article—the following command declarations are provided (see Table 3.7):

```
\Tiny  \tiny  \SMALL  \Small  \small
           \normalsize
\large  \Large  \LARGE  \huge  \Huge
```

The command `\SMALL` is also called `\scriptsize` and the command `\Small` is also called `\footnotesize`. The font size commands are listed in order of increasing—to be more precise, nondecreasing—size.

Command	Sample text
`\Tiny`	sample text
`\tiny`	sample text
`\SMALL` or `\scriptsize`	sample text
`\Small` or `\footnotesize`	sample text
`\small`	sample text
`\normalsize`	sample text
`\large`	sample text
`\Large`	sample text
`\LARGE`	sample text
`\huge`	sample text
`\Huge`	sample text

Table 3.7: Font size commands.

Two commands allow the user to increase or decrease font size: `\larger` moves up one size, `\smaller` moves down one. Both commands take an optional argument. For example, `\larger[2]` moves up 2 sizes.

3.6.7 *Orthogonality*

You are now familiar with the commands that change the font family, shape, series, and size. Each of these commands affects one and only one font attribute. For example, if you change the series, then the font family, shape, and size do not change. These commands act independently. In LaTeX terminology, the commands are *orthogonal.* From the user's point of view this behavior has an important consequence: *The order in which these commands are given does not matter.* Thus

```
\Large \itshape \bfseries
```

has the same effect as

```
\bfseries \itshape \Large
```

Note that LaTeX 2.09's two-letter commands (see Section 3.6.8) are not orthogonal.

Orthogonality also means that you can combine these font attributes in any way you like. For instance, the commands

```
\sffamily \slshape \bfseries \Large
```

instruct LaTeX to change the font family to sans serif, the shape to slanted, the series to bold, and the size to `\Large`. If the corresponding font is not available, LaTeX uses a font that is available, and issues a warning.

The font substitution algorithm (see Section 7.9.3 of *The* LaTeX *Companion,* 2nd edition [56] for details) may not provide the font you really want, so it is your responsibility to make sure that the necessary fonts are available. We discuss this topic further in Section 17.5.

3.6.8 *Obsolete two-letter commands*

Users of LaTeX 2.09 and AMS-LaTeX version 1.1 are accustomed to using the two-letter commands `\bf`, `\it`, `\rm`, `\sc`, `\sf`, `\sl`, and `\tt`. These commands are not part of LaTeX. They are, however, still defined in most document classes. The two-letter commands

1. switch to the document font family,

2. change to the requested shape.

There are a number of reasons not to use them. The two-letter commands

- are not part of LaTeX,
- require manual italic corrections,
- are not orthogonal (see Section 3.6.7).

`\slshape \bfseries` is the same as `\bfseries \slshape` (slanted bold), but `\sl\bf` is not the same as `\bf\sl`. Indeed, `{\sl\bf sample}` gives **sample** and `{\bf\sl sample}` produces *sample*.

3.6.9 Low-level commands

The font-characteristic changing commands we have discussed so far in this section are the *high-level* font commands. Each of these commands is implemented by LaTeX and the document class using *low-level* font commands. The low-level commands have been developed for document class and package writers. See Section 7.9 of *The* LaTeX *Companion*, 2nd edition [56].

There is one use of low-level commands you should keep in mind. When you choose a font size for your document or for some part thereof, you also determine the `\baselineskip`, the distance from the baseline of one line to the baseline of the next. Typically, a 10-point font size uses a 12 point `\baselineskip`. Occasionally, you may want to change the font size along with the `\baselineskip`. A command for accomplishing this is

```
\fontsize{9pt}{11pt}\selectfont
```

which changes the font size to 9 point and the `\baselineskip` to 11 point. To make this change for a single paragraph, you can type

```
{%special paragraph
\fontsize{9pt}{11pt}\selectfont

text

}%end special paragraph
```

Observe the blank line that follows `text` and marks the end of the paragraph; `\par` would accomplish the same thing.

3.7 Lines, paragraphs, and pages

When typesetting a document, LaTeX breaks the text into lines, paragraphs, and pages. Sometimes you may not like how LaTeX has chosen to lay out your text. There are ways to influence how LaTeX does its work and these are discussed in this section.

3.7.1 Lines

LaTeX typesets a document one paragraph at a time. It tries to split the paragraph into lines of equal width; for a preliminary discussion, see Section 1.4. If it fails to do so successfully and a line is too wide, you get an `overfull \hbox` message. Here is a typical example:

```
Overfull \hbox (15.38948pt too wide) in paragraph
             at lines 11--16
```

```
[]\OT1/cmr/m/n/10 In sev-eral sec-tions of the course
                 on ma-trix
the-ory, the strange term ``hamiltonian-
```

The log file records these messages. To place a visual warning in the typeset version of your document as well, use the draft document class option

```
\documentclass[draft]{amsart}
```

Lines that are too wide are be marked with a *slug* (a black box) in the margin. A slug is a vertical bar of width \overfullrule.

Do not worry about such messages while writing the document. If you are preparing the final version and receive a message for an overfull \hbox, the first line of defense is to see whether optional hyphens would help (see Section 3.4.9). Read the warning message carefully to see which words LaTeX cannot hyphenate properly. If adding optional hyphens does not help, a simple rephrasing of the problem sentence often does the trick.

Recall that there are 72.27 points in an inch (see Section 1.4). So if the message indicates a 1.55812 pt overflow, for instance, you can safely ignore it.

Tip If you do not want the 1.55812pt overflow reported whenever the document is typeset, you can enclose the offending paragraph (including the blank line indicating the end of the paragraph) between the lines

```
{\setlength{\hfuzz}{2pt}
```

and

```
}% end of \hfuzz=2pt
```

Choose an argument that is slightly more than the reported error (maybe 2pt). This does not affect the typeset output, but the warning message and the slug, if you are using the draft option, are suppressed.

Alternatively, enclose the offending paragraph including the blank line indicating the end of the paragraph in a setlength environment:

```
\begin{setlength}{\hfuzz}{2pt}
\end{setlength}
```

Breaking lines

There are two forms of the line breaking command:

- The \\ and \newline commands break the line at the point of insertion but do not stretch it.
- The \linebreak command breaks the line at the point of insertion and stretches the line to make it of the normal width.

The text following any of these commands starts at the beginning of the next line, without indentation. The \\ command is often used, but `\linebreak` is rarely seen. (See Section 17.6 for an application of the `\linebreak` command.) I illustrate the effect of these commands:

```
There are two forms of the line breaking command:

There are two forms\\ of the line breaking command:

There are two forms \newline of the line breaking command:

There are two forms \linebreak of the line breaking command:
```

typeset as

There are two forms of the line breaking command:
There are two forms
of the line breaking command:
There are two forms
of the line breaking command:
There are two forms of the line breaking command:

If you force a line break in the middle of a paragraph with the `\linebreak` command and LaTeX thinks that there is too little text left on the line to stretch it to full width, you get a message such as

```
Underfull \hbox (badness 4328) in paragraph
                                         at lines 8--12
```

The \\ command has two important variants:

- `\\[`*length*`]`, where *length* is the interline space you wish to specify after the line break, for instance, `12pt`, `.5in`, or `1.2cm`. Note how the units are abbreviated.
- `\\*`, which prohibits a page break following the line break.

The `\\*[`*length*`]` form combines the two variants. We illustrate the `\\[`*length*`]` command:

```
It is also semimodular.\\[15pt]  In particular,
```

which is typeset as

It is also semimodular.

In particular,

Since `\\` can be modified by $*$ or by `[ ]`, LaTeX may get confused if the line after a `\\` command starts with a $*$ or `[`. In such cases, type $*$ as `{*}` or `[` as `{[}`. For instance, to get

There are two sources of problems:
[a] The next line starts with `[`.

type

```
There are two sources of problems:\\
{[}a] The next line starts with \texttt{[}.
```

If you fail to type `{[}`, you get the message

```
! Missing number, treated as zero.
<to be read again>
                    a
l.16 [a]
         The next line starts with \texttt{[}.
```

Rule ■ `\\`

Without optional arguments, the `\\` command and the `\newline` command are the same *in text,* but not within environments or command arguments.

You can qualify the `\linebreak` command with an optional argument: `0` to `4`. The higher the argument, the more it forces the occurrence of a line break. `\linebreak[4]` is the same as `\linebreak`, while `\linebreak[0]` allows the line break but does not force it.

The `\nolinebreak` command plays the opposite role. `\nolinebreak[0]` = `\linebreak[0]`, and `\nolinebreak[4]` = `\nolinebreak`. `\nolinebreak` is seldom used since the tie (`~`) and the `\text` command (see Section 3.4.3) accomplish the same goal most of the time.

Double spacing

It is convenient to proofread documents double spaced. Sone journals even require submissions to be double spaced.

To typeset a document double spaced, include the command

```
\renewcommand{\baselinestretch}{1.5}
```

in its preamble.

Alternatively, use George D. Greenwade's `setspace`. Load this package with a

```
\usepackage{setspace}
```

command in the preamble of the document and specify

```
\doublespacing
```

in the preamble. This changes not just the line spacing but a number of other parameters to make your article look good.

See also Section 3.1.13 of *The LaTeX Companion,* 2nd edition [56].

3.7.2 Paragraphs

Paragraphs are separated by blank lines or by the `\par` command. LaTeX error messages always show paragraph breaks as `\par`. The `\par` form is also very useful in custom commands and environments (see Sections 14.1 and 14.2).

In some document classes, the first line of a paragraph is automatically indented. Indentation can be prevented with the `\noindent` command and can be forced with the `\indent` command.

Sometimes—for instance, in a schedule, glossary, or index—you may want a *hanging indent,* where the first line of a paragraph is not indented, and all the others are indented by a specified amount.

Hanging indents are created by specifying the amount of indentation specified by `\hangindent` and set with the `\setlength` command:

```
\setlength{\hangindent}{30pt}
\noindent
\textbf{sentence} a group of words terminated by
 a period, exclamation point, or question mark.

\setlength{\hangindent}{30pt}
\noindent
\textbf{paragraph} a group of sentences terminated by a
blank line or by the new paragraph command.
```

produces

sentence a group of words terminated by a period, exclamation point, or question mark.

paragraph a group of sentences terminated by a blank line or by the new paragraph command.

Notice that the `\setlength` command must be repeated for each paragraph.

Sometimes you may want to change the value of `\hangafter`, the length command that specifies the number of lines not to be indented. The default value is 1. To change it to 2, use the command

```
\setlength{\hangafter}{2}
```

For more about the `\setlength` command, see Section 14.5.2. *The LaTeX Companion,* 2nd edition [56] discusses in Section 3.1.4 the style parameters of a paragraph.

The preferred way to shape a paragraph or series of paragraphs is with a custom list environment (see Section 14.6).

3.7.3 Pages

There are two page breaking commands:

- \newpage, which breaks the page at the point of insertion but does not stretch the content
- \pagebreak, which breaks the page at the point of insertion and stretches the page's content to normal length

Text following either command starts at the beginning of the next page, indented.

As you can see, the page breaking commands are analogous to the line breaking commands discussed in Section 3.7.1. This analogy continues with the optional argument, 0 to 4:

\pagebreak[0] to \pagebreak[4]
\nopagebreak[0] to \nopagebreak[4]

There are also special commands for allowing or forbidding page breaks in multiline math displays (see Section 7.9).

When preparing the final version of a document (see Section 17.5), you may have to extend or shrink a page by a line or two to prevent it from breaking at an unsuitable line. You can do so with the \enlargethispage command. For instance,

```
\enlargethispage{\baselineskip}
```

adds one line to the page length. On the other hand,

```
\enlargethispage{-\baselineskip}
```

makes the page one line shorter.

```
\enlargethispage{10000pt}
```

makes the page very long.

The *-ed version, \enlargethispage*, squeezes the page as much as possible.

There are two more variants of the \newpage command. The

```
\clearpage
```

command does a \newpage and typesets all the figures and tables waiting to be processed (see Section 8.4.3). The variant

```
\cleardoublepage
```

is used with the twoside document class option (see Sections 9.5 and 10.1.2). It does a \clearpage and in addition makes the next printed page a right-hand, that is, odd-numbered, page, by inserting a blank page if necessary. If for your document class this does not work, use the package cleardoublepage.sty in the samples folder.

Section 17.6 discusses the use of some of these commands in the final preparation of books.

3.7.4 Multicolumn printing

Many document classes provide the `twocolumn` option for two-column typesetting (see Sections 9.5 and 10.1.2). In addition, there is a `\twocolumn` command which starts a new page by issuing a `\clearpage` and then typesets in two columns. An optional argument provides a two-column wide title. Use the `\onecolumn` command to switch back to a one-column format.

Frank Mittelbach's `multicol` package (see Section 10.3.1) provides the much more sophisticated `multicols` environment, which can start in the middle of a page, can handle more than two columns, and can be customized in a number of ways (see Section 3.5.3 of *The* LaTeX *Companion,* 2nd edition [56]).

3.8 Spaces

The judicious use of horizontal and vertical space is an important part of the formatting of a document. Fortunately, most of the spacing decisions are made by the document class, but LaTeX has a large number of commands that allow the user to insert horizontal and vertical spacing.

Remember that LaTeX ignores excess spaces, tabs, and end-of-line characters. If you need to add horizontal or vertical space, then you must choose from the commands in this section. Use them sparingly (see Section 2.6).

3.8.1 Horizontal spaces

In this section, we discuss fixed length horizontal space commands. Variable length horizontal space is discussed in Section 3.8.4.

When typing text, there are three commands that are often used to create horizontal space, shown between the bars in the display below:

```
\␣       ␣
\quad    ␣␣
\qquad   ␣␣␣␣
```

The `\quad` command creates a 1 em space and `\qquad` creates a 2 em space (see Section 3.8.3). The interword space created by `\␣` can both stretch and shrink. There are other commands that create smaller amounts of space. All the math spacing commands of Section 6.1.3—with the exception of `\mspace`—can be used in ordinary text (see Sections A.9 and B.6), but the `\hspace` and `\phantom` commands may be more appropriate.

The `\hspace` command takes a length as a parameter. The length may be negative. For example,

```
\textbar\hspace{12pt}\textbar      | |
\textbar\hspace{.5in}\textbar      |      |
\textbar\hspace{1.5cm}\textbar     |      |
```

or `\hspace{-40pt}`. The command `\hspace` is often used with a negative argument when placing illustrations.

The `\phantom{`*argument*`}` command produces a space the width and height of the space that would be occupied by its typeset argument

```
\textbar need space\textbar                  |need space|
\textbar\phantom{need space}\textbar         |          |
```

and

```
alpha \phantom{beta} gamma \phantom{delta}\\
\phantom{alpha} beta \phantom{gamma} delta
```

produces

```
alpha       gamma
      beta        delta
```

The `\phantom` command is very useful for fine tuning aligned math formulas (see Sections 6.1 and 7.5.3). The variant

`\hphantom{`*argument*`}`

creates a space with the horizontal dimension that would be occupied by its typeset *argument* and with zero height.

For instance, the last two lines of the dedication of this book were typed as follows in the `flushright` environment:

```
\textbf{Kate} (8), \phantom{\textbf{Jay} (3)}\\[8pt]
                            \textbf{Jay} (3)
```

See Section B.6 for a table of all horizontal text-spacing commands.

It is easy to see how we can emulate the `\pause` command of `beamer`—see Section 12.2.1—with the `\phantom` command.

Horizontal space variant

When LaTeX typesets a line, it removes all spaces from the beginning of the line, including the space created by `\hspace`, `\quad`, and other spacing commands. Using the `*`-ed variant of `\hspace`, `\hspace*`, prevents LaTeX from removing the space you have specified.

For example,

```
And text\\
\hspace{20pt}And text\\
\hspace*{20pt}And text
```

is typeset as

```
And text
And text
    And text
```

Use the \hspace* command for creating customized indentation. To indent a paragraph by 24 points, give the command

```
\noindent\hspace*{24pt}And text
```

which typesets as

        And text

To break a line and indent the next line by 24 points, give the command

```
And text\\
\hspace*{24pt}And text
```

which produces

And text
        And text

3.8.2 *Vertical spaces*

You can add some interline space with the command \\[*length*], as discussed in Section 3.7.1. You can also do it with the \vspace command, which works just like the \hspace command (see Section 3.8.1), except that it creates vertical space. Here are some examples:

```
\vspace{12pt}  \vspace{.5in}  \vspace{1.5cm}.
```

Standard amounts of vertical space are provided by the three commands

```
\smallskip  \medskip  \bigskip
```

The space these commands create depends on the document class and the font size. With the document class and font I am using for this book, they represent a vertical space of 3 points, 6 points, and 12 points, respectively. 12 points is the baseline skip (see Section 3.6.9) in standard LaTeX documents with the default 10pt option.

Rule ■ **Vertical space commands**
All vertical space commands add the vertical space *after* the typeset line in which the command appears.

To obtain

end of text.

        New paragraph after vertical space

type

```
end of text.

\vspace{12pt}

New paragraph after vertical space
```

The following example illustrates the unexpected way the vertical space is placed if the command that creates it does not start a new paragraph:

```
end of text.
\vspace{12pt}
The following example illustrates the unexpected way
the vertical space is placed if the
command that creates it does not start a new paragraph:
```

It typesets as

end of text. The following example illustrates the unexpected way the vertical

space is placed if the command that creates it does not start a new paragraph:

Vertical space variants

LaTeX removes vertical space from the beginning and end of each page, including space produced by `\vspace`. The space created by the variant `\vspace*` is not removed by LaTeX under any circumstances. Use this command, for instance, to start the typeset text (say, of a letter) not at the top of the page.

The `\phantom` command has also a vertical variant: `\vphantom`. The command `\vphantom{`*`argument`*`}` creates a vertical space with the vertical dimension that would be occupied by its typeset argument, *`argument`*.

3.8.3 Relative spaces

The length of a space is usually given in *absolute units:* `12pt` (points), `.5cm` (centimeters), `1.5in` (inches). Sometimes, *relative units*, em and ex, are more appropriate, units that are relative to the size of the letters in the current font. The unit 1 em is approximately the width of an M in the current font, while 1 ex is approximately the height of an x in the current font. These units are used in commands such as

`\hspace{12em}` and `\vspace{12ex}`

The `\quad` and `\qquad` commands (Section 3.8.1) produce 1 em and 2 em spaces.

3.8.4 Expanding spaces

Horizontal spaces

The `\hfill`, `\dotfill`, and `\hrulefill` commands fill all available space in the line with spaces, dots, or a horizontal line, respectively. If there are two of these commands on the same line, the space is divided equally between them. These commands can be used to center text, to fill lines with dots in a table of contents, and so on.

To obtain

2. Boxes...34
ABC   and   DEF
ABC____________________and____________________DEF

type

```
2. Boxes\dotfill 34\\
ABC\hfill and\hfill DEF\\
ABC\hrulefill and\hrulefill DEF
```

In a centered environment—such as a `\title` (see Section 9.2.1) or a `center` environment (see Section 4.3)—you can use `\hfill` to set a line flush right:

This is the title

First Draft

Author

To achieve this effect, type

```
\begin{center}
  This is the title\\
  \hfill First Draft\\
  Author
\end{center}
```

Vertical spaces

The vertical analogue of `\hfill` is `\vfill`. This command fills the page with vertical space so that the text before the command and the text after the command stretch to the upper and lower margin. You can play the same games with it as with `\hfill` in Section 3.8.4.

The command `\vfill` stands for `\vspace{\fill}`, so it is ignored at the beginning of a page. Use `\vspace*{\fill}` if you need it at the beginning of a page.

3.9 Boxes

Sometimes it can be useful to typeset text in an imaginary box, and treat that box as a single large character. A single-line box can be created with the `\text` or `\makebox` commands and a multiline box of a prescribed width can be created with the `\parbox` command or `minipage` environment.

3.9.1 Line boxes

The `\text` command provides a *line box* that typesets its argument without line breaks. As a result, you may find the argument extending into the margin. The resulting box is handled by LaTeX as if it were a single large character. For instance,

```
\text{database}
```

causes LaTeX to treat the eight characters of the word `database` as if they were one. This technique has a number of uses. It prevents LaTeX from breaking the argument (see Section 3.4.9). It also allows you to use the phrase in the argument in a formula (see Section 5.4.6).

The argument of `\text` is typeset in a size appropriate for its use, for example, as a subscript or superscript. See Section 5.4.6 for an example.

Line boxes—a refinement

The `\mbox` command is the short form of the `\makebox` command. Both `\mbox` and `\text` prevent breaking the argument, but `\mbox` does not change size in subscripts and superscripts.

The full form of the `\makebox` command is

```
\makebox[width][alignment]{text}
```

where the arguments are

- *`width`*, the (optional) width of the box. If `[`*`width`*`]` is omitted, the box is as wide as necessary to enclose its contents.
- *`alignment`*, (optionally) one of `c` (the default), `l`, `r`, or `s`. The text is centered by default, `l` sets the argument flush left, `r` right, and `s` stretches the text the full length of the box if there is blank space in the argument.
- *`text`*, the text in the box.

A *`width`* argument can be specified in inches (in), centimeters (cm), points (pt), em, or ex (see Sections 3.8.3 and 14.5.2).

The following examples,

```
\makebox{Short title.}End\\
\makebox[2in][l]{Short title.}End\\
\makebox[2in]{Short title.}End\\
\makebox[2in][r]{Short title.}End\\
\makebox[2in][s]{Short title.}End
```

typeset as

Short title.End
Short title.   End
          Short title.   End
                    Short title.End
Short   title.End

The optional width argument, *width*, can use four length commands:

```
\height  \depth  \totalheight  \width
```

These are the dimensions of the box that would be produced without the optional width argument.

Here is a simple example. The command

```
\makebox{hello}
```

makes a box of width `\width`. To typeset `hello` in a box three times the width, that is, in a box of width 3`\width`, use the command

```
\makebox[3\width]{hello}
```

So

```
start\makebox[3\width]{hello}end
```

typesets as

start   hello   end

The formal definition of these four length commands is the following:

- `\height` is the height of the box above the baseline
- `\depth` is the depth of the box below the baseline
- `\totalheight` is the sum of `\height` and `\depth`
- `\width` is the width of the box

There is an interesting variant of the `\makebox` command. The `\rlap` command makes a box and pretends that it is of width zero. For instance,

```
\newcommand{\circwithdot}
   {\mathbin{\rlap{$\mspace{2mu}\cdot$}\hbox{$\circ$}}}
```

defines the command `\circwithdot`, so you can type

```
$f\circwithdot\varphi$
```

which prints as $f \odot \varphi$. There is also an `\llap` command.

3.9.2 Frame boxes

Boxed text is very emphatic. For example, Do not touch! is typed as

```
\fbox{Do not touch!}
```

This is a *frame box,* hence the command \fbox or \framebox.

Boxed text cannot be broken, so if you want a frame around more than one line of text, you should put the text as the argument of a \parbox command or within a minipage environment (see Section 3.9.3), and then put that into the argument of an \fbox command. For instance,

```
\fbox{\parbox{3in}{Boxed text cannot be broken,
so if you want to frame more than one line
of text, place it in the argument of a
\bsl\texttt{parbox}
command or within a
\texttt{minipage} environment.}}
```

produces

Boxed text cannot be broken, so if you want to frame more than one line of text, place it in the argument of a \parbox command or within a minipage environment.

The \bsl command is defined in Section 14.1.1. See Section 6.7.2 for boxed formulas.

The \framebox command works exactly like \makebox, except that it draws a frame around the box.

```
\framebox[2in][l]{Short title}
```

produces

Short title

You can use this command to typeset the number 1 in a square box, as required by the title of Michael Doob's [12]:

TEX *Starting from* 1

```
\framebox{\makebox[\totalheight]{1}}
```

which typesets as

1

Note that

```
\framebox[\totalheight]{1}
```

typesets as

1

which is not a square box. Indeed, `\totalheight` is the height of 1, which becomes the width of the box. The total height of the box, however, is the height of the character 1 to which you have to add twice the `\fboxsep`, the separation between the contents of the box and the frame, defined as 3 points, and twice the `\fboxrule`, the width of the line, or rule, defined as 0.4 points. These lengths are in general also added to the width of the box, but not in this case, because we forced the width to equal the height of the character.

You can use the `\fbox` command to frame the name of an author:

```
\author{\fbox{author's name}}
```

3.9.3 *Paragraph boxes*

A paragraph box works like a paragraph. The text it contains is wrapped around into lines. The width of these lines is set by the user.

The `\parbox` command typesets the contents of its second argument as a paragraph with a line width supplied as the first argument. The resulting box is handled by LaTeX as a single large character. For example, to create a 3-inch wide column,

Fred Wehrung's new result shows the limitation of
E. T. Schmidt's construction, especially for large
lattices.

type

```
\parbox{3in}{Fred Wehrung's new result shows the
limitation of E.\,T. Schmidt's construction,
especially for large lattices.}
```

Paragraph boxes are especially useful when working within a `tabular` environment. See the subsection on refinements in Section 4.6 for examples of multiline entries.

The width of the paragraph box can be specified in inches (in), centimeters (cm), points (pt), or the relative measurements em and ex (see Section 3.8.3), among others (see Section 14.5.2 for a complete listing of measurement units).

Tip The \parbox command requires two arguments. Dropping the first argument results in a message such as

```
! Missing number, treated as zero.
<to be read again>
                   T
l.175
```

Dropping the second argument does not yield a message but the result is probably not what you intended. The next character is taken as the contents of the \parbox.

Paragraph box refinements

The "character" created by a \parbox command is placed on the line so that its vertical center is aligned with the center of the line. An optional first argument b or t forces the paragraph box to align along its bottom or top. For an example, see Section 4.6. The full syntax of \parbox is

```
\parbox[alignment][height][inner-alignment]{width}{text}
```

Just as for the \makebox command (see Section 3.9.1), the

\height   \depth   \totalheight   and   \width

commands may be used in the *height* argument instead of a numeric argument.

The *inner-alignment* argument is the vertical equivalent of the *alignment* argument for \makebox, determining the position of *text* within the box and it may be any one of t, b, c, or s, denoting top, bottom, centered, or stretched alignment, respectively. When the *inner-alignment* argument is not specified, it defaults to *alignment*.

Paragraph box as an environment

The minipage environment is very similar to the \parbox command. It typesets the text in its body using a line width supplied as an argument. It has an optional argument for bottom or top alignment, and the other \parbox refinements also apply. The difference is that the minipage environment can contain displayed text environments discussed in Chapter 4.

The minipage environment can also contain footnotes (see Section 3.5.2) that are displayed within the minipage. See Section 3.2.1 of *The* LaTeX *Companion,* 2nd edition [56] for complications that may arise therefrom.

3.9.4 Marginal comments

A variant of the paragraph box, the \marginpar command, allows you to add marginal comments. So

```
\marginpar{Do not use this often}
```

Do not use this often

produces the comment displayed in the margin.

The AMS warning in the book [31] (also displayed here below the marginal comment) is defined as

Ⓐ

```
\marginpar{{\Large%
\textcircled{\raisebox{.7pt}{\normalsize\textbf A}}}}
```

The \textcircled command is discussed in Section 3.4.7, while the \raisebox command is introduced in Section 3.9.6.

Rule ■ Marginal comments and math environments
Do not use marginal comments in equations or multiline math environments.

Tip Avoid using too many marginal comments on any given page—LaTeX may have to place some of them on the next page.

If the document is typeset two-sided, then the marginal comments are set in the outside margin. The form

```
\marginpar[left-comment]{right-comment}
```

uses the required argument *right-comment* when the marginal comment is set in the right margin and the optional argument *left-comment* when the marginal comment is set in the left margin.

The width of the paragraph box for marginal comments is stored in the length command \marginparwidth (see Section 14.5.2 for length commands). If you want to change it, use

```
\setlength{\marginparwidth}{new_width}
```

as in

```
\setlength{\marginparwidth}{90pt}
```

The default value of this width is set by the document class. If you want to know the present setting, type

```
\the\marginparwidth
```

in your document and typeset it, or, in interactive mode (see Sections 14.1.8 and C.3), type

```
*\showthe\marginparwidth
```

(* is the interactive prompt).

See Sections 3.2.8 and 4.1 of *The LaTeX Companion,* 2nd edition [56] for other style parameters pertaining to marginal notes.

3.9.5 Solid boxes

A solid filled box is created with a `\rule` command. The first argument is the width and the second is the height. For instance, to obtain

end of proof symbol: ■

type

```
end of proof symbol: \rule{1.6ex}{1.6ex}
```

In fact, this symbol is usually slightly lowered:

end of proof symbol: ■

This positioning is done with an optional first argument:

```
end of proof symbol: \rule[-.23ex]{1.6ex}{1.6ex}
```

Here is an example combining `\rule` with `\makebox` and `\hrulefill`:

```
1 inch:\quad\makebox[1in]{\rule{.4pt}{4pt}%
   \hrulefill\rule{.4pt}{4pt}}
```

which produces

1 inch: └────────────┘

Struts

Solid boxes of zero width are called *struts.* Struts are invisible, but they force LaTeX to make room for them, changing the vertical alignment of lines. Standard struts can also be added with the `\strut` or `\mathstrut` command. To see how struts work, compare

[ab] and [ab] and [ab]

typed as

```
\fbox{ab} and \fbox{\strut ab} and \fbox{$\mathstrut$ab}
```

Struts are especially useful for fine tuning tables (see Section 4.6) and formulas (see math struts in Section 6.5).

Rule ■ **Zero distance**

`0pt, 0in, 0cm, 0em` all stand for zero width. `0` by itself is not acceptable.

For example, `\rule{0}{1.6ex}` gives the message

```
! Illegal unit of measure (pt inserted).
<to be read again>
                    h
l.251 \rule{0}{1.6ex}
```

If the `\rule` command has no argument or only one, LaTeX generates a message. For instance, `\rule{1.6ex}` gives the message

```
! Paragraph ended before \@rule was complete.
```

or

```
! Missing number, treated as zero.
```

In the first message, the reference to `\@rule` suggests that the problem is with the `\rule` command. Checking the syntax of the `\rule` command, you find that an argument is missing. The second message is more informative, since there is, indeed, a missing number.

3.9.6 Fine tuning boxes

The command

```
\raisebox{displacement}{text}
```

typesets *text* in a box with a vertical *displacement*. If *displacement* is positive, the box is raised; if it is negative, the box is lowered.

The `\raisebox` command allows us to play games:

```
fine-\raisebox{.5ex}{tun}\raisebox{-.5ex}{ing}
```

produces fine-tuning.

The `\raisebox` command has two optional arguments:

```
\raisebox{0ex}[1.5ex][0.75ex]{text}
```

forces LaTeX to typeset *text* as if it extended 1.5 ex above and 0.75 ex below the line, resulting in a change in the interline space above and below the line. A simple version of this command, `\smash`, is discussed in Section 6.5.

In the AMS warning in the book [31] (shown on page 94), the `\raisebox` command is used to properly center the bold A in the circle:

```
\Large\textcircled{\raisebox{.7pt}{\normalsize\textbf A}}
```

CHAPTER

4

Text environments

There are three types of text environments in LaTeX:

1. Displayed text environments; text within such an environment usually is typeset with some vertical space around it
2. Text environments that create a "large symbol"
3. Style and size environments

We start by discussing a very important rule about blank lines in displayed text environments. Then we proceed in Section 4.2 to the most often used displayed text environments: lists. We continue with the style and size environments in Section 4.3.

The most important displayed text environments in math are proclamations or theorem-like structures, proclamations with style, and the `proof` environment, discussed in detail in Sections 4.4 and 4.5.

The `tabular` environment discussed in Section 4.6 produces a "large symbol", a table, which is of limited use in math. In Section 4.7, we discuss the `tabbing` environment, which may be used for computer code. The legacy environments `quote`,

G. Grätzer, *More Math Into* LaTeX, DOI 10.1007/978-3-319-23796-1_4

quotation, and verse are discussed in Section 4.8, along with the verbatim environment, which is often used to display LaTeX source in a typeset LaTeX document.

4.1 Some general rules for displayed text environments

As you know, blank lines play a special role in LaTeX, usually indicating a paragraph break. Since displayed text environments structure the printed display themselves, the rules about blank lines are relaxed somewhat. However, a blank line trailing an environment signifies a new paragraph for the text following the environment.

Rule ■ Blank lines in displayed text environments

1. Blank lines are ignored immediately after \begin{*name*} or immediately before \end{*name*} except in a verbatim environment.
2. A blank line after \end{*name*} forces the text that follows to start a new paragraph.
3. As a rule, you should not have a blank line before \begin{*name*}.
4. The line after any theorem or proof always begins a new paragraph, even if there is no blank line or \par command.

The page breaking commands in Section 3.7.3 apply to text environments, as does the line breaking command \\ discussed in Section 3.7.1.

4.2 List environments

LaTeX provides three list environments: enumerate, itemize, and description. LaTeX also provides a generic list environment that can be customized to fit your needs. See Section 14.6 on custom lists.

Most document classes redefine the spacing and some stylistic details of lists, especially since the list environments in the legacy document classes are not very pleasing. In this section, the list environments are formatted as they are by our standard document class, amsart. Throughout the rest of the book, lists are formatted as specified by this book's designer.

4.2.1 Numbered lists

A *numbered list* is created with the enumerate environment:

This space has the following properties:

(1) Grade 2 Cantor;
(2) Half-smooth Hausdorff;
(3) Metrizably smooth.

Therefore, we can apply the Main Theorem.

typed as

```
\noindent This space has the following properties:
\begin{enumerate}
   \item Grade 2 Cantor\label{Cantor};
   \item Half-smooth Hausdorff\label{Hausdorff};
   \item Metrizably smooth\label{smooth}.
\end{enumerate}
Therefore, we can apply the Main Theorem.
```

Each item is introduced with an `\item` command. The numbers LaTeX generates can be labeled and cross-referenced (Section 8.4.2). This construct can be used in theorems and definitions, for listing conditions or conclusions.

If you use `\item` in the form `\item[]`, you get an unnumbered item in the list, while `\item[a]` replaces the number of the item with a. This is another form of absolute referencing, see Section 8.4.2.

Tip Do not label absolute references. It may lead to problems that are hard to explain.

4.2.2 Bulleted lists

A *bulleted list* is created with the `itemize` environment:

We set out to accomplish a variety of goals:

- To introduce the concept of smooth functions.
- To show their usefulness in differentiation.
- To point out the efficacy of using smooth functions in Calculus.

is typed as

```
\noindent We set out to accomplish a variety of goals:
\begin{itemize}
   \item To introduce the concept of smooth functions.
   \item To show their usefulness in differentiation.
   \item To point out the efficacy of using smooth
      functions in Calculus.
\end{itemize}
```

4.2.3 Captioned lists

In a *captioned list* each item has a title (caption) specified by the optional argument of the `\item` command. Such lists are created with the `description` environment:

In this introduction, we describe the basic techniques:

Chopped lattice: a reduced form of a lattice;
Boolean triples: a powerful lattice construction;
Cubic extension: a subdirect power flattening the congruences.

is typed as

```
\noindent In this introduction, we describe
   the basic techniques:
\begin{description}
  \item[Chopped lattice] a reduced form of a lattice;
  \item[Boolean triples] a powerful lattice construction;
  \item[Cubic extensions] a subdirect power flattening
       the congruences.
\end{description}
```

4.2.4 A rule and combinations

There is only one rule you must remember.

Rule ■ List environments
An `\item` command must immediately follow
`\begin{enumerate}`, `\begin{itemize}`, or `\begin{description}`.

Of course, spaces and line breaks can separate them.

If you break this rule, you get a message. For instance,

```
\begin{description}
This is wrong!
  \item[Chopped lattice] a reduced lattice;
```

gives the message

```
! LaTeX Error: Something's wrong--perhaps a missing \item.

l.105   \item[Chopped lattice]
                                a reduced lattice;
```

If you see this message, remember the rule for list environments and check for text preceding the first `\item`.

You can nest up to four list environments; for instance,

(1) First item of Level 1.
  (a) First item of Level 2.
    (i) First item of Level 3.
      (A) First item of Level 4.
      (B) Second item of Level 4.
    (ii) Second item of Level 3.
  (b) Second item of Level 2.
(2) Second item of Level 1.

Referencing the second item of Level 4: 1(a)iB

which is typed as

```
\begin{enumerate}
   \item First item of Level 1.
   \begin{enumerate}
      \item First item of Level 2.
      \begin{enumerate}
         \item First item of Level 3.
         \begin{enumerate}
            \item First item of Level 4.
            \item Second item of Level 4.\label{level4}
          \end{enumerate}
          \item Second item of Level 3.
       \end{enumerate}
       \item Second item of Level 2.
   \end{enumerate}
   \item Second item of Level 1.
\end{enumerate}
Referencing the second item of Level 4: \ref{level4}
```

Note that the label `level4` collected all four of the counters (see Section 8.4.2).

You can also mix list environments:

(1) First item of Level 1.
- First item of Level 2.
    (a) First item of Level 3.
        – First item of Level 4.
        – Second item of Level 4.
    (b) Second item of Level 3.
- Second item of Level 2.

(2) Second item of Level 1.

Referencing the second item of Level 4: 1a

which is typed as

```
\begin{enumerate}
   \item First item of Level 1.
   \begin{itemize}
      \item First item of Level 2.
      \begin{enumerate}
         \item First item of Level 3.
         \begin{itemize}
            \item First item of Level 4.
            \item Second item of Level 4.\label{enums}
          \end{itemize}
          \item Second item of Level 3.
       \end{enumerate}
       \item Second item of Level 2.
   \end{itemize}
   \item Second item of Level 1.
\end{enumerate}
Referencing the second item of Level 4: \ref{enums}
```

Now the label enums collects only the two enumerate counters (see Section 8.4.2).

The indentations are, of course, not needed. I use them to keep track of the level of nesting.

In all three types of list environment, the `\item` command may be followed by an optional argument, which is displayed at the beginning of the typeset item:

```
\item[label]
```

Note that for `enumerate` and `itemize` the resulting typography may leave something to be desired.

Tip If the text following an \item command starts with an opening square bracket, [, then LaTeX thinks that \item has an optional argument. To prevent this problem from occurring, type [as {[}. Similarly, a closing square bracket,], *inside* the optional argument should be typed as {]}.

Tip You may want to use a list environment solely for the way the items are displayed, without any labels. You can achieve this effect by using \item[].

You can change the style of the numbers in an enumerate environment by redefining the counter as suggested in Section 14.5.1:

```
\renewcommand{\labelenumi}{{\normalfont (\roman{enumi})}}
```

The labels then are displayed as (i), (ii), and so on. This modification only works if you do not want to reference these items. If you want the \ref command to work properly, use David Carlisle's enumerate package (see Section 10.3.1). For an example of how to use Carlisle's environment, see Section 14.2.1. Section 3.3 of *The LaTeX Companion,* 2nd edition [56] explains how to customize the three list environments and discusses Bernd Schandl's paralist package, which provides a number of new list environments and makes customizing the three legacy list environments much easier. For custom lists, see Section 14.6.

4.3 Style and size environments

There are several text environments that allow you to set font characteristics. They have the same names as their corresponding command declarations:

```
        rmfamily  sffamily  ttfamily
upshape  itshape  em  slshape  scshape
               bfseries
```

For instance,

```
\begin{ttfamily}
   text
\end{ttfamily}
```

typesets *text* just like {\ttfamily *text*} would. Remember to use the command-declaration names for the environment names, that is, use rmfamily, not textrm and ttfamily, not texttt (see 3.6.2). There are also text environments for changing the font size, from tiny to Huge (see Section 3.6.6).

If you are getting overwhelmed by the large number of environments changing style and size, consult Tables 3.6 and 3.7 (see also Section B.3.2).

Horizontal alignment of a paragraph is controlled by the `flushleft`, `flushright`, and `center` environments. Within the `flushright` and `center` environments, it is customary to force new lines with the `\\` command, while in the `flushleft` environment, you normally allow LaTeX to wrap the lines.

These text environments can be used separately or in combination, as in

The **simplest**text environments set the printing style and size.
The commands and the environments have similar names.

typed as

```
\begin{flushright}
   The \begin{bfseries}simplest\end{bfseries}
   text environments set the
   printing style and size.\\
   The commands and the environments have similar names.
\end{flushright}
```

There are command declarations that correspond to these environments:

- `\centering` centers text
- `\raggedright` left aligns text
- `\raggedleft` right aligns text

The effect of one of these commands is almost the same as that of the corresponding environment except that the environment places additional vertical space before and after the displayed paragraphs. For such a command declaration to affect the way a paragraph is formatted, the scope must include the whole paragraph, including the blank line at the end of the paragraph, preferably indicated with a `\par` command.

The `\centering` command is used often with the `\includegraphics` command (see Section 8.4.3).

4.4 Proclamations (theorem-like structures)

Theorems, lemmas, definitions, and so forth are a major part of mathematical writing. In LaTeX, these constructs are typed in displayed text environments called *proclamations* or *theorem-like structures*.

In the `firstarticle.tex` sample article (see p. 4), there is only a single theorem.

In the `secondarticle.tex` sample article (see pp. 272–275), there are a number of different proclamations in a variety of styles, with varying degrees of emphasis. Proclamations with style are discussed in Section 4.4.2.

The two steps are required to use a proclamation:

Step 1 *Define* the proclamation with a \newtheorem command *in the preamble* of the document. For instance, the line

```
\newtheorem{theorem}{Theorem}
```

defines a theorem environment.

Step 2 *Invoke* the proclamation as an environment *in the body* of your document. Using the proclamation definition from Step 1, type

```
\begin{theorem}
   My first theorem.
\end{theorem}
```

to produce a theorem:

Theorem 1. *My first theorem.*

In the proclamation definition

```
\newtheorem{theorem}{Theorem}
```

the first argument, theorem, is the name of the environment that invokes the theorem. The second argument, Theorem, is the name that is used when the proclamation is typeset. LaTeX numbers the theorems automatically and typesets them with vertical space above and below. The phrase **Theorem 1.** appears, followed by the theorem itself, which may be emphasized. Of course, the formatting of the theorem depends on the document class and the proclamation style (see Section 4.4.2).

You may also specify an optional argument,

```
\begin{theorem}[The Fuchs-Schmidt Theorem]
   The statement of the theorem.
\end{theorem}
```

that appears as the name of the theorem:

Theorem 1 (The Fuchs-Schmidt Theorem). *The statement of the theorem.*

LaTeX is very fussy about how proclamations are defined. For example, in the introductory article firstarticle.tex (see Section 1.8), if the closing brace is dropped from the end of line 8,

```
\newtheorem{definition}{Definition
```

you get a message such as

```
Runaway argument?
{Definition \newtheorem {notation}{Notation}
! Paragraph ended before \@ynthm was complete.
<to be read again>
                   \par
l.10
```

Line 10 is the line after the `\newtheorem` commands. The message conveys the information that something is wrong in the paragraph before line 10.

If you forget an argument, as in

```
\newtheorem{definition}
```

LaTeX produces a message such as

```
! LaTeX Error: Missing \begin{document}.

l.9 \newtheorem{n
                  otation}{Notation}
```

In the message, the line

```
! LaTeX Error: Missing \begin{document}.
```

usually means that LaTeX became confused and believes that some text typed in the preamble should be moved past the line

```
\begin{document}
```

The mistake could be anywhere in the preamble above the line LaTeX indicates. If you encounter such a message, try to isolate the problem by commenting out parts of the preamble (see Section 3.5.1 and also Section 2.6).

Rule ■ Lists in proclamations

If a proclamation starts with a list environment, precede the list by `\hfill`.

If you do not, as in

```
\begin{definition}\label{D:prime}
   \begin{enumerate}
   \item $u$ is \emph{bold} if $u = x^2$.\label{mi1}
   \item $u$ is \emph{thin} if $u = \sqrt{x}$.\label{mi2}
   \end{enumerate}
\end{definition}
```

your typeset list starts on the first line of the proclamation:

Definition 1. (1) *u is* bold *if* $u = x^2$.
(2) *u is* thin *if* $u = \sqrt{x}$.

If you add the `\hfill` command,

```
\begin{definition}\hfill
\begin{enumerate}
```

the list in the definition typesets correctly:

Definition 1.
(1) *u is* bold *if* $u = x^2$.
(2) *u is* thin *if* $u = \sqrt{x}$.

Consecutive numbering

If you want to number two sets of proclamations consecutively, you can do so by first defining one proclamation, and then using its name as an optional argument of the second proclamation. For example, to number the lemmas and propositions in your paper consecutively, you type the following two lines in your preamble:

```
\newtheorem{lemma}{Lemma}
\newtheorem{proposition}[lemma]{Proposition}
```

Lemmas and propositions are then consecutively numbered as **Lemma 1**, **Proposition 2**, **Proposition 3**, and so on.

Let me emphasize: The optional argument of a proclamation definition must be the name of a proclamation that *has already been defined.*

Numbering within a section

The `\newtheorem` command may also have a different optional argument; it causes LaTeX to number the proclamations within sections. For example,

```
\newtheorem{lemma}{Lemma}[section]
```

numbers the lemmas in Section 1 as **Lemma 1.1** and **Lemma 1.2**. In Section 2, you have **Lemma 2.1** and **Lemma 2.2**, and so on.

Instead of `section`, you may use any sectioning command provided by the document class, such as `chapter`, `section`, and `subsection`.

Consecutive numbering and numbering within a section can be combined. For example,

```
\newtheorem{lemma}{Lemma}[section]
\newtheorem{proposition}[lemma]{Proposition}
```

sets up the `lemma` and `proposition` environments so that they are numbered consecutively within sections: **Lemma 1.1**, **Proposition 1.2**, **Proposition 1.3** and **Proposition 2.1**, **Lemma 2.2**, and so on.

4.4.1 The full syntax

The full form of `\newtheorem` is

```
\newtheorem{envname}[procCounter]{Name}[secCounter]
```

where the two optional arguments are mutually exclusive, and

`envname` is the name of the environment to be used in the body of the document. For instance, you may use `theorem` for the *`envname`* of a theorem, so that a theorem is typed inside a `theorem` environment. Of course, *`envname`* is just a label; you are free to choose any environment name, such as `thm` or `george` (as long as *the name is not in use as the name of another command or environment*). This argument is also the name of the counter LaTeX uses to number these text environments.

`procCounter` is an optional argument. It sets the new proclamation to use the counter of a previously defined proclamation and the two proclamations are consecutively numbered.

`Name` is the text that is typeset when the proclamation is invoked. So if `Theorem` is given as *`Name`*, then you get **Theorem 1**, **Theorem 2**, and so on in your document.

`secCounter` is an optional argument that causes the *`Name`* environments to be numbered within the appropriate sectioning units. So if `theorem` is the *`envname`* and `section` is the *`secCounter`*, then in Section 1 you have **Theorem 1.1**, **Theorem 1.2**, and so on. In Section 2 you get **Theorem 2.1**, **Theorem 2.2**, and so on. Proclamations may be numbered within subsections, sections, chapters, or any other sectioning unit automatically numbered by LaTeX.

4.4.2 Proclamations with style

You can choose one of three styles for your proclamations by preceding the definitions with the \theoremstyle{style} command, where *style* is one of the following:

- plain, the most emphatic
- definition
- remark, the least emphatic

There are a few extra options, including the \newtheorem* command, an unnumbered version of \newtheorem.

The following commands set the styles in the secondarticle.tex article. The typeset sample article (on pages 272–275) shows how the chosen styles affect the typeset proclamations.

```
\theoremstyle{plain}
\newtheorem{theorem}{Theorem}
\newtheorem{corollary}{Corollary}
\newtheorem*{main}{Main Theorem}
\newtheorem{lemma}{Lemma}
\newtheorem{proposition}{Proposition}

\theoremstyle{definition}
\newtheorem{definition}{Definition}

\theoremstyle{remark}
\newtheorem*{notation}{Notation}
```

A proclamation created by a \newtheorem command has the style of the last \theoremstyle command preceding it. The default style is plain.

Three examples

Here are three sets of proclamation definitions to illustrate different styles and numbering schemes.

Example 1

```
\theoremstyle{plain}
\newtheorem{theorem}{Theorem}
\newtheorem{lemma}[theorem]{Lemma}
\newtheorem{definition}[theorem]{Definition}
\newtheorem{corollary}[theorem]{Corollary}
```

In a document with this set of proclamation definitions you can use theorems, lemmas, definitions, and corollaries, typeset in the most emphatic (`plain`) style. They are all numbered consecutively: **Definition 1**, **Definition 2**, **Theorem 3**, **Corollary 4**, **Lemma 5**, **Lemma 6**, **Theorem 7**, and so on.

Example 2

```
\theoremstyle{plain}
\newtheorem{theorem}{Theorem}
\newtheorem*{main}{Main Theorem}
\newtheorem{definition}{Definition}[section]
\newtheorem{lemma}[definition]{Lemma}

\theoremstyle{definition}
\newtheorem*{Rule}{Rule}
```

In this document you may use theorems, definitions, and lemmas in the most emphatic (`plain`) style, and unnumbered rules in the less emphatic (`definition`) style. Definitions and lemmas are numbered consecutively within sections. You may also use the unnumbered Main Theorem. So, for example, you may have **Definition 1.1**, **Definition 1.2**, **Main Theorem**, **Rule**, **Lemma 1.3**, **Lemma 2.1**, **Theorem 1**, and so on.

Example 3

```
\theoremstyle{plain}
\newtheorem{theorem}{Theorem}
\newtheorem{corollary}{Corollary}
\newtheorem*{main}{Main Theorem}
\newtheorem{lemma}{Lemma}
\newtheorem{proposition}{Proposition}

\theoremstyle{definition}
\newtheorem{definition}{Definition}

\theoremstyle{remark}
\newtheorem*{notation}{Notation}
```

With these proclamation definitions you can use theorems, corollaries, lemmas, and propositions in the most emphatic (`plain`) style, and an unnumbered Main Theorem. You can have definitions in the less emphatic (`definition`) style. All are separately numbered. So in the document you may have **Definition 1**, **Definition 2**, **Main Theorem**, **Lemma 1**, **Proposition 1**, **Lemma 2**, **Theorem 1**, **Corollary 1**, and so on. You can also have Notations which are unnumbered and typeset in the least emphatic (`remark`) style.

Number swapping

Proclamations can be numbered on the left, as for instance, **3.2 Theorem.** To accomplish this, type the `\swapnumbers` command before the `\newtheorem` command corresponding to the proclamation definition you want to change. This command affects all of the proclamation definitions that follow it, so the proclamation definitions in the preamble should be in two groups. The regular ones should be listed first, followed by the `\swapnumbers` command, then all the proclamations that swap numbers.

Do not swap numbers unless the publisher demands it.

Custom theorem styles

You can define custom theorem styles with the `\newtheoremstyle` command. You should very seldom do this, the three theorem styles of the document class should suffice. For more detail, see [5].

4.5 Proof environments

A proof is the contents of a `proof` environment. For instance,

Proof. This is a proof, delimited by the q.e.d. symbol. □

typed as

```
\begin{proof}
This is a proof, delimited by the q.e.d.\ symbol.
\end{proof}
```

A proof is set off from the surrounding text with some vertical space. The end of the proof is marked with the symbol □ at the end of the line. There are a few examples of the `proof` environment in the `secondarticle.tex` sample article (pages 268–275).

We start with the same rule for proofs as we have for proclamations on page 106.

Rule ■ Lists in proofs

If a proof starts with a list environment, precede the list by `\hfill`.

If you want to suppress the symbol at the end of a proof, give the command

```
\begin{proof}
   ...
   \renewcommand{\qedsymbol}{}
\end{proof}
```

To suppress the end of the proof symbol in the whole article, give the

```
\renewcommand{\qedsymbol}{}
```

command in the preamble.

To substitute another phrase for *Proof,* such as *Necessity,* as in

Necessity. This is the proof of necessity. □

use the proof environment with an optional argument:

```
\begin{proof}[Necessity]
This is the proof of necessity.
\end{proof}
```

The optional argument may contain a reference, as in

```
\begin{proof}[Proof of Theorem~\ref{T:smooth}]
```

which might be typeset as

Proof of Theorem 5. This is the proof. □

It is easy to make the mistake of placing the optional argument after \begin:

```
\begin[Proof of Theorem~\ref{T:P*}]{proof}
```

You get a message

```
! LaTeX Error: Bad math environment delimiter.

l.91 \begin{equation}
                     \label{E:cong2}
```

which is not very helpful.

There is a problem with the placement of the q.e.d. symbol if the proof ends with a displayed formula (or a list environment). For instance,

```
\begin{proof}
Now the proof follows from the equation
\[
   a^2 = b^2 + c^2.
\]
\end{proof}
```

typesets as

Proof. Now the proof follows from the equation

$$a^2 = b^2 + c^2.$$

□

To correct the placement of the q.e.d. symbol, use the \qedhere command:

```
\begin{proof}
Now the proof follows from the equation
\[
   a^2 = b^2 + c^2.\qedhere
\]
\end{proof}
```

which typesets as

Proof. Now the proof follows from the equation

$$a^2 = b^2 + c^2. \qquad \square$$

4.6 Tabular environments

A tabular environment creates a table that LaTeX treats as a "large symbol". In particular, a table cannot be broken across pages.

Here is a simple table,

Name	1	2	3
Peter	2.45	34.12	1.00
John	0.00	12.89	3.71
David	2.00	1.85	0.71

, typeset inline. This looks awful, but it does make the point that the table is just a "large symbol". The table is typed as

```
\begin{tabular}{ | l | r | r | r | }
   \hline
   Name     & 1      & 2      & 3     \\ \hline
   Peter    &  2.45 & 34.12 & 1.00\\ \hline
   John     &  0.00 & 12.89 & 3.71\\ \hline
   David    &  2.00 & 1.85  & 0.71\\ \hline
\end{tabular}
```

Name	1	2	3
Peter	2.45	34.12	1.00
John	0.00	12.89	3.71
David	2.00	1.85	0.71

Table 4.1: Tabular table.

with no blank line before or after the environment.

This table can be horizontally centered with a `center` environment (see Section 4.3). It can also be placed within a `table` environment (see Section 8.4.3). This sets the table off from the surrounding text with vertical space and you can also use the float controls `b`, `t`, `h`, `p` to specify where the table should appear (see Section 8.4.3). This also allows you to define a caption, which can be placed before or after the table.

```
\begin{table}
   \begin{center}
     \begin{tabular}{ | l | r | r | r | }
       \hline
       Name     & 1      & 2      & 3    \\ \hline
       Peter    &  2.45 & 34.12 & 1.00\\ \hline
       John     &  0.00 & 12.89 & 3.71\\ \hline
       David    &  2.00 & 1.85  & 0.71\\ \hline
     \end{tabular}
     \caption{Tabular table.}\label{Ta:first}
   \end{center}
\end{table}
```

This table is displayed as Table 4.1. It can be listed in a list of tables (see Section 8.4.3) and the table number may be referenced using the command `\ref{Ta:first}`. Note that the label must be typed *between* the caption and the `\end{table}` command.

For another example, look at the two tables in the `fonttbl.tex` file in your `samples` folder. The first is typed as

```
\begin{tabular}{r|l|l|l|l|l|l|l|l|l|l}
 & 0 & 1 & 2 & 3 & 4 & 5 & 6 & 7 & 8 & 9\\ \hline

0& \symbol{0} &\symbol{1}&\symbol{2}&\symbol{3}&
\symbol{4}&\symbol{5}&\symbol{6}&\symbol{7}&
\symbol{8}&\symbol{9}\\ \hline
                    . . .

120& \symbol{120} &\symbol{121}&\symbol{123}&
\symbol{123}&\symbol{124}&\symbol{125}&\symbol{126}
```

```
&\symbol{127} && \\ \hline
\end{tabular}
```

The second table is the same except that the numbers run from 128 to 255. The typeset table is shown in Section 3.4.4.

Rule ■ `tabular` **environments**

1. `\begin{tabular}` requires an argument consisting of a character `l`, `r`, or `c`, meaning left, right, or center alignment, for each column, and optionally, the | symbols. Each | indicates a vertical line in the typeset table. Spaces in the argument are ignored but can be used for readability.

2. Columns are separated by ampersands (`&`) and rows are separated by `\\`.

3. `&` absorbs spaces on either side.

4. The `\hline` command creates a horizontal rule in the typeset table. It is placed either at the beginning of the table (after the `\begin` line) or it must follow a `\\` command.

5. If you use a horizontal line to finish the table, you must separate the last row of the table from the `\hline` command with the `\\` command.

6. `\begin{tabular}` takes an optional argument, `b` or `t`, to specify the bottom or the top vertical alignment of the table with the baseline. The default is center alignment.

Remember to put the optional argument `b` or `t` in square brackets, as in

```
\begin{tabular}[b]{ | l | r | r | r | }
```

If you forget to place an `\hline` command right after `\\` in the last row, you get a message such as

```
! Misplaced \noalign.
\hline ->\noalign
                   {\ifnum 0=`}\fi \hrule \@height
                   \arrayrulew...
1.9 ....00 & 1.85  & 0.71 \hline
```

More column-formatting commands

The required argument of the `tabular` environment may contain column-formatting commands of various types.

An `@`*-expression,* for instance, `@{.}`, replaces the space LaTeX normally inserts between two columns with its argument. For example,

```
\begin{tabular}{r @{.} l}
   3&78\\
   4&261\\
   4
\end{tabular}
```

creates a table with two columns separated by a decimal point. In effect, you get a single, decimal-aligned column:

```
3.78
4.261
4.
```

This example is an illustration. You should use David Carlisle's `dcolumn` package if you need a decimal-aligned column (see Section 10.3.1).

The width of a column depends on the entries in the column by default. You can specify a width by using the p column specifier:

`p{`*width*`}`

For instance, if you want the first column of Table 4.1 to be 1 inch wide, then type

```
\begin{tabular}{ | p{1in} | r | r | r | }\hline
   Name    &  1    & 2     & 3     \\ \hline
   Peter   &  2.45 & 34.12 & 1.00\\ \hline
   John    &  0.00 & 12.89 & 3.71\\ \hline
   David   &  2.00 & 1.85  & 0.71\\ \hline
\end{tabular}
```

which typesets as

Name	1	2	3
Peter	2.45	34.12	1.00
John	0.00	12.89	3.71
David	2.00	1.85	0.71

To center the items in the first column, precede *each* item with a `\centering` command (see Section 4.3). Note that the first column is actually somewhat over 1 inch wide, because of the extra space provided around the column boundaries.

The p column specifier can also be used for multiline entries.

Refinements

`\hline` draws a horizontal line the whole width of the table. `\cline{a-b}` draws a horizontal line from column *a* to column *b*. For instance,

`\cline{1-3}` or `\cline{4-4}`

Another useful command is `\multicolumn`, which is used to span more than one column, for example,

```
\multicolumn{3}{c}{\emph{absent}}
```

The first argument is the number of columns spanned by the entry, the second is the alignment (an optional vertical line designator | for this row only can also be included), and the third argument is the entry. Note that the entry for the spanned columns is in braces. An example is shown in Table 4.2, typed as follows:

```
\begin{table}[h!]
   \begin{center}
      \begin{tabular}{ | l | r | r | r | } \hline
         Name    & 1    & 2     & 3\\ \hline
         Peter   & 2.45 & 34.12 & 1.00\\ \hline
         John    & \multicolumn{3}{c |}{\emph{absent}}\\
         \hline
         David   & 2.00 & 1.85  & 0.71\\ \hline
      \end{tabular}
      \caption{Table with \bsl\texttt{multicolumn}.}
      \label{Ta:mc}
   \end{center}
\end{table}
```

The next example, shown in Table 4.3, uses the `\multicolumn` and `\cline` commands together:

```
\begin{table}[t]
   \begin{center}
      \begin{tabular}{ | c  c  | c | r | } \hline
         Name  & Month & Week & Amount\\ \hline
         Peter & Jan.  & 1    & 1.00\\ \cline{3-4}
               &       & 2    & 12.78\\ \cline{3-4}
               &       & 3    & 0.71\\ \cline{3-4}
               &       & 4    & 15.00\\ \cline{2-4}
               & \multicolumn{2}{| l}{Total} & 29.49\\
               \hline
         John  & Jan.  & 1    & 12.01\\ \cline{3-4}
               &       & 2    & 3.10\\ \cline{3-4}
```

```
                  &         & 3     & 10.10\\ \cline{3-4}
                  &         & 4     & 0.00\\ \cline{2-4}
                  & \multicolumn{2}{| l}{Total} & 25.21\\
                  \hline
              \multicolumn{3}{|l}{Grand Total} & 54.70\\
              \hline
          \end{tabular}
          \caption{Table with \bsl\texttt{multicolumn}
          and \bsl\texttt{cline}.}\label{Ta:multicol+cline}
      \end{center}
  \end{table}
```

The \parbox command (see Section 3.9.3) can be used to produce a single multi-line entry. Recall that the first argument of \parbox is the width of the box. A p{} width designator creates a column in which all entries can be multiline. As an example, to replace Grand Total by Grand Total for Peter and John, type the last line as

```
\multicolumn{3}{l}{ \parbox[b]{10em}{Grand Total\\
for Peter and John} } & 54.70\\ \hline
```

Note the use of the bottom alignment option (see Section 3.9.3). The last row of the

Name	1	2	3
Peter	2.45	34.12	1.00
John	*absent*		
David	2.00	1.85	0.71

Table 4.2: Table with \multicolumn.

Name	Month	Week	Amount
Peter	Jan.	1	1.00
		2	12.78
		3	0.71
		4	15.00
	Total		29.49
John	Jan.	1	12.01
		2	3.10
		3	10.10
		4	0.00
	Total		25.21
Grand Total			54.70

Table 4.3: Table with \multicolumn and \cline.

modified table prints

Grand Total for Peter and John	54.70

The spacing above Grand Total is not quite right. It can be adjusted with a strut (see Section 3.9.5),

```
\parbox[b]{10em}{\strut Grand Total\\
                for Peter and John:}
```

Finally, vertical spacing can be adjusted by redefining `\arraystretch`. For instance, in the table

	Area	**Students**
5th Grade:	63.4 m^2	22
6th Grade:	62.0 m^2	19
Overall:	62.6 m^2	20

typed as

```
\begin{center}
  \begin{tabular}{|r|c|c|}\hline
       & \textbf{Area}  & \textbf{Students}\\ \hline
    \textbf{5th Grade}: & 63.4 m\textsuperscript{2} &22\\
    \hline
    \textbf{6th Grade}: & 62.0 m\textsuperscript{2} &19\\
    \hline
    \textbf{Overall}: & 62.6 m\textsuperscript{2} &20\\
    \hline
   \end{tabular}
\end{center}
```

you may find that the rows are too crowded. The vertical spacing may be adjusted by adding the line

```
\renewcommand{\arraystretch}{1.25}
```

to the `tabular` environment. To limit its scope, add it after

```
\begin{center}
```

The adjusted table is typeset as

	Area	**Students**
5th Grade:	63.4 m^2	22
6th Grade:	62.0 m^2	19
Overall:	62.6 m^2	20

In some tables, horizontal and vertical lines do not always intersect as desired. Fine control over these intersections is provided by the `hhline` package (see Section 10.3.1).

Chapter 5 of *The LaTeX Companion,* 2nd edition [56] deals with tabular material, discussing many extensions, including multipage tables, decimal-point alignment, footnotes in tables, tables within tables, and so on.

4.6.1 Table styles

LaTeX can draw double horizontal and vertical lines in tables with ease. As a result, there are far too many double lines in LaTeX tables, resulting in cluttered and confusing tables. *The Chicago Manual of Style,* 15th edition [11] has almost 80 pages on tables. For simple tables it advocates a simple style, as shown in Table 4.4. Notice

- the generous space above and below the column heads, which has been achieved with the command `\rule[-8pt]{0pt}{22pt} \rule[-8pt]{0pt}{22pt}`
- some extra space above the first line of data, which has been achieved with the command `\rule{0pt}{14pt}`
- the columns of equal width, which has been achieved with `p{70pt}` commands,
- no vertical lines.

Most tables in this book have been designed according to this style using Simon Fear's `booktabs` package (see Section D.1).

Table 4.4: Smokers and Nonsmokers, by Sex.

	Smoke	Don't Smoke	Total
Males	1,258	2,104	3,362
Females	1,194	2,752	3.946
Total	2,452	4,856	7,308

4.7 Tabbing environments

Although of limited use for mathematical typesetting, the `tabbing` environment can be useful for typing algorithms, computer programs, and so forth. LaTeX calculates the width of a column in the `tabular` environment based on the widest entry (see Section 4.6). The tabbing environment allows you to control the width of the columns.

The \\ command is the line separator, tab stops are set by \= and are remembered by LaTeX in the order they are given, and \> moves to the next tab position.

You can easily reset tab positions. For instance, if you are past the second tab position by using \> twice, and there is a third tab position, the \= command resets it.

Lines of comments may be inserted with the `\kill` command, see the examples below, or with the `%` character. The difference is that a line with `\kill` can be used to set tab stops, whereas a commented out line cannot.

A simple example:

```
PrintTime
     Block[timing],
          timing = Timing[expr];
          Print[ timing[[1]] ];
     ]
End[]
```

typed as

```
{\ttfamily
\begin{tabbing}
   Print\=Time\\
   \>Block\=[timing,\\
   \>\>timing = Timing[expr];\\
   (careful with initialization)\kill
   \>\>Print[ timing[[1]] ];\\
   \>]\\
   End[]
\end{tabbing}
}% end \ttfamily
```

An alternative method is to use a line to set the tab stops, and then `\kill` the line so it does not print:

```
{\ttfamily
\begin{tabbing}
   \hspace*{.25in}\=\hspace{2ex}\=\hspace{2ex}\=
```

```
        \hspace{2ex}\kill
   \>  $k := 1$\\
   \>  $l_k := 0$; $r_k := 1$\\
   \>  loop\\
   \>  \> $m_k := (l_k + r_k)/2$\\
   \>  \> if $w < m_k$ then\\
   \>  \>  \> $b_k := 0$; $r_k := m_k$\\
   \>  \> else if $w > m_k$ then\\
   \>  \>  \> $b_k := 1$; $l_k := m_k$\\
   \>  \> end if\\
   \>  \> $k := k + 1$\\
   \> end loop
\end{tabbing}
}% end \ttfamily
```

which typesets as

$k := 1$
$l_k := 0$; $r_k := 1$
loop
  $m_k := (l_k + r_k)/2$
  if $w < m_k$ then
    $b_k := 0$; $r_k := m_k$
  else if $w > m_k$ then
    $b_k := 1$; $l_k := m_k$
  end if
  $k := k + 1$
end loop

Some simple rules:

- There is no \\ command on a line containing the `\kill` command.
- You may set the tabs in a `\kill` line with `\hspace` commands.
- The \> command moves to the next tab stop, even if the text you have already typed extends past that stop, which can result in overprinting.
- The `tabbing` environment has to be typeset with typewriter style font—note the `\ttfamily` command.

To illustrate the third rule, type

```
\begin{tabbing}
   This is short.\=\\
   This is much longer, \> and jumps back.
\end{tabbing}
```

which typesets as

This is short.
This is much longer, and jumps back.

If you do not follow the fourth rule, be careful with your tabbing. You do not really have to use typewriter style font—just beware of the pitfalls.

There are a number of packages that help type programming code. I mention here only two: `listings` and `program`. For more information, please consult Chapter 5 of *The LaTeX Companion,* 2nd edition [56].

4.8 Miscellaneous displayed text environments

There are four more displayed text environments, of limited use in math:
`quote`, `quotation`, `verse`, and `verbatim`.
We also discuss an inline version of the `verbatim` environment, the `\verb` command.

Quotes

The `quote` environment is used for short (one paragraph) quotations:

> It's not that I'm afraid to die. I just don't want to be there when it happens. *Woody Allen*
>
> Literature is news that STAYS news. *Ezra Pound*

which is typed as:

```
\begin{quote}
   It's not that I'm afraid to die. I just don't
   want to be there when it happens.
   \emph{Woody Allen}

   Literature is news that STAYS news.
   \emph{Ezra Pound}
\end{quote}
```

Note that multiple quotes are separated by blank lines.

Quotations

In the quotation environment, blank lines mark new paragraphs:

KATH: Can he be present at the birth of his child?

ED: It's all any reasonable child can expect if the dad is present at the conception.

Joe Orton

is typed as

```
\begin{quotation}
   KATH: Can he be present at the birth of his child?

   ED: It's all any reasonable child can expect
   if the dad is present at the conception.
   \begin{flushright}
      \emph{Joe Orton}
   \end{flushright}
\end{quotation}
```

Verses

A verse environment,

I think that I shall never see
A poem lovely as a tree.

Poems are made by fools like me,
But only God can make a tree.

Joyce Kilmer

is typed as

```
\begin{verse}
   I think that I shall never see\\
   A poem lovely as a tree.

   Poems are made by fools like me,\\
   But only God can make a tree.

   \begin{flushright}
      \emph{Joyce Kilmer}
```

```
    \end{flushright}
\end{verse}
```

Lines are separated by \\ and stanzas by blank lines. Long lines are typeset with hanging indent.

Verbatim typesetting

Finally, there is the `verbatim` text environment. You may need it if you write *about* LaTeX or some other computer programming language or if you have to include portions of a source file or user input in your typeset work. Most of the displayed source in this book was written in a `verbatim` environment. For instance, you may have to write to a journal about an article you are proofreading:

Formula (2) in Section 3 should be typed as follows:

```
\begin{equation}
D = \{ x_0 \mid x_0 \Rightarrow a_1 \} \tag{2}
\end{equation}
```

Please make the necessary corrections.

The problem is that if you just type

```
Formula (2) in Section 3 should be typed as follows:
\begin{equation}
    D = \{ x_0 \mid x_0 \Rightarrow a_1 \} \tag{2}
\end{equation}
Please make the necessary corrections.
```

it typesets as

Formula (2) in Section 3 should be typed as follows:

$$D = \{ x_0 \mid x_0 \Rightarrow a_1 \} \tag{2}$$

Please make the necessary corrections.

To get the proper typeset form, type it as follows:

```
Formula (2) in Section 3 should be typed as follows:
\begin{verbatim}
\begin{equation}
D = \{ x_0 \mid x_0 \Rightarrow a_1 \} \tag{2}
```

```
\end{equation}
\end{verbatim}
Please make the necessary corrections.
```

Rule ■ `verbatim` **text environments**

A `verbatim` environment cannot be placed within

- Another `verbatim` environment
- The argument of a command
- The closing line, `\end{verbatim}`, must be on a line by itself.

A violation of the first rule results in unmatched environment delimiters. You get an error message such as

```
! \begin{document} ended by \end{verbatim}.
```

A violation of the second rule gives an error message such as

```
! Argument of \@xverbatim has an extra }.
```

Tip There are two traps to avoid when using the `verbatim` environment.

1. If the `\end{verbatim}` line starts with spaces, a blank line is added to the typeset version.
2. Any characters following `\end{verbatim}` on the same line are dropped and you get a LaTeX warning.

To illustrate the first trap, type the last two lines of the previous example as follows:

```
␣\end{verbatim}
Please make the necessary corrections.
```

Then you find an unintended blank line before the last line.

The second trap can be seen if you type the last line of the above example as

```
\end{verbatim} Please make the necessary corrections.
```

When typeset, Please make the necessary corrections. does not appear, and you receive a warning

```
LaTeX Warning: Characters dropped after
 '\end{verbatim}' on input line 17.
```

The verbatim package provides several improved versions of the verbatim environment (see Section 10.3.1). To use this package, include the command

```
\usepackage{verbatim}
```

in the preamble. In fact, the rules discussed in this section are those of the verbatim package.

The verbatim environment has some interesting variants and a number of them are discussed in Section 3.4 of *The LaTeX Companion,* 2nd edition [56]. For instance, the alltt package, which is part of the standard LaTeX distribution (see Section 10.3) is used to type the command syntax in this book. See the full syntax of \newtheorem on page 108 for an example.

Verbatim typesetting inline

The verbatim environment also has an inline version called \verb. Here is an example:

```
Some European e-mail addresses contain \%;
recall that you have to type \verb+\%+ to get \%.
```

which prints

Some European e-mail addresses contain %; recall that you have to type \% to get %.

The character following the \verb command is a delimiter. In this example, I have used +. The argument starts with the character following the delimiter, and it is terminated by the next occurrence of the delimiter. In this example, the argument is \%.

Choose the delimiter character carefully. For instance, if you want to typeset

```
$\sin(\pi/2 + \alpha)$
```

verbatim, and you type

```
\verb+$\sin(\pi/2 + \alpha)$+
```

then you get the message

```
! Missing $ inserted.
<inserted text>
                $
l.5 \verb+$\sin(\pi/2 + \alpha
                                    )$+
```

Indeed, the argument of \verb is $\sin(\pi/2 because the second + terminates the \verb command. Then LaTeX tries to typeset \alpha)$+, but cannot because it is not in math mode. Use another character, such as !, in place of +:

```
\verb!$\sin(\pi/2 + \alpha)$!
```

Rule ■ verb command

- The entire \verb command must be on a single line of your source file.
- There can be no space between the \verb command and the delimiter.
- The \verb command cannot appear in the argument of another command.
- The \verb command cannot be used within an aligned math environment.
- Do not use * as a delimiter.

If you violate the first rule, as in

```
\verb!$\sin(\pi/2 +
\alpha)$!
```

you get the message

```
! LaTeX Error: \verb command ended by end of line.

1.6 \verb!$\sin(\pi/2 +
```

The \verb command has a *-ed version which prints spaces as ␣ symbols. For example, \today␣the is typed as \verb*+\today the+.

The \verb command can perform the function of the verbatim environment. The last message, which was displayed in a verbatim environment, may be typed as follows:

```
you get the message\\[8pt]
\verb|! LaTeX Error: \verb command ended by end of line.|\\
\verb| |\\
\verb|1.6 \verb!$\sin(\pi/2 +|\\[8pt]
```

Rule ■ Simulating verbatim with verb

1. End the line before the verbatim environment with \\[8pt].
2. Each line *xxx* of the verbatim environment is placed in the construct:

   \verb|*xxx*|

   If | occurs in *xxx*, then choose a different delimiter.

3. The last line *yyy* of the verbatim environment is placed in the construct:

   `\verb|`*yyy*`|\\[8pt]`

   If `|` occurs in *yyy*, then choose a different delimiter.

However, simulating `verbatim` with `verb` takes away the flexibility `verbatim` provides in displaying the page.

CHAPTER 5

Typing math

LaTeX was designed for typesetting math. I address this topic in detail.

A math formula can be typeset *inline*, as part of the current paragraph, or *displayed*, on a separate line or lines with vertical space before and after the formula.

In this and the next chapter we discuss formulas that are set inline or displayed on a *single line*. In Chapter 7 we address *multiline* math formulas.

We start with a discussion of LaTeX's basic math environments (Section 5.1), spacing rules in math (Section 5.2), and continue with the `equation` environment (Section 5.3). The basic constructs of a formula—arithmetic (including subscripts and superscripts), binomial coefficients, ellipses, integrals, roots, and text—are discussed in detail in Section 5.4. From the basic constructs of that section, you can build very complicated formulas, one step at a time. The process is illustrated in Section 5.9.

Delimiters, operators, and math accents are dealt with in Sections 5.5–5.7. In Section 5.8, we discuss three types of stretchable horizontal lines that can be used above or below a formula: braces, bars, and arrows. There are also stretchable arrow math symbols.

Section 5.10 is our *Formula Gallery,* in which you find a large number of illustrations, some straightforward, some more imaginative, of the math constructs introduced in the preceding sections.

G. Grätzer, *More Math Into* LaTeX, DOI 10.1007/978-3-319-23796-1_5

5.1 *Math environments*

A formula in a LaTeX document can be typeset *inline,* like the congruence $a \equiv b \pod{\theta}$ or the integral $\int_{-\infty}^{\infty} e^{-x^2}\, dx = \sqrt{\pi}$, or *displayed*, as in

$$a \equiv b \pod{\theta}$$

or

$$\int_{-\infty}^{\infty} e^{-x^2}\, dx = \sqrt{\pi}$$

Notice how changing these two formulas from inline to displayed affects their appearance.

Inline and displayed math formulas are typeset using the *math environments* `math` and `displaymath`, respectively. Because math formulas occur so frequently, LaTeX has abbreviations: the special braces `\(` and `\)` or `$` are used for the `math` environment, and `\[` and `\]` for the `displaymath` environment. So our inline example may be typed as

```
$a \equiv b \pod{\theta}$
```

or

```
\( a \equiv b \pod{\theta} \)
```

or

```
\begin{math}
   a \equiv b \pod{\theta}
\end{math}
```

The displayed example can be typed as

```
 \[
   \int_{-\infty}^{\infty} e^{-x^{2}} \, dx = \sqrt{\pi}
 \]
```

or

```
\begin{displaymath}
   \int_{-\infty}^{\infty} e^{-x^{2}} \, dx = \sqrt{\pi}
\end{displaymath}
```

Using `$` as a delimiter for a `math` environment is a bit of an anomaly, since the same character is used as both an opening and closing delimiter. This dual purpose use makes it more difficult for LaTeX to diagnose an incorrect use of `$`. For instance,

```
Let $a be a real number, and let $f$ be a function.
```

would be interpreted by LaTeX as follows:

- `Let` is ordinary text
- `$a be a real number, and let $` is math
- `f` is interpreted as ordinary text
- `$ be a function.` is thought to be a `math` environment (opened by $) that should be closed by the next $ in the paragraph

Because the paragraph ends with no more dollar signs appearing, you get the message

```
!! Missing $ inserted.
<inserted text>
                $
l.29
```

and giving you the line number of the end of the paragraph. This message tells you that LaTeX would place a $ at the end of the paragraph when it proceeds with the typesetting. Press Return; LaTeX produces the following:

Let *abearealnumber*, *andlet*f*beafunction.*

The text that ended up in a math environment is run together because math environments ignore spaces (see Section 5.2).

If you use `\(` and `\)` as special braces for the `math` environment, LaTeX handles the same mistake more elegantly:

```
Let \( a be a real number, and let \( f \) be a function.
```

gives the message

```
! LaTeX Error: Bad math environment delimiter.

l.25 Let \( a be a real number, and let \(
                                f \) be a function.
```

LaTeX realizes that the first `\(` opens a `math` environment, so the second `\(` must be in error. In this case, the line number in the message is correct.

Throughout this book, like nearly everyone else, I use $ to delimit inline math.

TeX uses $$ to open and close a displayed math environment. In LaTeX, this may occasionally cause problems. Don't do it! Try the `fleqn` document class option of `amsart` (see Section 9.5) as an example of what can go wrong.

Rule ■ Math environments
No blank lines are permitted in a `math` or `displaymath` environment.

If you violate this rule, LaTeX generates a message,

```
! Missing $ inserted.
<inserted text>
                $
...
l.7
```

where the line number points inside the environment.

Multiline math environments, such as the examples in Sections 1.7.3–1.7.4, are discussed in Chapter 7.

5.2 *Spacing rules*

In text, the most important spacing rule is that any number of spaces in the source file equals one space in the typeset document. The spacing rule for math mode is even more straightforward.

Rule ■ Spacing in math
LaTeX ignores spaces in math.

In other words, all spacing in math mode is provided by LaTeX. For instance,

```
$a+b=c$
```

and

```
$a + b = c$
```

are both typeset as $a + b = c$.

There are two exceptions to this rule:

1. A space indicating the end of a command name is recognized. For instance, in

   ```
   $a \quad b$
   ```

   LaTeX does not ignore the space between `\quad` and `b`.

2. If you switch back to text mode inside a math formula with a `\text` command (see Section 5.4.6), then the text spacing rules apply in the argument of such a command.

As you see, LaTeX provides controls for spaces in typeset math. The spaces you type in math do not affect the typeset document. But keep this tip in mind.

Tip Format your source file so that it is easy to read.

When typing a source file, the following is good practice:

- Place `\[` and `\]` on lines by themselves.
- Leave spaces before and after binary operations and binary relations, including the equal sign.
- Indent—by three spaces, for example—the contents of environments so they stand out.
- Keep a formula on a single line of the source file, if you can.

Develop your own style of typing math, and stick with it.

Tip The spacing after a comma is different in math and text. Do not leave a trailing comma in inline math.

So do not type

```
If $a = b,$ then
```

but move the comma out.

5.3 Equations

An *equation* is a numbered formula displayed on a single typeset line.

Equations are typed in an `equation` environment. The `equation` environment and `displaymath` environment are exactly the same except that the `equation` environment assigns a number to the displayed formula

$$\int_{-\infty}^{\infty} e^{-x^{2}} \, dx = \sqrt{\pi} \tag{1}$$

This example is typed as

```
\begin{equation}\label{E:int}
   \int_{-\infty}^{\infty} e^{-x^{2}} \, dx = \sqrt{\pi}
\end{equation}
```

The \label command in the equation environment is optional. If you use a \label command, the number assigned to the equation can be referenced with the \ref command. So

```
see~(\ref{E:int})
```

typesets as see (1). Even better, use the \eqref command, which places the parentheses automatically:

```
see~\eqref{E:int}
```

also typesets as see (1). In fact, the \eqref command does more: It typesets the reference *upright,* even in italicized or slanted text. For more information about cross-referencing, see Section 1.7.2.

Analogously, the \upn command forces the use of upright characters for digits, punctuations, parentheses, etc. LaTeX numbers equations consecutively. As a rule, equations are numbered consecutively throughout articles, whereas in books, numbering starts from 1 at the start of each chapter. You may also choose to have equations numbered within each section—(1.1), (1.2), ..., in Section 1; (2.1), (2.2), ..., in Section 2; and so on—by including the command

```
\numberwithin{equation}{section}
```

in the preamble of your document (see Section 8.2). "Manual control" of numbering is discussed in Section 14.5.1, group numbering in Section 7.4.4.

The *-ed form of the equation environment suppresses numbering, so it is equivalent to the displaymath environment (or the special braces \[and \]).

Rule ■ Equation environment

1. No blank lines are permitted within an equation or equation* environment.

2. No blank line before the environment.

If you typeset

```
\begin{equation}\label{E:int}
   \int_{-\infty}^{\infty} e^{-x^{2}} \, dx = \sqrt{\pi}

\end{equation}
```

LaTeX generates the familiar, but misleading, message

```
! Missing $ inserted.
```

5.4 Basic constructs

A formula is built by combining various basic constructs. This section discusses the following constructs:

- Arithmetic operations – Subscripts and superscripts
- Binomial coefficients
- Ellipses
- Integrals
- Roots
- Text
- Hebrew and Greek letters

Read carefully the basic constructs *important for your work*. Additional constructs are discussed in subsequent sections of this chapter.

5.4.1 Arithmetic operations

The *arithmetic operations* are typed pretty much as you would expect. To get $a + b$, $a - b$, $-a$, a/b, and ab, type

```
$a + b$, $a - b$, $-a$, $a / b$, $a b$
```

There are two other forms of multiplication and one of division: $a \cdot b$, $a \times b$, and $a \div b$. They are typed as follows:

```
$a \cdot b$,  $a \times b$,  $a \div b$
```

In displayed formulas, *fractions* are usually typed with the `\frac` command. To get

$$\frac{1 + 2x}{x + y + xy}$$

type

```
\[
   \frac{1 + 2x}{x + y + xy}
\]
```

You can use display-style fractions inline with `\dfrac`, and inline-style fractions in displayed math environments with `\tfrac`; for example, $\dfrac{3 + a^{2}}{4 + b}$ is typed as

```
$\dfrac{3 + a^{2}}{4 + b}$
```

and

$$\tfrac{3+a^2}{4+b}$$

is typed as

```
\[
  \tfrac{3 + a^{2}}{4 + b}
\]
```

The `\dfrac` command is often used in matrices whose entries would look too small with the `\frac` command. See Formula 20 in the *Formula Gallery* (Section 5.10) for an example, and Section 6.7.1 for other fraction variants.

Subscripts and superscripts

Subscripts are typed with `_` and *superscripts* with `^`. Remember to enclose the subscripted or superscripted expression in braces:

```
\[
   a_{1},\ a_{i_{1}},\ a^{2},\  a^{b^{c}},\ a^{i_{1}},\
      a_{i} + 1,\ a_{i + 1},\ a_{1}^{2},\ a^{2}_{1}
\]
```

typesets as

$$a_{1},\ a_{i_{1}},\ a^{2},\ a^{b^{c}},\ a^{i_{1}},\ a_{i} + 1,\ a_{i + 1},\ a_{1}^{2},\ a^{2}_{1}$$

For $a^{b^{c}}$, type `$a^{b^{c}}$`, not `$a^{b}^{c}$`. If you type the latter, you get the message

```
! Double superscript.
```

Similarly, $a_{b_{c}}$ is typed as `$a_{b_{c}}$`, not as `$a_{b}_{c}$`.

In many instances, the braces for the subscripts and superscripts could be omitted, but you should type them anyway.

Tip You may safely omit the braces for a subscript or superscript that is a single digit or letter, as in `$a_1$` and `$(a + b)^x$`, which are typeset as a_1 and $(a+b)^x$. Be careful, however, if you have to edit `$a_1$` to make it a_{12}, then the braces can no longer be omitted, you must type `$a_{12}$` to obtain a_{12} because `$a_12$` typesets as a_12.

There is one symbol that is automatically superscripted in math mode, the prime, that is, `'`. To get $f'(x)$, type `$f'(x)$`. However, to get f'^2 you must type

```
$f^{\prime 2}$
```

Typing `${f'}^{2}$` results in ${f'}^{2}$, with the 2 too high; typing it as `$f'^{2}$` causes a double superscript error. Sometimes you may want a symbol to appear superscripted or subscripted by itself, as in the phrase

use the symbol † to indicate the dualspace

typed as

```
use the symbol ${}^{\dagger}$ to indicate the dualspace
```

where `{}` is the *empty group*. The empty group can be used to separate symbols, to terminate commands, or as the base for subscripting and superscripting.

The `\sb` and `\sp` commands also typeset subscripts and superscripts, respectively, as in

```
$a\sb{1} - a\sp{x + y}$
```

which produces $a_1 - a^{x+y}$. These commands are seldom used, however, except in the `alltt` environment (see Section 10.3) and in the *Mathematical Reviews* of the AMS.

For multiline subscripts and superscripts, see Section 5.6.4.

5.4.2 *Binomial coefficients*

Binomials are typeset with the `\binom` command. Here are two examples shown inline, $\binom{a}{b+c}$ and $\binom{\frac{n^2-1}{2}}{n+1}$, and displayed:

$$\binom{a}{b + c} \text{ and } \binom{\frac{n^{2} - 1}{2}}{n + 1}$$

The latter is typed as

```
\[
   \binom{a}{b + c} \text{ and }
   \binom{\frac{n^{2} - 1}{2}}{n + 1}
\]
```

You can use display-style binomials inline with `\dbinom`, and inline-style binomials in displayed math environments with `\tbinom`. For example, $\dbinom{a}{b + c}$ is typed as `$\dbinom{a}{b + c}$`. See Section 6.7.1 for other variants.

5.4.3 *Ellipses*

There are two types of *ellipsis* in math, the *low* or *on-the-line ellipsis*, as in

$$F(x_1, x_2, \dots, x_n)$$

and the *centered ellipsis*, as in

$$x_1 + x_2 + \cdots + x_n$$

These two formulas are typed as

```
\[
   F(x_{1}, x_{2}, \dots, x_{n})
\]
```

and

```
\[
   x_{1} + x_{2} + \dots + x_{n}
\]
```

LaTeX uses the symbol following a `\dots` command to decide whether to use a low or centered ellipsis. If it fails to make the right decision as in

$$\alpha(x_1 + x_2 + \ldots)$$

typed as

```
\[
   \alpha(x_{1} + x_{2} + \dots)
\]
```

help LaTeX by giving the command `\ldots` for low and `\cdots` for centered ellipsis. So to get the last formula right, type

```
\[
   \alpha(x_{1} + x_{2} + \cdots)
\]
```

and it typesets correctly:

$$\alpha(x_1 + x_2 + \cdots)$$

There are five more variants of the `\dots` command:

- `\dotsc` for an ellipsis followed by a comma
- `\dotsb` for an ellipsis followed by a binary operation or relation
- `\dotsm` for an ellipsis followed by multiplication
- `\dotsi` for an ellipsis with integrals
- `\dotso` for an "other" ellipsis

These commands not only force the ellipsis to be low or centered, but also adjust the spacing.

See Section 7.7.1 for an example of *vertical dots* with the `\vdots` command and *diagonal dots* with the `\ddots` command.

5.4.4 Integrals

You have already seen the formula $\int_{-\infty}^{\infty} e^{-x^2}\,dx = \sqrt{\pi}$ in both inline and displayed forms in the first section of this chapter. The lower limit is typeset as a subscript and the upper limit is typeset as a superscript. To force the limits below and above the integral symbol, use the `\limits` command. The `\nolimits` command does the reverse. To typeset $\int\limits_{-\infty}^{\infty} e^{-x^{2}} \, dx = \sqrt{\pi}$, type

```
$\int\limits_{-\infty}^{\infty} e^{-x^{2}} \, dx = \sqrt{\pi}$
```

See Section 9.5 for a discussion of the `intlimits` document class option.

There are five commands to produce variants of the basic integral symbol:

```
\oint  \iint  \iiint  \iiiint  \idotsint
```

which typeset as

$$\oint \quad \iint \quad \iiint \quad \iiiint \quad \idotsint$$

For complicated bounds, use the `\substack` command or the `subarray` environment (see Section 5.6.4).

5.4.5 Roots

The `\sqrt` command produces a square root, for instance,

`$\sqrt{5}$`	typesets as	$\sqrt{5}$
`$\sqrt{a + 2b + c^{2}}$`	typesets as	$\sqrt{a + 2b + c^{2}}$

Here is a more interesting example:

$$\sqrt{1 + \sqrt{1 + \frac{1}{2}\sqrt{1 + \frac{1}{3}\sqrt{1 + \frac{1}{4}\sqrt{1 + \cdots}}}}}$$

typed as

```
\[
   \sqrt{1 + \sqrt{1 + \frac{1}{2}\sqrt{1 + \frac{1}{3}
    \sqrt{1 + \frac{1}{4}\sqrt{1 + \cdots}}}}}
\]
```

For n-th roots other than the square root, that is, $n \neq 2$, specify n with an optional argument. To get $\sqrt[3]{5}$, type `$\sqrt[3]{5}$`.

Root refinement

In $\sqrt[g]{5}$, typed as `$\sqrt[g]{5}$`, the placement of g is not very pleasing. LaTeX provides two additional commands to allow you to adjust the position of g:

`\leftroot` moves g *to the left*—or *to the right* with a negative argument

`\uproot` moves g *up*—or *down* with a negative argument

You may prefer one of the following variants:

$\sqrt[\leftroot{2}\uproot{2} g]{5}$ typed as `$\sqrt[\leftroot{2} \uproot{2} g]{5}$`
$\sqrt[\uproot{2} g]{5}$ typed as `$\sqrt[\uproot{2} g]{5}$`

Experiment with `\leftroot` and `\uproot` to find the best spacing.

Note that LaTeX is very finicky with this optional argument. Typing a space after `[`, as in `$\sqrt[ \uproot{2} g]{5}$`, gives the message

```
! Package amsmath Error: Invalid use of \uproot.
```

There may also be problems with vertical spacing under the root symbol (see Section 6.5).

5.4.6 *Text in math*

LaTeX allows you to include text in formulas with the `\text` command. The formula

$$A = \{ x \mid x \in X_{i}, \text{ for some } i \in I \}$$

is typed as

```
\[
   A = \{ x \mid x \in X_{i}, \text{ for some }
        i \in I \}
\]
```

Note that you have to leave space before `for` and after `some` inside the argument of `\text`. The argument of the `\text` command is always typeset in a single line.

Sometimes it is more convenient to go into math mode within the argument of a `\text` command rather than end the `\text` and start another, as in

$$A = \{ x \mid \text{for x large} \}$$

which may be typed as

```
\[
   A = \{ x \mid \text{for $x$ large} \}
\]
```

The \text command correctly sizes its argument to match the context. The formula

$$a_{\text{left}} + 2 = a_{\text{right}}$$

is typed as

```
\[
   a_{\text{left}} + 2 = a_{\text{right}}
\]
```

Note that \text typesets its argument *in the size and shape* of the surrounding text. If you want the text in a formula to be typeset in the document font family (see Section 3.6.2) independent of the surrounding text, use

```
\textnormal{ ... }
```

or

```
{\normalfont ...}
```

For instance, if you have a constant a_{right}, then in a theorem:

Theorem 1. *The constant* a_{right} *is recursive in a.*

The subscript is wrong. To get it right, type the constant as

```
$a_{\normalfont\text{right}}$
```

Now the theorem typesets as

Theorem 1. *The constant* a_{right} *is recursive in a.*

Any of the text font commands with arguments (see Section 3.6.3) can also be used in math formulas. For instance, \textbf uses the size and shape of the surrounding text to typeset its argument in bold (extended).

Tip If a is subscripted l for left, as in a_{l}, type it as `$a_\text{l}$`. Unfortunately, many papers use a_l, typed as `$a_l$`. The rule is simple:

if in a_{l}, the character l is text, type it as `$a_\text{l}$`;

if in a_l, the character l is a variable, type it as `$a_l$`.

5.4.7 Hebrew and Greek letters

Math uses only four Hebrew letters: $\aleph$, $\beth$, $\daleth$, $\gimel$, typed as

```
\com{aleph}, \com{beth}, \com{daleth}, \com{gimel}
```

The 26 Greek letters come in lower case and some also in upper case. There is no upper case α, because it is the same as the Latin letter A. Seven lower case Greek letter also come in a variant. For instance, the variant of ϕ is φ. Tables 5.1 and 5.2 list them all; see also Section A.1.

Type	Typeset	Type	Typeset	Type	Typeset
\alpha	α	\iota	ι	\sigma	σ
\beta	β	\kappa	κ	\tau	τ
\gamma	γ	\lambda	λ	\upsilon	υ
\delta	δ	\mu	μ	\phi	ϕ
\epsilon	ϵ	\nu	ν	\chi	χ
\zeta	ζ	\xi	ξ	\psi	ψ
\eta	η	\pi	π	\omega	ω
\theta	θ	\rho	ρ		
\varepsilon	ε	\varpi	ϖ	\varsigma	ς
\vartheta	ϑ	\varrho	ϱ	\varphi	φ
	\digamma	$\digamma$	\varkappa	$\varkappa$	

Table 5.1: Lowercase Greek letters

Type	Typeset	Type	Typeset	Type	Typeset
\Gamma	Γ	\Xi	Ξ	\Phi	Φ
\Delta	Δ	\Pi	Π	\Psi	Ψ
\Theta	Θ	\Sigma	Σ	\Omega	Ω
\Lambda	Λ	\Upsilon	Υ		
\varGamma	$\varGamma$	\varXi	$\varXi$	\varPhi	$\varPhi$
\varDelta	$\varDelta$	\varPi	$\varPi$	\varPsi	$\varPsi$
\varTheta	$\varTheta$	\varSigma	$\varSigma$	\varOmega	$\varOmega$
\varLambda	$\varLambda$	\varUpsilon	$\varUpsilon$		

Table 5.2: Uppercase Greek letters

5.5 Delimiters

Delimiters are used to enclose some subformulas. In the following formula we use two delimiters: parentheses and square brackets: `$[(a*b)+(c*d)]^2$`; this typesets as $[(a * b) + (c * d)]^2$. LaTeX knows that parentheses and square brackets are delimiters, and spaces them accordingly.

The standard delimiters are shown in Table 5.3. Note that delimiters are math symbols with special spacing rules and you can use them in any way you please, not only in pairs. LaTeX does not stop you from typing `\uparrow(x]`, which typesets as $\uparrow (x]$.

Observe the difference in spacing between $||a||$ and $\|a\|$. The first, $||a||$, was typed incorrectly as `$|| a ||$`. As a result, the vertical bars are too far apart. The second was typed correctly using the appropriate delimiter commands: `$\| a \|$`. Here they are again side-by-side, enlarged:

$$||a|| \quad \|a\|$$

<table>
<tr><th>Name</th><th>Type</th><th>Typeset</th></tr>
<tr><td>left parenthesis</td><td><code>(</code></td><td>$($</td></tr>
<tr><td>right parenthesis</td><td><code>)</code></td><td>$)$</td></tr>
<tr><td>left bracket</td><td><code>[</code> or <code>\lbrack</code></td><td>$[$</td></tr>
<tr><td>right bracket</td><td><code>]</code> or <code>\rbrack</code></td><td>$]$</td></tr>
<tr><td>left brace</td><td><code>\{</code> or <code>\lbrace</code></td><td>$\{$</td></tr>
<tr><td>right brace</td><td><code>\}</code> or <code>\rbrace</code></td><td>$\}$</td></tr>
<tr><td>backslash</td><td><code>\backslash</code></td><td>$\backslash$</td></tr>
<tr><td>forward slash</td><td><code>/</code></td><td>$/$</td></tr>
<tr><td>left angle bracket</td><td><code>\langle</code></td><td>$\langle$</td></tr>
<tr><td>right angle bracket</td><td><code>\rangle</code></td><td>$\rangle$</td></tr>
<tr><td>vertical line</td><td><code>|</code> or <code>\vert</code></td><td>$\vert$</td></tr>
<tr><td>double vertical line</td><td><code>\|</code> or <code>\Vert</code></td><td>$\Vert$</td></tr>
<tr><td>left floor</td><td><code>\lfloor</code></td><td>$\lfloor$</td></tr>
<tr><td>right floor</td><td><code>\rfloor</code></td><td>$\rfloor$</td></tr>
<tr><td>left ceiling</td><td><code>\lceil</code></td><td>$\lceil$</td></tr>
<tr><td>right ceiling</td><td><code>\rceil</code></td><td>$\rceil$</td></tr>
<tr><td>upward</td><td><code>\uparrow</code></td><td>$\uparrow$</td></tr>
<tr><td>double upward</td><td><code>\Uparrow</code></td><td>$\Uparrow$</td></tr>
<tr><td>downward</td><td><code>\downarrow</code></td><td>$\downarrow$</td></tr>
<tr><td>double downward</td><td><code>\Downarrow</code></td><td>$\Downarrow$</td></tr>
<tr><td>up-and-down</td><td><code>\updownarrow</code></td><td>$\updownarrow$</td></tr>
<tr><td>double up-and-down</td><td><code>\Updownarrow</code></td><td>$\Updownarrow$</td></tr>
<tr><td>upper-left corner</td><td><code>\ulcorner</code></td><td>$\ulcorner$</td></tr>
<tr><td>upper-right corner</td><td><code>\urcorner</code></td><td>$\urcorner$</td></tr>
<tr><td>lower-left corner</td><td><code>\llcorner</code></td><td>$\llcorner$</td></tr>
<tr><td>lower-right corner</td><td><code>\lrcorner</code></td><td>$\lrcorner$</td></tr>
</table>

Table 5.3: Standard delimiters.

5.5.1 Stretching delimiters

All delimiters, except the four "corners", can stretch to enclose the subformula. For example,

$$\left(\frac{1}{2} \right)^{\alpha}$$

is typed as

```
\[
   \left( \frac{1}{2} \right)^{\alpha}
\]
```

The `\left` and `\right` commands instruct LaTeX to stretch the parentheses. The general construction is

`\left` *delim1* and `\right` *delim2*

where *delim1* and *delim2* are chosen from Table 5.3. They are usually, but not always, a matching pair—see the examples below. LaTeX inspects the formula between the `\left` and `\right` commands and decides what size delimiters to use. The `\left` and `\right` commands *must be paired* in order for LaTeX to know the extent of the material to be vertically measured. However, we repeat, the delimiters need not be the same.

If you want to stretch a single delimiter, you have to pair it with a *blank delimiter*, represented by the `\left.` and `\right.` commands. Here are some examples of stretching delimiters:

$$\left| \frac{a + b}{2} \right|, \quad \left\| A^{2} \right\|, \quad \left(\frac{a}{2}, b \right], \quad \left. F(x) \right|_{a}^{b}$$

typed as

```
\[
   \left| \frac{a + b}{2} \right|, \quad
   \left\| A^{2} \right\|,         \quad
   \left( \frac{a}{2}, b \right],  \quad
   \left. F(x) \right|_{a}^{b}
\]
```

There are also two convenient abbreviations:

`\left<` for `\left\langle`
`\right>` for `\right\rangle`

The `\left` and `\right` commands have one more use. For the delimiters `|`, `\|`, and all the arrows, the same symbol represents the left and right delimiters, which can sometimes cause problems as in Example 2 in Section 6.1.4. In such cases, you should use the `\left` and `\right` commands to tell LaTeX whether the delimiter is a left or a right delimiter. LaTeX also provides the `\lvert` and `\rvert` for `|` as left and right delimiter, and `\lVert` and `\rVert` for `\|`.

5.5.2 Delimiters that do not stretch

LaTeX provides the \big, \Big, \bigg, and \Bigg commands to produce delimiters of larger sizes. These delimiters do not stretch. For example,

```
\[
   (\quad \big(\quad \Big(\quad \bigg(\quad \Bigg(
\]
```

typesets as

$$(\quad \big(\quad \Big(\quad \bigg(\quad \Bigg($$

LaTeX also provides the more specific

\bigl, \Bigl, \biggl, \Biggl, \bigr, \Bigr, \biggr, and \Biggr

commands to produce larger left and right delimiters.

For integral evaluation, you can choose one of the following:

$$F(x) |^{b}_{a} \quad F(x) \bigr|^{b}_{a} \quad F(x) \Bigr|^{b}_{a}$$

typed as

```
\[
   F(x) |^{b}_{a}        \quad
   F(x) \bigr|^{b}_{a}  \quad
   F(x) \Bigr|^{b}_{a}
\]
```

5.5.3 Limitations of stretching

In a number of situations the stretching computed by LaTeX is not ideal, so you should use a larger sized non-stretching variant. Here are some typical examples:

1. Large operators

```
\[
   \left[ \sum_i a_i \right]^{1/p} \quad
      \biggl[ \sum_i a_i \biggr]^{1/p}
\]
```

typesets as

$$\left[\sum_i a_i \right]^{1/p} \quad \biggl[\sum_i a_i \biggr]^{1/p}$$

You may prefer the second version with \biggl[and \biggr].

2. Groupings

```
\[
  \left( (a_1 b_1) - (a_2 b_2) \right)
  \left( (a_2 b_1) + (a_1 b_2) \right)
  \quad
  \bigl( (a_1 b_1) - (a_2 b_2) \bigr)
  \bigl( (a_2 b_1) + (a_1 b_2) \bigr)
\]
```

typesets as

$$\left((a_1 b_1) - (a_2 b_2) \right) \left((a_2 b_1) + (a_1 b_2) \right) \quad \bigl((a_1 b_1) - (a_2 b_2) \bigr) \bigl((a_2 b_1) + (a_1 b_2) \bigr)$$

You may prefer the clearer groupings provided by `\bigl(` and `\bigr)`.

3. Inline formulas The delimiters produced by `\left` and `\right` use too much interline space in $\left\lvert \frac{b'}{d'} \right\rvert$, typed as

```
\left\lvert \frac{b'}{d'} \right\rvert
```

Use `\bigl` and `\bigr` to produce delimiters that fit within the normal line spacing: $\bigl\lvert \frac{b'}{d'} \bigr\rvert$, typed as

```
\bigl\lvert \frac{b'}{d'} \bigr\rvert
```

5.5.4 Delimiters as binary relations

The symbol | can be used as a delimiter, as in $|x + y|$, and also as a binary relation, as in $\{ x \in \mathcal{R} \mid x^{2} \leq 2 \}$. As a binary relation it is typed as `\mid`. The previous formula is typed as

```
$\{ x \in \mathcal{R} \mid x^{2} \leq 2 \}$
```

`\bigm` and `\biggm` produce larger variants, with spacing on either side like binary relations. For example,

$$\left\{ x \biggm| \int_{0}^x t^{2}\, dt\leq 5 \right\}$$

is typed as

```
\[
 \left\{ x \biggm|\int_{0}^x t^{2}\, dt\leq 5 \right\}
\]
```

5.6 Operators

You cannot just type `sin x` to typeset the sine function in math mode. Indeed,

```
$sin x$
```

produces *sinx* instead of sin x, as you intended. Type this function as

```
$\sin x$
```

The `\sin` command prints sin with the proper style and spacing. LaTeX calls `\sin` an *operator* or log-*like function*.

5.6.1 Operator tables

There are two types of operators:

1. *Operators without limits*, such as `\sin`
2. *Operators with limits*, such as `\lim`, that take a subscript in inline mode and a "limit" in displayed math mode. For example, $\lim_{x \to 0} f(x) = 1$ is typed as

   ```
   $\lim_{x \to 0}  f(x) = 1$
   ```

   The same formula displayed,

   $$\lim_{x \to 0} f(x) = 1$$

   is typed as

   ```
   \[
      \lim_{x \to 0} f(x) = 1
   \]
   ```

The operators are listed in Tables 5.4 and 5.5 (see also Section A.6). The entries in the last two rows of Table 5.5 can be illustrated by

$$\varliminf_{x \to 0} \quad \varlimsup_{x \to 0} \quad \varinjlim_{x \to 0} \quad \varprojlim_{x \to 0}$$

Type	Typeset	Type	Typeset	Type	Typeset	Type	Typeset
`\arccos`	arccos	`\cot`	cot	`\hom`	hom	`\sin`	sin
`\arcsin`	arcsin	`\coth`	coth	`\ker`	ker	`\sinh`	sinh
`\arctan`	arctan	`\csc`	csc	`\lg`	lg	`\tan`	tan
`\arg`	arg	`\deg`	deg	`\ln`	ln	`\tanh`	tanh
`\cos`	cos	`\dim`	dim	`\log`	log		
`\cosh`	cosh	`\exp`	exp	`\sec`	sec		

Table 5.4: Operators without limits.

Type	Typeset	Type	Typeset
`\det`	$\det$	`\limsup`	$\limsup$
`\gcd`	$\gcd$	`\max`	$\max$
`\inf`	$\inf$	`\min`	$\min$
`\lim`	$\lim$	`\Pr`	$\Pr$
`\liminf`	$\liminf$	`\sup`	$\sup$
`\injlim`	$\injlim$	`\projlim`	$\projlim$
`\varliminf`	$\varliminf$	`\varlimsup`	$\varlimsup$
`\varinjlim`	$\varinjlim$	`\varprojlim`	$\varprojlim$

Table 5.5: Operators with limits.

which are typed as

```
\[
   \varliminf_{x \to 0} \quad  \varlimsup_{x \to 0}  \quad
   \varinjlim_{x \to 0} \quad  \varprojlim_{x \to 0}
\]
```

The following examples illustrate some more entries from Table 5.5:

$$\injlim_{x \to 0} \quad \liminf_{x \to 0} \quad \limsup_{x \to 0} \quad \projlim_{x \to 0}$$

These operators were typed as

```
\[
   \injlim_{x \to 0} \quad \liminf_{x \to 0} \quad
   \limsup_{x \to 0} \quad \projlim_{x \to 0}
\]
```

You can force the limits in a displayed formula into the subscript position with the `\nolimits` command. For example, the formulas

$$\injlim\nolimits_{x \to 0} \quad \liminf\nolimits_{x \to 0} \quad \limsup\nolimits_{x \to 0} \quad \projlim\nolimits_{x \to 0}$$

are typed as

```
\[
   \injlim\nolimits_{x \to 0} \quad
   \liminf\nolimits_{x \to 0} \quad
   \limsup\nolimits_{x \to 0} \quad
   \projlim\nolimits_{x \to 0}
\]
```

5.6.2 Congruences

\mod is a special operator used for congruences. Congruences are usually typeset using the \pmod or \pod variant. There is also the \bmod command, which is used as a binary operation. All four commands are shown in Table 5.6.

See Sections 14.1.2 and 14.1.9 for a discussion of related custom commands.

Type	Typeset
`$a \equiv v \mod{\theta}$`	$a \equiv v \mod \theta$
`$a \bmod b$`	$a \bmod b$
`$a \equiv v \pmod{\theta}$`	$a \equiv v \pmod{\theta}$
`$a \equiv v \pod{\theta}$`	$a \equiv v \pod{\theta}$

Table 5.6: Congruences.

5.6.3 Large operators

Here is a sum typeset inline, $\sum_{i=1}^{n} x_i^2$, and displayed,

$$\sum_{i=1}^{n} x_i^2$$

In the latter form, the sum symbol is larger. Operators that behave in this way are called *large operators*. Table 5.7 gives a complete list of large operators.

You can use the \nolimits command if you wish to show the limits of large operators as subscripts and superscripts in a displayed math environment.

The formula

$$\bigsqcup\nolimits_{ \mathfrak{m} } X = a$$

is typed as

```
\[
   \bigsqcup\nolimits_{ \mathfrak{m} } X = a
\]
```

You can use the \limits command if you wish to show the limits of large operators below and above the operator symbol in an inline math environment. For example, $\bigsqcup\limits_{ \mathfrak{m} } X = a$ is typed as

```
$\bigsqcup\limits_{ \mathfrak{m} } X = a$
```

Sums and products are very important constructs. The examples

$$\frac{z^d - z_0^d}{z - z_0} = \sum_{k=1}^{d} z_0^{k-1} z^{d-k} \quad \text{and} \quad (T^n)'(x_0) = \prod_{k=0}^{n-1} T'(x_k)$$

are typed as

Type	Inline	Displayed
`\int_{a}^{b}`	$\int_{a}^{b}$	$\displaystyle\int_{a}^{b}$
`\oint_{a}^{b}`	$\oint_{a}^{b}$	$\displaystyle\oint_{a}^{b}$
`\iint_{a}^{b}`	$\iint_{a}^{b}$	$\displaystyle\iint_{a}^{b}$
`\iiint_{a}^{b}`	$\iiint_{a}^{b}$	$\displaystyle\iiint_{a}^{b}$
`\iiiint_{a}^{b}`	$\iiiint_{a}^{b}$	$\displaystyle\iiiint_{a}^{b}$
`\idotsint_{a}^{b}`	$\idotsint_{a}^{b}$	$\displaystyle\idotsint_{a}^{b}$
`\prod_{i=1}^{n}`	$\prod_{i=1}^{n}$	$\displaystyle\prod_{i=1}^{n}$
`\coprod_{i=1}^{n}`	$\coprod_{i=1}^{n}$	$\displaystyle\coprod_{i=1}^{n}$
`\bigcap_{i=1}^{n}`	$\bigcap_{i=1}^{n}$	$\displaystyle\bigcap_{i=1}^{n}$
`\bigcup_{i=1}^{n}`	$\bigcup_{i=1}^{n}$	$\displaystyle\bigcup_{i=1}^{n}$
`\bigwedge_{i=1}^{n}`	$\bigwedge_{i=1}^{n}$	$\displaystyle\bigwedge_{i=1}^{n}$
`\bigvee_{i=1}^{n}`	$\bigvee_{i=1}^{n}$	$\displaystyle\bigvee_{i=1}^{n}$
`\bigsqcup_{i=1}^{n}`	$\bigsqcup_{i=1}^{n}$	$\displaystyle\bigsqcup_{i=1}^{n}$
`\biguplus_{i=1}^{n}`	$\biguplus_{i=1}^{n}$	$\displaystyle\biguplus_{i=1}^{n}$
`\bigotimes_{i=1}^{n}`	$\bigotimes_{i=1}^{n}$	$\displaystyle\bigotimes_{i=1}^{n}$
`\bigoplus_{i=1}^{n}`	$\bigoplus_{i=1}^{n}$	$\displaystyle\bigoplus_{i=1}^{n}$
`\bigodot_{i=1}^{n}`	$\bigodot_{i=1}^{n}$	$\displaystyle\bigodot_{i=1}^{n}$
`\sum_{i=1}^{n}`	$\sum_{i=1}^{n}$	$\displaystyle\sum_{i=1}^{n}$

Table 5.7: Large operators.

```
\[
   \frac{z^{d} - z_{0}^{d}}
        {z - z_{0}} =
   \sum_{k = 1}^{d} z_{0}^{k - 1} z^{d - k}
   \text{\quad and\quad}
   (T^{n})'(x_{0}) = \prod_{k=0}^{n - 1} T'(x_{k})
\]
```

5.6.4 Multiline subscripts and superscripts

The \substack command provides multiline limits for large operators. For instance,

$$\sum_{\substack{i<n\\ i \text{ even}}} x_i^2$$

is typed as

```
\[
   \sum_{ \substack{ i < n\\
                     i \text{ even} } }
    x_{i}^{2}
\]
```

There is only one rule to remember. Use the line separator command \\. You can use the \substack command wherever subscripts or superscripts are used.

The lines are centered by \substack, so if you want them set flush left, as in

$$\sum_{\begin{subarray}{l} i<n\\ i \text{ even}\end{subarray}} x_i^2$$

then use the subarray environment with the argument l:

```
\[
   \sum_{ \begin{subarray}{l}
            i < n\\
            i \text{ even}
          \end{subarray} }
     x_{i}^{2}
\]
```

See Section 14.1.6 for another example.

5.7 Math accents

The accents used in text (see Section 3.4.7) cannot be used in math formulas. For accents in formulas a separate set of commands is provided. All math accents are shown in Table 5.8 (see also Section A.8). The `amsxtra` package is needed for the accents in the second column. To use them, make sure to place in the preamble the line

```
\usepackage{amsxtra}
```

You can also use double accents, such as

		amsxtra	
Type	Typeset	Type	Typeset
`\acute{a}`	$\acute{a}$		
`\bar{a}`	$\bar{a}$		
`\breve{a}`	$\breve{a}$	`\spbreve`	$\breve{}$
`\check{a}`	$\check{a}$	`\spcheck`	$\vee$
`\dot{a}`	$\dot{a}$	`\spdot`	$\cdot$
`\ddot{a}`	$\ddot{a}$	`\spddot`	$\cdot\cdot$
`\dddot{a}`	$\dddot{a}$	`\spdddot`	$\cdots$
`\ddddot{a}`	$\ddddot{a}$		
`\grave{a}`	$\grave{a}$		
`\hat{a}`	$\hat{a}$		
`\widehat{a}`	$\widehat{a}$	`\sphat`	$\hat{}$
`\mathring{a}`	$\mathring{a}$		
`\tilde{a}`	$\tilde{a}$		
`\widetilde{a}`	$\widetilde{a}$	`\sptilde`	$\sim$
`\vec{a}`	$\vec{a}$		

Table 5.8: Math accents

```
\[
  \hat{\hat{A}}
\]
```

which typesets as $\hat{\hat{A}}$.

The two "wide" varieties, `\widehat` and `\widetilde`, expand to fit the symbols (their arguments) covered: $\widehat{A}$, $\widehat{ab}$, $\widehat{iii}$, $\widehat{aiai}$, $\widehat{iiiii}$, and $\widetilde{A}$, $\widetilde{ab}$, $\widetilde{iii}$, $\widetilde{aiai}$, $\widetilde{iiiii}$ (the last example is typed as `$\widetilde{iiiii}$`). If the base is too wide, the accent is centered:

$$\widehat{ABCDE}$$

The "sp" commands, provided by the **amsxtra** package, are used for superscripts, as illustrated in Table 5.8. If you use a lot of accented characters, you should appreciate custom commands (see Section 14.1.1).

Notice the difference between $\bar{a}$ and $\overline{a}$, typed as

```
$\bar{a}$ $\overline{a}$
```

For other examples of the `\overline` command, see Section 5.8.2.

To use an arbitrary symbol as an accent or to create "underaccents", use Javier Bezos' `accents` package.

5.8 Stretchable horizontal lines

LaTeX provides three types of stretchable horizontal lines that appear above or below a formula, braces, bars, and arrows. There are also stretchable arrow math symbols.

5.8.1 Horizontal braces

The `\overbrace` command places a brace of variable size above its argument, as in

$$\overbrace{a + b + \dots + z}$$

which is typed as

```
\[
  \overbrace{a + b + \dots + z}
\]
```

A superscript adds a label to the brace, as in

$$\overbrace{a + a + \dots + a}^{n}$$

which is typed as

```
\[
  \overbrace{a + a + \dots + a}^{n}
\]
```

The `\underbrace` command works similarly, placing a brace below its argument. A subscript adds a label to the brace, as in

$$\underbrace{a + a + \dots + a}_{n}$$

which is typed as

```
\[
   \underbrace{a + a + \dots + a}_{n}
\]
```

The following example combines these two commands:

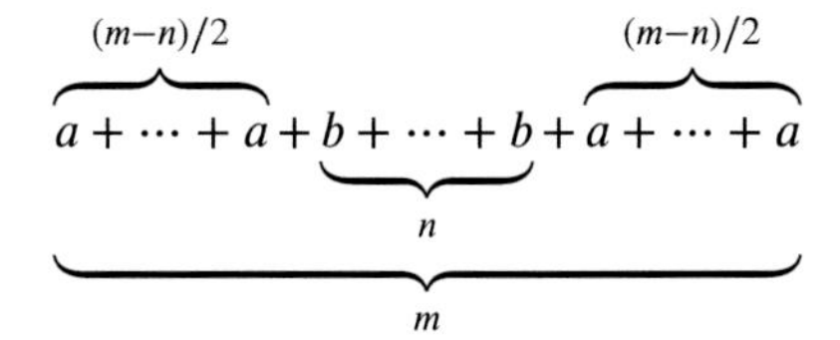

This example is typed as

```
\[
   \underbrace{
      \overbrace{a + \dots + a}^{(m - n)/2}
      + \underbrace{b + \dots + b}_{n}
      + \overbrace{a + \dots + a}^{(m - n)/2}
    }_{m}
\]
```

5.8.2 *Overlines and underlines*

The `\overline` and `\underline` commands draw lines above or below a formula. For example,

$$\overline{\overline{X} \cup \overline{\overline{X}}} = \overline{\overline{X}}$$

is typed as

```
\[
   \overline{ \overline{X} \cup \overline{\overline{X}} }
      = \overline{ \overline{X} }
\]
```

Similarly, you can place arrows above and below an expression:

$$\overleftarrow{a} \quad \overrightarrow{aa}$$
$$\overleftrightarrow{aaa} \quad \underleftarrow{aaaa} \quad \underrightarrow{aaaaa} \quad \underleftrightarrow{aaaaaa}$$

which is typed as

```
\begin{gather*}
   \overleftarrow{a}          \quad \overrightarrow{aa}\\
   \overleftrightarrow{aaa} \quad \underleftarrow{aaaa}\quad
   \underrightarrow{aaaaa}  \quad \underleftrightarrow{aaaaaa}
\end{gather*}
```

5.8.3 Stretchable arrow math symbols

There are two stretchable arrow math symbols that extend to accommodate a formula above or below the arrows with the `\xleftarrow` and `\xrightarrow` commands. The formula on top is given as the argument (possibly empty) and the formula below is an optional argument.

$$A \xrightarrow{\text{1-1}} B \xleftarrow[\alpha\to\beta]{\text{onto}} C \xleftarrow[\gamma]{} D \xleftarrow{} E$$

is typed as

```
\[
   A \xrightarrow{\text{1-1}} B \xleftarrow[\alpha\to\beta]
      {\text{onto}} C \xleftarrow[\gamma]{} D \xleftarrow{} E
\]
```

There are other stretchable arrow math symbols described in Section 7.8, but they can only be used in commutative diagrams. See Section 13.4 for the Ti*k*Z arrows and the `tikz-cd` package arrows.

5.9 Building a formula step-by-step

It is easy to build complex formulas from the components described in this section. Try the formula

$$\sum_{i=1}^{\left[\frac{n}{2}\right]} \binom{x_{i,i+1}^{i^2}}{\left[\frac{i+3}{3}\right]} \frac{\sqrt{\mu(i)^{\frac{3}{2}}(i^2-1)}}{\sqrt[3]{\rho(i)-2}+\sqrt[3]{\rho(i)-1}}$$

We build this formula in several steps. Create a new file in your `work` folder. Name it `formula.tex`, type in the following lines, and save it:

```
%File: formula.tex
\documentclass{sample}
\begin{document}
\end{document}
```

At present, the file has an empty `document` environment. Type each part of the formula as an inline or displayed formula within this environment, so that you can typeset the document and check for errors.

Step 1 We start with $\left[\frac{n}{2}\right]$. Type the following line into `formula.tex`:

```
$\left[ \frac{n}{2} \right]$
```

and test it by typesetting the document.

Step 2 Now you can do the sum

$$\sum_{i=1}^{\left[\frac{n}{2}\right]}$$

For the superscript, you can copy and paste the formula created in Step 1 (without the dollar signs), so that you have

```
\[
   \sum_{i = 1}^{ \left[ \frac{n}{2} \right] }
\]
```

Step 3 Next, do the two formulas in the binomial

$$x_{i,i+1}^{i^2} \qquad \left[\frac{i+3}{3}\right]$$

Type them as separate formulas in `formula.tex`:

```
\[
   x_{i, i + 1}^{i^{2}}\qquad\left[ \frac{i + 3}{3} \right]
\]
```

Step 4 Now it is easy to do the binomial. Piece together the following formula by copying and pasting the previous formulas and dropping the `\qquad` command:

```
\[
   \binom{x_{i,i + 1}^{i^{2}}}{\left[\frac{i + 3}{3}\right]}
\]
```

which typesets as

$$\binom{x_{i,i+1}^{i^2}}{\left[\frac{i+3}{3}\right]}$$

Step 5 Next, type the formula under the square root, $\mu(i)^{\frac{3}{2}}(i^2 - 1)$:

`$\mu(i)^{ \frac{3}{2} } (i^{2} - 1)$`

and then the square root, $\sqrt{\mu(i)^{\frac{3}{2}}(i^2 - 1)}$:

`$\sqrt{ \mu(i)^{ \frac{3}{2} } (i^{2} - 1) }$`

Step 6 The two cube roots, $\sqrt[3]{\rho(i) - 2}$ and $\sqrt[3]{\rho(i) - 1}$, are easy to type:

`$\sqrt[3]{ \rho(i) - 2 }$` `$\sqrt[3]{ \rho(i) - 1 }$`

Step 7 Now the fraction

$$\frac{\sqrt{\mu(i)^{\frac{3}{2}}(i^{2}-1)}}{\sqrt[3]{\rho(i)-2}+\sqrt[3]{\rho(i)-1}}$$

is typed, copied, and pasted together as

```
\[
  \frac{ \sqrt{ \mu(i)^{ \frac{3}{2}} (i^{2} -1) } }
       { \sqrt[3]{\rho(i) - 2} + \sqrt[3]{\rho(i) - 1} }
\]
```

Step 8 Finally, the whole formula,

$$\sum_{i=1}^{\left[\frac{n}{2}\right]}\binom{x_{i,i+1}^{i^{2}}}{\left[\frac{i+3}{3}\right]}\frac{\sqrt{\mu(i)^{\frac{3}{2}}(i^{2}-1)}}{\sqrt[3]{\rho(i)-2}+\sqrt[3]{\rho(i)-1}}$$

is formed by copying and pasting the pieces together, leaving only one pair of displayed math delimiters:

```
\[
  \sum_{i = 1}^{ \left[ \frac{n}{2} \right] }
    \binom{ x_{i, i + 1}^{i^{2}} }
          { \left[ \frac{i + 3}{3} \right] }
    \frac{ \sqrt{ \mu(i)^{ \frac{3}{2}} (i^{2} - 1) } }
         {\sqrt[3]{\rho(i)-2} + \sqrt[3]{\rho(i) - 1}}
\]
```

Note the use of

- Hierarchical indentation, to keep track of the structure of the formula
- Spacing to help highlight the braces—some text editors help you balance braces
- Separate lines for the various pieces of formulas that are more than a line long

It is to your advantage to *keep your source file readable.* LaTeX does not care how its input is formatted, and would happily accept the following:

```
\[\sum_{i=1}^{\left[\frac{n}{2}\right]}\binom{x_{i,i+1}
^{i^{2}}}{\left[\frac{i+3}{3}\right]}\frac{\sqrt{\mu(i)
^{\frac{3}{2}}(i^{2}-1)}}{\sqrt[3]{\rho(i)-2}+\sqrt[3]
{\rho(i)-1}}\]
```

But this haphazard style not only makes it more difficult for your coauthors or editor to work with your source file, it also makes finding mistakes difficult. Try to find the error in the next version:

```
\[\sum_{i=1}^{\left[\frac{n}{2}\right]}
\binom{x_{i,i+1}^{i^{2}}}{\left[\frac{i+3}{3}\right]}
\frac{\sqrt{\mu(i)^{\frac{3}{2}}}(i^{2}-1)}}{\sqrt[3]
{\rho(i)-2}+\sqrt[3]{\rho(i)-1}}\]
```

Answer: `\frac{3}{2` should be followed by `}}` and not by `}}}`.

5.10 Formula Gallery

In this section I present a collection of formulas—some simple, some complex—that illustrate the power of LaTeX.

Some of these examples require the `amssymb` package, so it is a good idea to include the line

```
\usepackage{amssymb,latexsym}
```

following the `\documentclass` line of any article.

Formula 1 A set-valued function

$$x \mapsto \{ c \in C \mid c \leq x \}$$

```
\[
  x \mapsto \{ c \in C \mid c \leq x \}
\]
```

Formula 2

$$\left| \bigcup (I_{j} \mid j \in J) \right| < \mathfrak{m}$$

```
\[
  \left| \bigcup ( I_{j} \mid j \in J ) \right| < \mathfrak{m}
\]
```

We use the delimiters `\left|` and `\right|`, see Section 5.5.1. The Fraktur $\mathfrak{m}$ is introduced in Section 6.4.2.

Formula 3 Note that you have to add spacing both before and after the text fragment in the following example. The argument of `\text` is typeset in text mode, so spaces are recognized.

$$A = \{ x \in X \mid x \in X_{i}, \text{ for some $i \in I$} \}$$

```
\[
  A = \{ x \in X \mid x \in X_{i},
           \text{ for some $i \in I$} \}
\]
```

Formula 4 Space to show logical structure:

$$\langle a_1, a_2 \rangle \leq \langle a'_1, a'_2 \rangle \qquad \text{iff} \qquad a_1 < a'_1 \quad \text{or} \quad a_1 = a'_1 \text{ and } a_2 \leq a'_2$$

```
\[
   \langle a_{1}, a_{2} \rangle \leq
   \langle a'_{1}, a'_{2}\rangle \qquad \text{if{f}}
   \qquad a_{1} < a'_{1} \quad \text{or}
   \quad a_{1} = a'_{1} \text{ and } a_{2} \leq a'_{2}
\]
```

Note that in `if{f}` (in the argument of the first `\text`) the second `f` is enclosed in braces to avoid the use of the ligature—the merging of the two f's. For the proper way of typesetting iff without a ligature, see Section 3.4.6.

Formula 5 Here are some examples of Greek letters:

$$\Gamma_{u'} = \{\gamma \mid \gamma < 2\chi,\ B_{\alpha} \nsubseteq u',\ B_{\gamma} \subseteq u' \}$$

```
\[
   \Gamma_{u'} = \{\gamma \mid \gamma < 2\chi,\ B_{\alpha}
      \nsubseteq u', \ B_{\gamma} \subseteq u' \}
\]
```

See Section A.1 for a complete listing of Greek letters. We use the command `\␣` to properly space the formula. This command can be used both in text and in math.

Formula 6 `\mathbb` allows you to use the blackboard bold math alphabet, which only provides capital letters:

$$A = B^{2} \times \mathbb{Z}$$

```
\[
   A = B^{2} \times \mathbb{Z}
\]
```

Formula 7 `\left[` and `\right]` provide stretched delimiters:

$$y^C \equiv z \vee \bigvee_{ i \in C } \left[s_{i}^{C} \right] \pmod{ \Phi }$$

```
\[
   y^C \equiv z \vee \bigvee_{ i \in C } \left[ s_{i}^{C}
     \right] \pmod{ \Phi }
\]
```

Notice how the superscript is set directly above the subscript in s_i^C.

Formula 8 A complicated congruence:

$$y \vee \bigvee ([B_{\gamma}] \mid \gamma \in \Gamma) \equiv z \vee \bigvee ([B_{\gamma}] \mid \gamma \in \Gamma) \pmod{ \Phi^{x} }$$

```
\[
  y \vee \bigvee ( [B_{\gamma}] \mid \gamma
   \in \Gamma ) \equiv z \vee \bigvee ( [B_{\gamma}]
   \mid \gamma \in \Gamma ) \pmod{ \Phi^{x} }
\]
```

Formula 9 Use `\nolimits` to force the "limit" of the large operator to display as a subscript (see Section 5.6.3):

$$f(\mathbf{x}) = \bigvee\nolimits_{\!\mathfrak{m}} \left(\bigwedge\nolimits_{\mathfrak{m}} (x_{j} \mid j \in I_{i}) \mid i < \aleph_{\alpha} \right)$$

```
\[
  f(\mathbf{x}) =
      \bigvee\nolimits_{\!\mathfrak{m}}
      \left(
     \bigwedge\nolimits_{\mathfrak{m}}
      ( x_{j} \mid j \in I_{i} )
      \mid i < \aleph_{\alpha}
      \right)
\]
```

Notice that I inserted a negative space (`\!`) to bring the $\mathfrak{m}$ a little closer to the big join symbol $\bigvee$.

Formula 10 The `\left.` command gives a blank left delimiter, which is needed to balance the `\right|` command:

$$\left. \widehat{F}(x) \right|_{a}^{b} = \widehat{F}(b) - \widehat{F}(a)$$

```
\[
  \left. \widehat{F}(x) \right|_{a}^{b}
   = \widehat{F}(b) - \widehat{F}(a)
\]
```

Formula 11 The `\underset` and `\overset` commands build new symbols (see Section 6.3.1):

$$u \underset{\alpha}{+} v \overset{1}{\thicksim} w \overset{2}{\thicksim} z$$

```
\[
  u \underset{\alpha}{+} v \overset{1}{\thicksim} w
   \overset{2}{\thicksim} z
\]
```

Note that the new symbols $\overset{1}{\thicksim}$ and $\overset{2}{\thicksim}$ are binary relations and $\underset{\alpha}{+}$ is a binary operation.

Formula 12 Small size bold **def**:

$$f(x) \overset{\mathbf{def}}{=} x^{2} - 1$$

```
\[
  f(x) \overset{ \mathbf{def} }{ = } x^{2} - 1
\]
```

Formula 13 Math accents run amok:

$$\overbrace{a^{\vee} + b^{\vee} + \dots + z^{\vee}}^{\breve{\breve{n}}}$$

```
\[
  \overbrace{a\spcheck + b\spcheck + \dots + z\spcheck}^
  {\breve{\breve{n}}}
\]
```

Recall that for the `\sp` commands you need the `amsxtra` package.

Formula 14

$$\begin{vmatrix} a + b + c & uv\\ a + b & c + d \end{vmatrix} = 7$$

```
\[
  \begin{vmatrix}
    a + b + c & uv\\
    a + b & c + d
  \end{vmatrix}
  = 7
\]
```

$$\begin{Vmatrix} a + b + c & uv\\ a + b & c + d \end{Vmatrix} = 7$$

```
\[
  \begin{Vmatrix}
    a + b + c & uv\\
    a + b & c + d
  \end{Vmatrix}
  = 7
\]
```

Formula 15

$$\boldsymbol{\alpha}^2\sum_{j \in \mathbf{N}} b_{ij} \hat{y}_{j} = \sum_{j \in \mathbf{N}} b^{(\lambda)}_{ij}\hat{y}_{j} + (b_{ii} - \lambda_{i}) \hat{y}_{i} \hat{y}$$

```
\[
   \boldsymbol{\alpha}^2\sum_{j \in \mathbf{N}} b_{ij}
    \hat{y}_{j} = \sum_{j \in \mathbf{N}}
    b^{(\lambda)}_{ij}\hat{y}_{j}
    + (b_{ii} - \lambda_{i}) \hat{y}_{i} \hat{y}
\]
```

`\mathbf{N}` makes a bold **N** and `\boldsymbol{\alpha}` produces a bold $\boldsymbol{\alpha}$ (see Section 6.4.2).

Formula 16 To produce the formula

$$\biggl(\prod^n_{ j = 1} \hat{ x }_{j} \biggr) H_{c} = \frac{1}{2}\hat{ k }_{ij} \det \widehat{ \mathbf{K} }(i|i)$$

try typing

```
\[
   \left( \prod^n_{j = 1} \hat{ x }_{j} \right) H_{c}=
   \frac{1}{2} \hat{k}_{ij} \det \hat{ \mathbf{K} }(i|i)
\]
```

which typesets as

$$\left(\prod^n_{j = 1} \hat{ x }_{j} \right) H_{c}= \frac{1}{2} \hat{k}_{ij} \det \hat{ \mathbf{K} }(i|i)$$

This is not quite right. You can correct the overly large parentheses by using the `\biggl` and `\biggr` commands in place of `\left(` and `\right)`, respectively (see Section 5.5.2). Adjust the small hat over **K** by using `\widehat`:

```
\[
   \biggl( \prod^n_{ j = 1} \hat{ x }_{j} \biggr)
   H_{c} = \frac{1}{2}\hat{ k }_{ij}
   \det \widehat{ \mathbf{K} }(i|i)
\]
```

which gives you the desired formula.

Formula 17 In this formula, I have used `\overline{I}` to get $\overline{I}$. You could, instead, use `\bar{I}`, which is typeset as $\bar{I}$.

$$\det \mathbf{K}(t = 1, t_1, \dots, t_n) = \sum_{I \in \mathbf{n}} (-1)^{|I|} \prod_{i \in I} t_i \prod_{j \in I} (D_j + \lambda_j t_j) \det \mathbf{A}^{(\lambda)} (\overline{I} | \overline{I}) = 0$$

```
\[
  \det \mathbf{K} (t = 1, t_{1}, \dots, t_{n}) =
  \sum_{I \in \mathbf{n} }(-1)^{|I|} \prod_{i \in I}t_{i}
  \prod_{j \in I} (D_{j} + \lambda_{j} t_{j})
  \det \mathbf{A}^{(\lambda)}
  (\overline{I} | \overline{I}) = 0
\]
```

Formula 18 The command \| provides the ‖ math symbol in this formula:

$$\lim_{(v,v')\to(0,0)} \frac{H(z+v) - H(z+v') - BH(z)(v-v')}{\| v - v' \|} = 0$$

```
\[
   \lim_{(v, v') \to (0, 0)}
    \frac{H(z + v) - H(z + v') - BH(z)(v - v')}
         {\| v - v' \|} = 0
\]
```

Formula 19 This formula uses the calligraphic math alphabet (introduced in Section 6.4.2):

$$\int_{\mathcal{D}} | \overline{\partial u} |^{2} \Phi_{0}(z) e^{\alpha |z|^2} \geq c_{4} \alpha \int_{\mathcal{D}} |u|^{2}\Phi_{0} e^{\alpha |z|^{2}} + c_{5} \delta^{-2} \int_{A} |u|^{2} \Phi_{0} e^{\alpha |z|^{2}}$$

```
\[
   \int_{\mathcal{D}} | \overline{\partial u} |^{2}
    \Phi_{0}(z) e^{\alpha |z|^2}
    \geq c_{4} \alpha \int_{\mathcal{D}} |u|^{2}\Phi_{0}
    e^{\alpha |z|^{2}}
    + c_{5} \delta^{-2} \int_{A} |u|^{2}
    \Phi_{0} e^{\alpha |z|^{2}}
\]
```

Formula 20 The `\hdotsfor` command sets dots that span multiple columns in a matrix. The `\dfrac` command is the displayed variant of the `\frac` command (see Section 5.4.1), used here because the matrix entries with `\frac` would look too small.

$$\mathbf{A} = \begin{pmatrix} \dfrac{\varphi \cdot X_{n,1}}{\varphi_{1} \times \varepsilon_{1}} & (x + \varepsilon_{2})^{2} & \cdots & (x + \varepsilon_{n-1})^{n-1} & (x + \varepsilon_{n})^{n} \\ \dfrac{\varphi \cdot X_{n,1}}{\varphi_{2} \times \varepsilon_{1}} & \dfrac{\varphi \cdot X_{n,2}}{\varphi_{2} \times \varepsilon_{2}} & \cdots & (x + \varepsilon_{n-1})^{n-1} & (x + \varepsilon_{n})^{n} \\ \hdotsfor{5} \\ \dfrac{\varphi \cdot X_{n,1}}{\varphi_{n} \times \varepsilon_{1}} & \dfrac{\varphi \cdot X_{n,2}}{\varphi_{n} \times \varepsilon_{2}} & \cdots & \dfrac{\varphi \cdot X_{n,n-1}}{\varphi_{n} \times \varepsilon_{n-1}} & \dfrac{\varphi \cdot X_{n,n}}{\varphi_{n} \times \varepsilon_{n}} \end{pmatrix} + \mathbf{I}_{n}$$

```
\[
  \mathbf{A} =
  \begin{pmatrix}
    \dfrac{\varphi \cdot X_{n, 1}} {\varphi_{1} \times
      \varepsilon_{1}} & (x + \varepsilon_{2})^{2}
      & \cdots & (x + \varepsilon_{n - 1})^{n - 1}
      & (x + \varepsilon_{n})^{n}\\[10pt]
    \dfrac{\varphi \cdot X_{n, 1}} {\varphi_{2} \times
      \varepsilon_{1}} & \dfrac{\varphi \cdot X_{n, 2}}
      {\varphi_{2} \times \varepsilon_{2}} & \cdots &
      (x + \varepsilon_{n - 1})^{n - 1}
      & (x + \varepsilon_{n})^{n}\\
    \hdotsfor{5}\\
    \dfrac{\varphi \cdot X_{n, 1}} {\varphi_{n} \times
      \varepsilon_{1}} & \dfrac{\varphi \cdot X_{n, 2}}
      {\varphi_{n} \times \varepsilon_{2}} & \cdots
      & \dfrac{\varphi \cdot X_{n, n - 1}} {\varphi_{n}
      \times \varepsilon_{n - 1}} &
      \dfrac{\varphi\cdot X_{n, n}}
            {\varphi_{n} \times \varepsilon_{n}}
  \end{pmatrix}
   + \mathbf{I}_{n}
\]
```

Recall the discussion of `\dots` vs. `\cdots` and `\ldots` in Section 5.4.3. In this formula, we have to use `\cdots`. Matrices are discussed in detail in Section 7.7.1.

Note the use of the command `\\[10pt]`. If you use `\\` instead, the first and second lines of the matrix are set too close.

I show you in Section 14.1.2 how to rewrite this formula to make it shorter and more readable.

CHAPTER

6

More math

In the previous chapter, we discussed the basic building blocks of a formula and how to put them together to form more complex formulas. This chapter starts out by going one step lower, to the characters that make up a formula. We discuss math symbols and math alphabets.

LaTeX was designed for typesetting math, so it is not surprising that it contains a very large number of math symbols. Appendix A lists them all. Section 6.2 introduces the STIX symbols, some 2,000 of them. Section 6.1 classifies and describes them. Section 6.3 discusses how to build new symbols from existing ones. Math alphabets and symbols are discussed in Section 6.4. Horizontal spacing commands in math are described in Section 6.5.

LaTeX provides a variety of ways to number and tag equations. These techniques are described in Section 6.6. We conclude the chapter with two minor topics: generalized fractions (Section 6.7.1) and boxed formulas (Section 6.7.2).

G. Grätzer, *More Math Into* LaTeX, DOI 10.1007/978-3-319-23796-1_6

6.1 Spacing of symbols

LaTeX provides a large variety of math symbols: Greek characters (α), binary operations ($\circ$), binary relations ($\leq$), negated binary relations ($\nleq$), arrows ($\nearrow$), delimiters ($\{$), and so on. All the math symbols provided by LaTeX are listed in the tables of Appendix A. Consider the formula

$$A = \{ x \in X \mid x \beta \geq x y > (x + 1)^{2} - \alpha \}$$

which is typed as

```
\[
   A = \{ x \in X \mid x \beta \geq x y
     > (x + 1)^{2} - \alpha \}
\]
```

The spacing of the symbols in the formula varies. In $x\beta$, the two symbols are very close. In $x \in X$, there is some space around the $\in$, and in $x + 1$, there is somewhat less space around the $+$.

6.1.1 Classification

LaTeX classifies symbols into several categories or *types* and spaces them accordingly. In the formula

$$A = \{ x \in X \mid x \beta \geq x y > (x + 1)^{2} - \alpha \}$$

we find

- Ordinary math symbols: A, x, X, β, and so on
- Binary relations: $=$, $\in$, $|$, $\geq$, and $>$
- Binary operations: $+$ and $-$
- Delimiters: $\{$, $\}$, $($, and $)$

As a rule, you do not have to be concerned with whether or not a given symbol in a formula, say $\times$, is a binary operation. LaTeX knows and spaces the typeset symbol correctly.

6.1.2 Three exceptions

There are three symbols with more than one classification:

$$+ \quad - \quad |$$

$+$ or $-$ could be either a binary operation, for instance, $a - b$, or a sign, for instance, $-b$.

Rule ■ + and −

+ or − are binary operations when preceded and followed by a symbol or an empty group, { }.

So, for instance, in

$$\begin{alignat*}{2}(A + BC)x &{}+{} &C &y = 0,\\ Ex &{}+{} &(F + G)&y = 23.\end{alignat*}$$

which is typed as (see the `alignat*` environment in Section 7.5.4)

```
\begin{alignat*}{2}
   (A + B C)x &{}+{} &C       &y = 0,\\
           Ex &{}+{} &(F + G)&y = 23.
\end{alignat*}
```

we use the empty groups, { }, to tell LaTeX that the second + in line 1 and the first + in line 2 of the formula are binary operations. If we leave out the empty groups, and type instead

```
\begin{alignat*}{2}
   (A + B C)x &+ &C      &y = 0,\\
           Ex &+ &(F + G)&y = 23.
\end{alignat*}
```

we get

$$\begin{alignat*}{2}(A + BC)x &+ &C &y = 0,\\ Ex &+ &(F + G)&y = 23.\end{alignat*}$$

Another illustration is given later in this section using the `\phantom` command. This problem often arises in split formulas, for example if the formula is split just before a + or −, you should start the next line with `{}+` or `{}-`. See Section 7.3 for examples.

The | symbol can play several different roles in a math formula, so LaTeX provides separate commands to specify the symbol's meaning.

Rule ■ The four roles of the | symbol

- `|` ordinary math symbol
- `\mid` binary relation
- `\left|` left delimiter
- `\right|` right delimiter

Note the differences between the spacing in $a|b$, typed as `$a | b$`, and in $a \mid b$, typed as `$a \mid b$`.

Name	Width	Short	Long
1 mu (math unit)	ı		\mspace{1mu}
thinspace	⌴	\,	\thinspace
medspace	⌴	\:	\medspace
thickspace	⌴	\;	\thickspace
interword space	⌴	\␣	
1 em	⌴		\quad
2 em	⌴		\qquad
Negative space			
1 mu	ı		\mspace{-1mu}
thinspace	⌴	\!	\negthinspace
medspace	⌴		\negmedspace
thickspace	⌴		\negthickspace

Table 6.1: Math spacing commands.

6.1.3 *Spacing commands*

There are some situations where LaTeX cannot typeset a formula properly and you have to add spacing commands. Luckily, LaTeX provides a variety of spacing commands, listed in Table 6.1. The \neg commands remove space by “reversing the print head”.

The \quad and \qquad commands are often used to adjust aligned formulas (see Chapter 7) or to add space before text in a math formula. The size of \quad (= 1 em) and \qquad (= 2 em) depends on the current font.

The \, and \! commands are the most useful for fine tuning math formulas, see some examples in the *Formula Gallery* and in the next section. The \mspace command and the math unit *mu* provides you with even finer control. 18 mu = 1 em, defined in Section 3.8.3. For example, \mspace{3mu} adds a space that is 1/6 em long. There is an interesting use of mu on page 90.

6.1.4 *Examples*

We present some examples of fine tuning. One more example can be found in Section 6.3.1.

Example 1 In Section 1.6, we type the formula $\int_{0}^{\pi} \sin x \, dx = 2$ as

```
$\int_{0}^{\pi} \sin x \, dx = 2$
```

Notice the thinspace spacing command \, between \sin x and dx. Without the command, LaTeX would have crowded sin x and dx: $\int_{0}^{\pi} \sin x dx = 2$.

Example 2 $| - f(x)|$, typed as `$|-f(x)|$`, is spaced incorrectly. – becomes a binary operation by the + and – rule. To get the correct spacing, as in $\left|-f(x)\right|$, type `$\left|-f(x)\right|$`. This form tells LaTeX that the first | is a left delimiter, by the | rule, and therefore – is the unary minus sign, not the binary subtraction operation.

Example 3 In $\sqrt{5}\text{side}$, typed as

```
$\sqrt{5} \text{side}$
```

$\sqrt{5}$ is too close to side. So type it as

```
$\sqrt{5} \, \text{side}$
```

which typesets as $\sqrt{5} \, \text{side}$.

Example 4 In $\sin x / \log n$, the division symbol / is too far from $\log n$, so type

```
$\sin x / \! \log n$
```

which prints $\sin x / \! \log n$.

Example 5 In $f(1 / \sqrt{n})$, typed as

```
$f(1 / \sqrt{n})$
```

the square root almost touches the closing parenthesis. To correct it, type

```
$f(1 / \sqrt{n}\,)$
```

which typesets as $f(1 / \sqrt{n}\,)$.

There is one more symbol with special spacing: the `\colon` command, used for formulas such as $f \colon A \to B$ (typed as `$f \colon A \to B$`). Observe that `$f: A \to B$` typesets as $f : A \to B$. The spacing is awful. See Section 6.3.3 on how to declare the type of a symbol.

6.1.5 The `phantom` command

The `\phantom{`*argument*`}` command (introduced for text in Section 3.8.1) produces a space in a formula equivalent to the space that would be occupied by its typeset argument. This command is one of the most powerful tools available to us for fine tuning alignments. Here are two simple illustrations:

$$A = \begin{pmatrix} 1 & 3 & 1 \\ 2 & 1 & 1 \\ -2 & 2 & -1 \end{pmatrix}$$

typed as

```
\[
  A = \begin{pmatrix}
        \phantom{-}1 & \phantom{-}3 & \phantom{-}1\\
        \phantom{-}2 & \phantom{-}1 & \phantom{-}1\\
                 -2 & \phantom{-}2 &            -1\\
      \end{pmatrix}
\]
```

and

$$\begin{aligned} a+b+c&+d\phantom{{}+e}=0,\\ c&+d+e=5. \end{aligned}$$

typed as

```
\begin{align*}
     a + b + c & + d \phantom{ {}+e } = 0,\\
             c & + d + e             = 5.
   \end{align*}
```

Note that yields incorrect spacing by the + and − rule:

$$\begin{aligned} a+b+c&+d=0,\\ c&+d+e=5. \end{aligned}$$

See Section 7.6.2 for an additional example.

6.2 The STIX math symbols

6.2.1 Swinging it

In a recent paper of mine (see arXiv: 1312.2537), I introduce the concept of a *swing*: a prime interval $\mathfrak{p}$ swings to another one, $\mathfrak{q}$, as exemplified by this diagram:

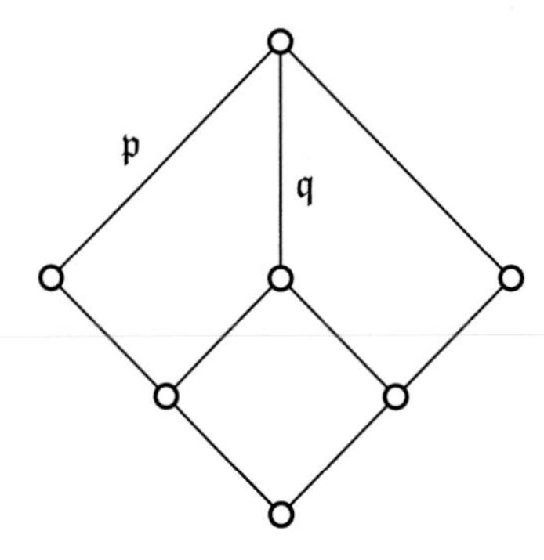

LaTeX provides a nice curved arrow, \curvearrowright, typeset as ↷; unfortunately, it is upside down (you don't swing that way). Utilizing the graphicx package, I solved my problem by defining

```
\newcommand{\swing}
  {\mathbin{{\rotatebox{180}{$\curvearrowleft$}}}}
```

which turns \curvearrowleft, ↶, 180 degrees: $\mathfrak{p}$ ↶ $\mathfrak{q}$. Raise it with \raisebox: $\mathfrak{p}$ ↶ $\mathfrak{q}$, typed as

```
\mathfrak p \swingraised \mathfrak q
```

where

```
\newcommand{\swingraised}{\mathbin{\raisebox{3.0pt}
  {\rotatebox{160}{$\curvearrowleft$}}}}
```

A better way to solve the problem is by utilizing the 2,000 or so new math symbols offered by STIX.

6.2.2 *The STIX project*

"The mission of the Scientific and Technical Information Exchange (STIX) font creation project is the preparation of a comprehensive set of fonts that serve the scientific and engineering community in the process from manuscript creation through final publication, both in electronic and print formats." The symbols were completed by 2006. The LaTeX Version 1.1.1 was released for my birthday in 2014.

6.2.3 *Installation and usage*

If you have a TeX installation from 2014, then you should already have the STIX fonts installed. To test whether you have the STIX fonts installed, try to typeset the following tiny article:

```
\documentclass{article}
\usepackage{stix}
\begin{document}
Some text, and a math formula $\ccwundercurvearrow$.
\end{document}
```

This should typeset as

> Some text, and a math formula ⤻.

To use the STIX fonts, load the STIX package, as in the example. Note that the STIX package has to be loaded *ahead of* the AMS packages. Since the amsart document class loads many AMS packages, we have to use the unusual construct:

```
\RequirePackage{stix}
\documentclass{amsart}
\begin{document}
Some text, and a math formula $\ccwundercurvearrow$.
\end{document}
```

This typesets the same as the previous example.

If your installation is not up-to-date, it is simpler to update it than to try to install the STIX fonts yourself. You can find the STIX fonts at

`https://sourceforge.net/projects/stixfonts`

Download the package; you get the folder `STIXv2`. In `STIXv2`, you find the folder `Fonts`. It contains

1. the fonts (inside the `Fonts` folder, there is a subfolder `fonts`, which contains a subfolder `opentype`, which contains a subfolder `public`; this contains a subfolder `stix`, containing the five font files);
2. the style file `stix.sty` (inside the `Fonts` folder, there is a subfolder `tex`, which contains a subfolder `latex`; this contains a subfolder `stix`, containing a lot of files, including the style file);
3. and the documentation `stix.pdf` (inside the `Fonts` folder, there is a subfolder `doc`, which contains a subfolder `fonts`; this contains a subfolder `stix`, containing several files, including the document file).

To install the fonts, follow the steps appropriate for your operating system.

All the math symbols are listed in 19 pages of `STIX.pdf`; we located this document in the third item above, see also in the `samples` folder. See Figure 6.1, for the second of these 19 pages. These pages show all the symbols and the commands necessary to produce them. So to get, type `\varcarriagereturn`. An * indicates that there is no bold version of the symbol. You can find `STIX.pdf` is the `samples` folder.

How to find a symbol in the 19 page listing? With perseverance. The symbols are divided into 12 parts; the longest is Relations, about seven pages! Within a part, they are listed by their hexadecimal number.

6.3 Building new symbols

No matter how many math symbols LaTeX provides, users always seem to want more. LaTeX gives you excellent tools to build new symbols from existing ones.

6.3.1 Stacking symbols

To place any symbol above, or below, any other, for instance, $\overset{u}{\sim}$, use the `\overset` command. It takes two arguments—the first argument is set in a smaller size above the

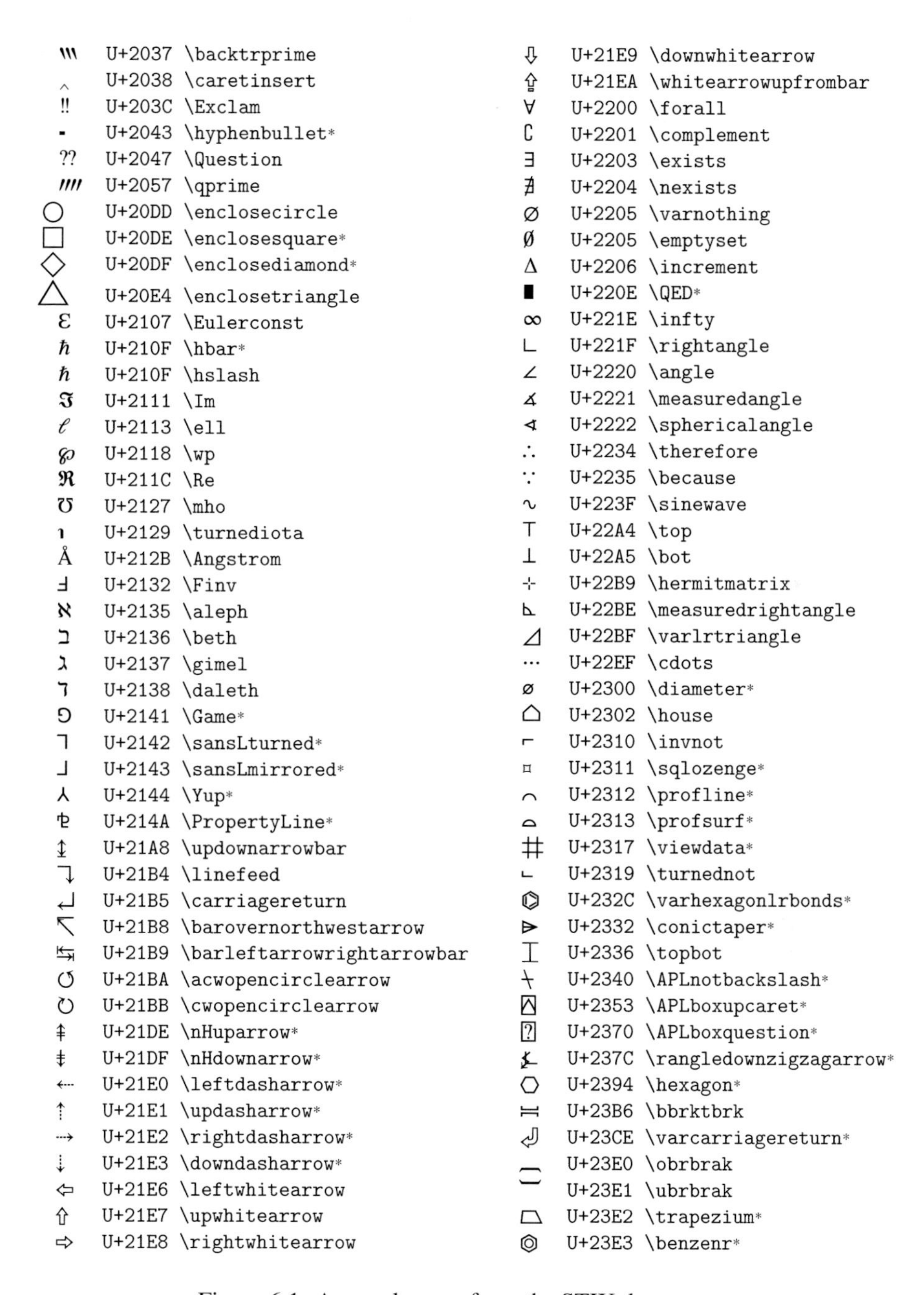

‷	U+2037	\backtrprime
‸	U+2038	\caretinsert
‼	U+203C	\Exclam
⁃	U+2043	\hyphenbullet*
⁇	U+2047	\Question
⁗	U+2057	\qprime
⃝	U+20DD	\enclosecircle
⃞	U+20DE	\enclosesquare*
⃟	U+20DF	\enclosediamond*
⃤	U+20E4	\enclosetriangle
ℇ	U+2107	\Eulerconst
ℏ	U+210F	\hbar*
ℏ	U+210F	\hslash
ℑ	U+2111	\Im
ℓ	U+2113	\ell
℘	U+2118	\wp
ℜ	U+211C	\Re
℧	U+2127	\mho
℩	U+2129	\turnediota
Å	U+212B	\Angstrom
Ⅎ	U+2132	\Finv
ℵ	U+2135	\aleph
ℶ	U+2136	\beth
ℷ	U+2137	\gimel
ℸ	U+2138	\daleth
⅁	U+2141	\Game*
⅂	U+2142	\sansLturned*
⅃	U+2143	\sansLmirrored*
⅄	U+2144	\Yup*
⅊	U+214A	\PropertyLine*
↨	U+21A8	\updownarrowbar
↴	U+21B4	\linefeed
↵	U+21B5	\carriagereturn
↸	U+21B8	\barovernorthwestarrow
↹	U+21B9	\barleftarrowrightarrowbar
↺	U+21BA	\acwopencirclearrow
↻	U+21BB	\cwopencirclearrow
⇞	U+21DE	\nHuparrow*
⇟	U+21DF	\nHdownarrow*
⇠	U+21E0	\leftdasharrow*
⇡	U+21E1	\updasharrow*
⇢	U+21E2	\rightdasharrow*
⇣	U+21E3	\downdasharrow*
⇦	U+21E6	\leftwhitearrow
⇧	U+21E7	\upwhitearrow
⇨	U+21E8	\rightwhitearrow
⇩	U+21E9	\downwhitearrow
⇪	U+21EA	\whitearrowupfrombar
∀	U+2200	\forall
∁	U+2201	\complement
∃	U+2203	\exists
∄	U+2204	\nexists
∅	U+2205	\varnothing
∅	U+2205	\emptyset
∆	U+2206	\increment
∎	U+220E	\QED*
∞	U+221E	\infty
∟	U+221F	\rightangle
∠	U+2220	\angle
∡	U+2221	\measuredangle
∢	U+2222	\sphericalangle
∴	U+2234	\therefore
∵	U+2235	\because
∿	U+223F	\sinewave
⊤	U+22A4	\top
⊥	U+22A5	\bot
⊹	U+22B9	\hermitmatrix
⊾	U+22BE	\measuredrightangle
⊿	U+22BF	\varlrtriangle
⋯	U+22EF	\cdots
⌀	U+2300	\diameter*
⌂	U+2302	\house
⌐	U+2310	\invnot
⌑	U+2311	\sqlozenge*
⌒	U+2312	\profline*
⌓	U+2313	\profsurf*
⌗	U+2317	\viewdata*
⌙	U+2319	\turnednot
⌬	U+232C	\varhexagonlrbonds*
⌲	U+2332	\conictaper*
⌶	U+2336	\topbot
⍀	U+2340	\APLnotbackslash*
⍓	U+2353	\APLboxupcaret*
⍰	U+2370	\APLboxquestion*
⍼	U+237C	\rangledownzigzagarrow*
⎔	U+2394	\hexagon*
⎶	U+23B6	\bbrktbrk
⏎	U+23CE	\varcarriagereturn*
⏠	U+23E0	\obrbrak
⏡	U+23E1	\ubrbrak
⏢	U+23E2	\trapezium*
⏣	U+23E3	\benzenr*

Figure 6.1: A sample page from the STIX document.

second argument. The spacing rules of the symbol in the second argument remain valid, i.e., the type remains the same. Since $\sim$ is a binary relation, so is $\overset{\text{u}}{\sim}$. The \underset command is the same except that the first argument is set under the second argument. For example,

$$\overset{\alpha}{a} \qquad \underset{\boldsymbol{\cdot}}{X} \qquad \overset{\alpha}{a_{i}} \qquad \overset{\alpha}{a}_{i}$$

are typed as

```
\[
   \overset{\alpha}{a}                  \qquad
   \underset{\boldsymbol{\cdot}}{X}  \qquad
   \overset{\alpha}{ a_{i} }            \qquad
   \overset{\alpha}{a}_{i}
\]
```

For the \boldsymbol command, see Section 6.4.3. Note that in the third example, $\overset{\alpha}{a_{i}}$, the α seems to be sitting too far to the right but the fourth example corrects that.

You can also use these commands with binary relations, as in

$$f(x) \overset{\text{def}}{=} x^{2} - 1$$

which is typed as

```
\[
   f(x) \overset{ \text{def} }{=} x^{2} - 1
\]
```

Since $=$ is a binary relation, $\overset{\text{def}}{=}$ becomes a binary relation, as shown by the spacing on either side. Here is another example,

$$\frac{a}{b} \overset{u}{+} \frac{c}{d} \overset{l}{+} \frac{e}{f}$$

which is typed as

```
\[
   \frac{a}{b} \overset{u}{+} \frac{c}{d}
    \overset{l}{+} \frac{e}{f}
\]
```

Note that $\overset{u}{+}$ and $\overset{l}{+}$ are properly spaced as binary operations.

As we discuss in Section 5.4.6, the safer definitions for these examples are

```
\[
   f(x) \overset{ \normalfont\text{def} }{=} x^{2} - 1
\]
```

and

```
\[
   \frac{a}{b} \overset{\normalfont u}{+} \frac{c}{d}
    \overset{\normalfont l}{+} \frac{e}{f}
\]
```

6.3.2 Negating and side-setting symbols

You can *negate* with the \not command; for instance, $a \not\in b$ and $a \not= b$ are typed as `$a \not\in b$` and `$a \not= b$`, respectively. It is preferable, however, to use the negated symbols $\notin$, typed as `$\notin$`, and $\ne$, typed as `$\ne$`. See the negated binary relations table in Section A.2. For instance, "a does not divide b", $a \nmid b$, should be typed as `$a \nmid b$`, not as `$a \not\mid b$`, which typesets as $a \not\mid b$. In Section 6.3.3, you learn how to improve $a \not\mid b$ to $a \mathrel{\not|} b$, typed as `$a \mathrel{\not|} b$`. However, `$a \nmid b$` is still best.

LaTeX provides the \sideset command to set symbols at the corners of large operators other than the "corners" (the last four delimiters in Table 5.3). This command takes three arguments:

```
\sideset{ _{ll}^{ul} }{ _{lr}^{ur} }{large_op}
```

where *ll* stands for the symbol to be placed at the lower left, *ul* for upper left, *lr* for lower right, and *ur* for upper right; *large_op* is a large operator. These two examples,

$$\sideset{}{_{a}^{c}}{\prod}\text{ and } \sideset{^{e}}{}{\prod}$$

are typed as

```
\[
  \sideset{}{_{a}^{c}}{\prod}\text{ and } \sideset{^{e}}{}{\prod}
\]
```

Note that the two first arguments are compulsory, although one or the other may be empty, while the third argument must contain the large operator.

Here is a more meaningful example:

```
\[
   \sideset{}{'}{\sum}_{\substack{ i < 10\\ j < 10 } } x_{i}z_{j}
\]
```

it is typeset as

$$\sideset{}{'}{\sum}_{\substack{ i < 10\\ j < 10 } } x_{i}z_{j}$$

In this example, note that prime ($'$) is an automatically superscripted symbol (see Section 5.4.1), so you do not have to type `^'` in the second argument. Typing `\sum'` would not work, since LaTeX would place the prime above the sum symbol.

Thus, `\sideset` helps in mixing sub- and superscripts in "limit" positions with others in "nolimit" positions, allowing for a total of six positions in displayed operators with limits. Try

$$\sideset{_{a}^{c}}{_{e}^{i}}{\prod}_{n}^{r}$$

typed as

```
\[
  \sideset{_{a}^{c}}{_{e}^{i}}{\prod}_{n}^{r}
\]
```

6.3.3 Changing the type of a symbol

Some symbols are binary relations and some are binary operations (see Section 6.1). In fact, you can force any symbol to behave like either type. The `\mathbin` command declares its argument to be a binary operation. For example,

```
\mathbin{\alpha}
```

makes this instance of `\alpha` behave like a binary operation, as in $a \mathbin{\alpha} b$, typed as

```
$a \mathbin{\alpha} b$
```

You can use the `\mathrel` command to make a symbol behave like a binary relation, as in the formula $a \mathrel{\alpha} b$, typed as

```
$a \mathrel{ \alpha } b$
```

You can see
$a \mathbin{\alpha} b$ (`$a \mathbin{\alpha} b$`)
$a \mathrel{\alpha} b$ (`$a \mathrel{\alpha} b$`)
that a binary relation provides a bit more space than a binary operation. There is an interesting use of `\mathbin` on page 90.

In Section 14.1.6, we discussed the `\DeclareMathOperator` command and its $*$-ed version, to declare a symbol—or any text or formula—a math operator.

6.4 Math alphabets and symbols

The classification of math symbols in the context of spacing was discussed in Section 6.1. The symbols in a formula can also be classified as *characters from math alphabets* and *math symbols*. In the formula

$$A = \{ x \in X \mid x\beta \geq xy > (x+1)^2 - \alpha \}$$

the following characters come from math alphabets:

$$A \quad x \quad X \quad y \quad 1 \quad 2$$

whereas these characters are math symbols:

$$= \quad \{ \quad \in \quad | \quad \beta \quad \geq \quad > \quad (\quad + \quad) \quad - \quad \alpha \quad \}$$

6.4.1 Math alphabets

The letters and digits typed in a math formula come from *math alphabets*. LaTeX's default math alphabet—the one you get if you do not ask for something else—is Computer Modern math italic for *letters*. In the formula $x^2 \vee y_3 = \alpha$, the characters x and y come from this math alphabet. The default math alphabet for *digits* is Computer Modern roman and the digits 2 and 3 in this formula are typeset in Computer Modern roman.

LaTeX has a number of commands to switch type style in math. The two most important commands select the bold and italic versions:

Command	Math alphabet	Produces
`\mathbf{a}`	math bold	$\mathbf{2\ Greek\ gammas,}\ \gamma\ \mathbf{and\ \Gamma}$
`\mathit{a}`	math italic	$\mathit{2\ Greek\ gammas,}\ \gamma\ \mathit{and\ \Gamma}$

These commands change the style of letters, numbers, and upper case Greek characters. But beware of the pitfalls. For instance, in `\mathit{left-side}` the hyphen typesets as a minus: $\mathit{left-side}$.

There are four more commands that switch math alphabets:

Command	Math Alphabet	Produces
`\mathsf{a}`	math sans serif	$\mathsf{2\ Greek\ gammas,}\ \gamma\ \mathsf{and\ \Gamma}$
`\mathrm{a}`	math roman	$\mathrm{2\ Greek\ gammas,}\ \gamma\ \mathrm{and\ \Gamma}$
`\mathtt{a}`	math typewriter	$\mathtt{2\ Greek\ gammas,}\ \gamma\ \mathtt{and\ \Gamma}$
`\mathnormal{a}`	math italic	$\mathnormal{2\ Greek\ gammas,}\ \gamma\ \mathnormal{and\ \Gamma}$

Math roman is used in formulas for operator names, such as sin in $\sin x$, and for text. For operator names, you should use the `\DeclareMathOperator` command or the `*`-ed version, which sets the name of the operator in math roman, and also provides the proper spacing (see Section 6.3.3). For text, you should use the `\text` command (see Section 5.4.6).

The `\mathnormal` command switches to the default math alphabet; it is seldom used in practice.

The Computer Modern fonts include a math bold italic alphabet, but LaTeX does not provide a command to access it.

Rule ■ **Math alphabets vs. text alphabets**
Do not use text alphabets in a math formula, except in the argument of a `\text` command.

It may not be easy to see the difference, but some things will not look right or may not align properly.

6.4.2 Math symbol alphabets

You may have noticed that α was not classified as belonging to an alphabet in the example at the beginning of this section. Indeed, α is treated by LaTeX as a math symbol rather than as a member of a math alphabet. You cannot italicize or slant it, nor is there a sans serif version. There is a bold version, but you must use the `\boldsymbol` command to produce it. For instance, $\boldsymbol{\alpha}_{\boldsymbol{\beta}}$, is typed as

```
$\boldsymbol{\alpha}_{\boldsymbol{\beta}}$
```

Note that $\boldsymbol{\beta}$ appears in a small size in $\boldsymbol{\alpha}_{\boldsymbol{\beta}}$.

Four "alphabets of symbols" are built into LaTeX.

Greek The examples α, β, Γ are typed as

```
$\alpha, \beta, \Gamma$
```

See Section A.1 for the symbol tables.

Calligraphic an uppercase-only alphabet invoked with the `\mathcal` command. The examples $\mathcal{A}, \mathcal{C}, \mathcal{E}$ are typed as

```
$\mathcal{A}, \mathcal{C}, \mathcal{E}$
```

Euler Fraktur invoked by the `\mathfrak` command. The examples $\mathfrak{n}, \mathfrak{N}, \mathfrak{p}, \mathfrak{P}$ are typed as

```
$\mathfrak{n}, \mathfrak{N}, \mathfrak{p}, \mathfrak{P}$
```

Blackboard bold uppercase-only math alphabet, invoked with `\mathbb`. The examples $\mathbb{A}, \mathbb{B}, \mathbb{C}$ are typed as

```
$\mathbb{A}, \mathbb{B}, \mathbb{C}$
```

6.4.3 *Bold math symbols*

In math, most characteristics of a font are specified by LaTeX. One exception is boldface. To make a *letter* bold from a math alphabet within a formula, use the `\mathbf` command. For instance, in

we choose the vector $\mathbf{v}$

the bold $\mathbf{v}$ is produced by `$\mathbf{v}$`.

To obtain bold math *symbols*, use the `\boldsymbol` command. For example, the bold symbols

$$\boldsymbol{5} \quad \boldsymbol{\alpha} \quad \boldsymbol{\Lambda} \quad \boldsymbol{\mathcal{A}} \quad \boldsymbol{\to} \quad \boldsymbol{A}$$

are typed as

```
\[
  \boldsymbol{5}              \quad \boldsymbol{\alpha}
  \quad \boldsymbol{\Lambda}\quad\boldsymbol{\mathcal{A}}
  \quad \boldsymbol{\to}     \quad \boldsymbol{A}
\]
```

Note that `\boldsymbol{A}` typesets as $\boldsymbol{A}$, a bold math italic A. To get an upright $\mathbf{A}$, type `$\mathbf{A}$`. The digit 5 did not really need `\boldsymbol`; `\mathbf{5}` gives the same result.

To make an entire formula bold, use the `\mathversion{bold}` command, as in

```
{\mathversion{bold} $a \equiv c \pod{\theta}$}
```

which typesets as $\boldsymbol{a \equiv c \pod{\theta}}$. Note that the `\mathversion{bold}` command is given *before the formula.*

To typeset $\boldsymbol{\mathcal{AMS}}$, type

```
$\boldsymbol{ \mathcal{A} } \boldsymbol{ \mathcal{M} }
\boldsymbol{ \mathcal{S} }$
```

or

```
$\boldsymbol{ \mathcal{AMS} }$
```

or

```
{\mathversion{bold} $\mathcal{AMS}$}
```

Within the scope of `\mathversion{bold}`, you can undo its effect with

```
\mathversion{normal}
```

Not all symbols have bold variants. For example, if you type

```
$\sum \quad \boldsymbol{\sum}$
```

you get $\sum \quad \sum$, two identical symbols. If you want to obtain a bold version, use the *poor man's bold* invoked by the \pmb command. This command typesets the symbol three times very close to one another producing a bold symbol of some quality. Note that \pmb does destroy the type of the symbol, \pmb{\sum} is no longer spaced like a large operator. To make it into a large operator, declare in the preamble

```
\DeclareMathOperator{\boldsum}{\pmb{\sum}}
```

and

```
\DeclareMathOperator*{\boldsumlim}{\pmb{\sum}}
```

Compare the following four variants of sum:

$$\sum_{i=1}^{n} i^{2} \quad \textstyle\sum_{i=1}^{n} i^{2} \quad \textstyle\sum_{i=1}^{n} i^{2} \quad \displaystyle\sum_{i=1}^{n} i^{2}$$

The first sum is typed (in displayed math mode) as

```
\sum_{i = 1}^{n} i^{2}
```

The second uses poor man's bold, but does not declare the result to be a large operator:

```
\pmb{\sum}_{i = 1}^{n} i^{2}
```

The third uses the math operator declared:

```
\boldsum_{i = 1}^{n} i^{2}
```

The fourth uses the math operator with limit declared:

```
\boldsumlim_{i = 1}^{n} i^{2}
```

6.4.4 Size changes

There are four math font sizes, invoked by the command declarations

- \displaystyle, normal size for displayed formulas
- \textstyle, normal size for inline formulas
- \scriptstyle, normal size for subscripted and superscripted symbols
- \scriptscriptstyle, normal size for doubly subscripted and superscripted symbols

These commands control a number of style parameters in addition to the size. Compare the two fractions

$$\frac{1}{\displaystyle 2 + \frac{1}{3}} \quad \frac{1}{ 2 + \frac{1}{3} }$$

which are typed as

```
\[
   \frac{1}{\displaystyle 2 + \frac{1}{3}} \quad
      \frac{1}{ 2 + \frac{1}{3} }
\]
```

6.4.5 Continued fractions

In addition to the \frac, \dfrac, and \tfrac commands (see Section 5.4.1), LaTeX makes typesetting continued fractions even easier by providing the \cfrac command. The \cfrac command takes an optional argument, l or r, to place the numerator on the left or on the right. For example,

$$\cfrac{1}{ 2 + \cfrac{1}{3 + \cdots} } \qquad \cfrac[l]{1}{2 + \cfrac[l]{1}{3 + \cdots}}$$

is typed as

```
\[
  \cfrac{1}{ 2 + \cfrac{1}{3 + \cdots} } \qquad
     \cfrac[l]{1}{2 + \cfrac[l]{1}{3 + \cdots}}
\]
```

6.5 Vertical spacing

As a rule, all horizontal and vertical spacing in a math formula is done by LaTeX. Nevertheless, you often need to adjust horizontal spacing (see Section 6.1). There is seldom a need to adjust vertical spacing, but there are a few exceptions.

The formula $\sqrt{a} + \sqrt{b}$ does not look quite right, because the square roots are not uniform. You can correct this with \mathstrut commands, which inserts an invisible vertical space:

```
$\sqrt{\mathstrut a} + \sqrt{\mathstrut b}$
```

typesets as $\sqrt{\mathstrut a} + \sqrt{\mathstrut b}$. See Section 3.9.5 for struts in general.

Another way to handle this situation is with the \vphantom (vertical phantom) command, which measures the height of its argument and places a math strut of that height into the formula. So

```
$\sqrt{\vphantom{b} a} + \sqrt{b}$
```

also prints uniform square roots, $\sqrt{\vphantom{b} a} + \sqrt{b}$. The \vphantom method is more versatile than the previous one.

Here is a more complicated example from a recent research article:

$$\Theta_i = \bigcup \big(\Theta (\overline{a \wedge b}, \overline{\vphantom{b}a} \wedge \overline{b}) \mid a,\ b \in B_i \big) \vee \bigcup \big(\Theta(\overline{a \vee b}, \overline{\vphantom{b}a} \vee \overline{b}) \mid a,\ b \in B_i \big),$$

typed as

```
\[
   \Theta_i = \bigcup \big( \Theta (\overline{a \wedge b},
   \overline{\vphantom{b}a} \wedge \overline{b})
   \mid a,\ b \in B_i \big)
   \vee \bigcup \big( \Theta(\overline{a \vee b},
   \overline{\vphantom{b}a} \vee \overline{b} )
   \mid a,\ b \in B_i \big),
 \]
```

Another useful command for vertical spacing is the \smash command. It directs LaTeX to pretend that its argument does not protrude above or below the line in which it is typeset.

For instance, the two lines of this admonition:

> It is *very important* that you memorize the integral $\frac{1}{\int f(x)\, dx} = 2g(x) + C$, which will appear on the next test.

are too far apart because LaTeX had to make room for the fraction. However, in this instance, the extra vertical space is not necessary because the second line is very short. To correct this, place the formula in the argument of a \smash command:

```
It is \emph{very important} that you memorize the
integral $\smash{\frac{1}{\int f(x) \, dx}} = 2g(x) + C$,
which will appear on the next test.
```

LaTeX produces the following:

> It is *very important* that you memorize the integral $\smash{\frac{1}{\int f(x)\, dx}} = 2g(x) + C$, which will appear on the next test.

An optional argument to the \smash command controls which part of the formula is ignored, t to smash the top and b to smash the bottom.

6.6 Tagging and grouping

You can attach a name to an equation using the `\tag` command. In the `equation` or `equation*` environments,

```
\tag{name}
```

attaches the tag *name* to the equation—*name* is typeset as text. The tag replaces the number.

Recall that the numbering of an equation is *relative,* that is, the number assigned to an equation is relative to the placement of the equation with respect to other equations in the document. An equation tag, on the other hand, is *absolute*—the tag remains the same even if the equation is moved.

If there is a tag, the `equation` and the `equation*` environments are equivalent. For example,

$$\int_{-\infty}^{\infty} e^{-x^{2}} \, dx = \sqrt{\pi} \tag{Int}$$

may be typed as

```
\begin{equation*}
   \int_{-\infty}^{\infty} e^{-x^{2}} \, dx
   = \sqrt{\pi}\tag{Int}
\end{equation*}
```

or

```
\begin{equation}
   \int_{-\infty}^{\infty} e^{-x^{2}} \, dx
   = \sqrt{\pi}\tag{Int}
\end{equation}
```

or

```
\[
   \int_{-\infty}^{\infty} e^{-x^{2}} \, dx
   = \sqrt{\pi}\tag{Int}
\]
```

The `\tag*` command is the same as `\tag` except that it does not automatically enclose the tag in parentheses. To get

$$\int_{-\infty}^{\infty} e^{-x^{2}} \, dx = \sqrt{\pi} \tag*{A–B}$$

type

```
\begin{equation}
  \int_{-\infty}^{\infty} e^{-x^{2}} \, dx = \sqrt{\pi}
   \tag*{A--B}
\end{equation}
```

Tagging allows numbered variants of equations. For instance, the equation

$$A^{[2]} \diamond B^{[2]} \cong (A \diamond B)^{[2]} \tag{1}$$

may need a variant:

$$A^{\langle 2 \rangle} \diamond B^{\langle 2\rangle} \equiv (A \diamond B)^{\langle 2 \rangle} \tag{1$'$}$$

If the label of the first equation is `E:first`, then the second equation may be typed as follows:

```
\begin{equation}\tag{\ref{E:first}$'$}
   A^{\langle 2 \rangle} \diamond B^{\langle 2\rangle}
   \equiv (A \diamond B)^{\langle 2 \rangle}
\end{equation}
```

Such a tag is absolute in the sense that it does not change if the equation is moved. But if it references a label and the number generated by LaTeX for the label changes, the tag changes.

In contrast, *grouping* applies to a group of *adjacent* equations. Suppose the last equation was numbered (1) and the next group of equations is to be referred to as (2), with individual equations numbered as (2a), (2b), and so on. Enclosing these equations in a `subequations` environment accomplishes this goal. For instance,

$$A^{[2]} \diamond B^{[2]} \cong (A \diamond B)^{[2]} \tag{1a}$$

and its variant

$$A^{\langle 2 \rangle} \diamond B^{\langle 2\rangle} \equiv (A \diamond B)^{\langle 2 \rangle} \tag{1b}$$

are typed as

```
\begin{subequations}\label{E:joint}
   \begin{equation}\label{E:original}
      A^{[2]} \diamond B^{[2]} \cong (A \diamond B)^{[2]}
   \end{equation}

   \begin{equation}\label{E:modified}
      A^{\langle 2 \rangle} \diamond B^{\langle 2\rangle}
      \equiv (A \diamond B)^{\langle 2\rangle}
   \end{equation}
\end{subequations}
```

Referring to these equations, you find that

- `\eqref{E:joint}` resolves to (1)
- `\eqref{E:original}` resolves to (1a)
- `\eqref{E:modified}` resolves to (1b)

Note that in this example, references to the second and third labels produce numbers, (1a) and (1b), that also appear in the typeset version. The group label, `E:joint`, references the entire group, but (1) does not appear in the typeset version unless referenced.

A `subequations` environment can contain the multiline math constructs discussed in Chapter 7 (see Section 7.4.4).

6.7 Miscellaneous

6.7.1 Generalized fractions

The generalized fraction command provides the facility to typeset many variants of fractions and binomials, such as $\genfrac{}{}{1pt}{1}{a+b}{c}$ and $\genfrac{]}{[}{0pt}{1}{a+b}{c}$. The syntax is

```
\genfrac{left-delim}{right-delim}{thickness}{mathstyle}
        {numerator}{denominator}
```

where

- *left-delim* is the left delimiter for the formula (default: none)
- *right-delim* is the right delimiter for the formula (default: none)
- *thickness* is the thickness of the fraction line, in the form x`pt` (default: the normal weight, 0.4pt), for instance, `12pt` for 12 point width
- *mathstyle* is one of
  - `0` for `\displaystyle`
  - `1` for `\textstyle`
  - `2` for `\scriptstyle`
  - `3` for `\scriptscriptstyle`
  - Default: Depends on the context. If the formula is being set in display style, then the default is `0`, and so on
- *numerator* is the numerator
- *denominator* is the denominator

All arguments must be specified. The empty argument, {}, gives the default value.

Examples

1. `\frac{numerator}{denominator}`
   is the same as

   `\genfrac{}{}{}{}{numerator}{denominator}`

2. `\dfrac{numerator}{denominator}`
   is the same as

   `\genfrac{}{}{}{0}{numerator}{denominator}`

3. `\tfrac{numerator}{denominator}`
   is the same as

   `\genfrac{}{}{}{1}{numerator}{denominator}`

4. `\binom{numerator}{denominator}`
   is the same as

   `\genfrac{(}{)}{0pt}{}{numerator}{denominator}`

5. Here are some more examples:

$$\frac{a+b}{c} \quad \genfrac{}{}{1pt}{}{a+b}{c} \quad \genfrac{}{}{1.5pt}{}{a+b}{c} \quad \genfrac{}{}{2pt}{}{a+b}{c} \quad \genfrac{[}{]}{0pt}{}{a+b}{c} \quad \genfrac{]}{[}{0pt}{}{a+b}{c}$$

   typed as

```
\[
   \frac{a + b}{c} \quad
   \genfrac{}{}{1pt}{}{a + b}{c}   \quad
   \genfrac{}{}{1.5pt}{}{a + b}{c} \quad
   \genfrac{}{}{2pt}{}{a + b}{c}   \quad
   \genfrac{[}{]}{0pt}{}{a + b}{c} \quad
   \genfrac{]}{[}{0pt}{}{a + b}{c}
\]
```

$$\frac{a+b}{c} \quad \genfrac{}{}{0.4pt}{}{a+b}{c}$$

   typed as

```
\[
   \frac{a + b}{c} \quad
   \genfrac{}{}{0.4pt}{}{a + b}{c}   \quad
\]
```

   You can choose the delimiters from Table 5.3.

If a `\genfrac` construct is used repeatedly, you should name it. See Section 14.1 for custom commands.

6.7.2 Boxed formulas

The \boxed command puts its argument in a box, as in

$$\boxed{\int_{-\infty}^{\infty} e^{-x^{2}}\, dx = \sqrt{\pi}} \tag{2}$$

typed as

```
\begin{equation}
   \boxed{ \int_{-\infty}^{\infty} e^{-x^{2}}\, dx
   = \sqrt{\pi} }
\end{equation}
```

The \boxed command can also be used in the argument of a \text command. Note that

```
\fbox{Hello world}
```

and

```
$\boxed{\text{Hello world}}$
```

produce the same $\boxed{\text{Hello world}}$.

Morten Høgholm's mathtools package contains many variants of boxes.

CHAPTER

7

Multiline math displays

7.1 Visual Guide

LaTeX is about typesetting math. It knows a lot about typesetting inline formulas, but not much about how to display a multiline formula to best reflect its meaning in a visually pleasing way. So you have to decide the visual structure of a multiline formula and then use the tools provided by LaTeX to code and typeset it.

For many mathematical documents the three constructs of Chapter 1 suffice: *simple* and *annotated* alignments, and the *cases* construct. To help you choose the appropriate tool for more complicated constructs, we start by introducing the basic concepts and constructions with the *Visual Guide* of Figure 7.1.

7.1.1 Columns

Multiline math formulas are displayed in *columns*. The columns are either *adjusted*, that is, centered, or set flush left or right, or *aligned*, that is, an alignment point is designated for each column and for each line. Moreover, the columns are either separated by the *intercolumn space* or adjacent with no separation.

G. Grätzer, *More Math Into* LaTeX, DOI 10.1007/978-3-319-23796-1_7

Adjusted environments

$$\begin{gather} x_1x_2 + x_1^2x_2^2 + x_3 \\ x_1x_3 + x_1^2x_3^2 + x_2 \\ x_1x_2x_3 \end{gather}$$

gather
one column, centered

$$\begin{multline} (x_1x_2x_3x_4x_5x_6)^2 \\ + (x_1x_2x_3x_4x_5 + x_1x_3x_4x_5x_6 + x_1x_2x_4x_5x_6 + x_1x_2x_3x_5x_6)^2 \\ + (x_1x_2x_3x_4 + x_1x_2x_3x_5 + x_1x_2x_4x_5 + x_1x_3x_4x_5)^2 \end{multline}$$

multline
flush left, centered, flush right

Adjusted subsidiary environments

$$\begin{pmatrix} 1 & 0 & \dots & 0 \\ 0 & 1 & \dots & 0 \\ \vdots & \vdots & \ddots & \vdots \\ 0 & 0 & \dots & 1 \end{pmatrix}$$

matrix
multicolumn, centered

$$\begin{array}{cccc} a+b+c & uv & x-y & 27 \\ a+b & u+v & z & 134 \end{array}$$

array
multicolumn
each column adjusted independently

$$f(x) = \begin{cases} -x^2, & \text{if } x < 0; \\ \alpha + x, & \text{if } 0 \leq x \leq 1; \\ x^2, & \text{otherwise.} \end{cases}$$

cases
columns flush left

Aligned environments

$$\begin{align} f(x) &= x + yz & g(x) &= x + y + z \\ h(x) &= xy + xz + yz & k(x) &= (x + y)(x + z)(y + z) \end{align}$$

align
multicolumn, aligned

$$\begin{flalign} f(x) &= x + yz & g(x) &= x + y + z \\ h(x) &= xy + xz + yz & k(x) &= (x + y)(x + z)(y + z) \end{flalign}$$

flalign
multicolumn, aligned

$$\begin{alignat}{4} a_{11}x_1 &+ a_{12}x_2 &&+ a_{13}x_3 && &&= y_1 \tag{17} \\ a_{21}x_1 &+ a_{22}x_2 && && + a_{24}x_4 &&= y_2 \tag{18} \\ a_{31}x_1 & &&+ a_{33}x_3 && + a_{34}x_4 &&= y_3 \tag{19} \end{alignat}$$

alignat
multicolumn, aligned

Aligned subsidiary environment

$$\begin{split} 0 &= \langle \dots, 0, \dots, \overset{i}{d}, \dots, 0, \dots \rangle \wedge \langle \dots, 0, \dots, \overset{j}{a}, \dots, 0, \dots \rangle \\ &\equiv \langle \dots, 0, \dots, \overset{j}{a}, \dots, 0, \dots \rangle \pmod{\Theta} \end{split} \tag{3.4}$$

split
one column, aligned

Figure 7.1: The *Visual Guide* for multiline math formulas.

One column

As in Chapter 1, we start with a simple align:

$$\begin{aligned} r^2 &= s^2 + t^2,\\ 2u + 1 &= v + w^\alpha \end{aligned}$$

This is a single column, aligned at the $=$ signs, and coded with the `align` environment (see Section 1.7.3).

Two columns

The annotated align, coded with the `align` environment (see Section 1.7.3),

$$\begin{aligned} x &= x \wedge (y \vee z) && \text{(by distributivity)}\\ &= (x \wedge y) \vee (x \wedge z) && \text{(by condition (M))}\\ &= y \vee z \end{aligned}$$

has two columns. The first column is aligned like our example of simple align, but the second column is aligned flush left. There is a sizeable intercolumn space.

7.1.2 *Subsidiary math environments*

The cases example in Section 1.7.4:

$$f(x) = \begin{cases} -x^2, & \text{if } x < 0;\\ \alpha + x, & \text{if } 0 \leq x \leq 1;\\ x^2, & \text{otherwise.} \end{cases}$$

introduces a new concept. The part of the formula to the right of $=$ is a multiline construct. This is an example of a *subsidiary math environment* that can only be used *inside another math environment*. It creates a "large math symbol", in this case

$$\begin{cases} -x^2, & \text{if } x < 0;\\ \alpha + x, & \text{if } 0 \leq x \leq 1;\\ x^2, & \text{otherwise.} \end{cases}$$

So the cases example:

$$f(x) = \text{large math symbol}$$

is a single line displayed formula, where "large math symbol" is replaced by the `cases` construct.

7.1.3 *Adjusted columns*

An *adjusted column* is either set *centered*, or *flush left*, or *flush right*. This may happen by default, built into the environment, or so specified in the code.

For instance, in the displayed formula

$$\begin{gathered} x_1x_2 + x_1^2x_2^2 + x_3, \\ x_1x_3 + x_1^2x_3^2 + x_2 \end{gathered}$$

typeset with the gather environment, by default all the lines are centered.

On the other hand, in

$$\left(\begin{array}{lcr} 1 & 100 & 115 \\ 201 & 0 & 1 \end{array} \right)$$

coded with the array subsidiary math environment, the first column is flush left, the second centered, the third flush right.

7.1.4 Aligned columns

Aligned columns, on the other hand, are only of one kind, aligned by you. For instance,

$$\begin{alignedat}{2} f(x) &= x + yz & \qquad g(x) &= x + y + z \\ h(x) &= xy + xz + yz & \qquad k(x) &= (x + y)(x + z)(y + z) \end{alignedat}$$

is coded with the alignat environment. It has two aligned columns, both aligned at the = signs.

7.1.5 Touring the Visual Guide

Figure 7.1, the *Visual Guide*, shows thumbnail pictures of the various kinds of multiline math environments and subsidiary math environments.

The first part of the *Visual Guide* illustrates gather and multline. The gather environment is a one-column, centered math environment—discussed in Section 7.2—which is used to display a *number of formulas* collected into one multiline formula. In contrast, multline—discussed in Section 7.3—displays *one long formula* in a number of lines. The first line is set flush left, the last line set flush right, and the rest (if any) of the lines are centered.

The third part of the *Visual Guide* illustrates the align environment and two of its variants, alignat and flalign, discussed in Section 7.5.

Three adjusted subsidiary math environments—matrix, cases, and array—are illustrated in second part of the *Visual Guide* and presented in Section 7.7.

The aligned subsidiary math environments aligned and gathered look just like the align and gather environments, so they are not illustrated in the *Visual Guide*. The aligned and gathered environments—along with \itemref—are discussed in Section 7.6, along with the split subsidiary math environment; this last one is illustrated in the last part of the *Visual Guide*.

7.2 Gathering formulas

The gather environment groups a number of one-line formulas, each centered on a separate line:

$$x_1 x_2 + x_1^2 x_2^2 + x_3, \tag{1}$$

$$x_1 x_3 + x_1^2 x_3^2 + x_2, \tag{2}$$

$$x_1 x_2 x_3. \tag{3}$$

Formulas (1)–(3) are typed as follows:

```
\begin{gather}
 x_{1} x_{2}+x_{1}^{2} x_{2}^{2} + x_{3},\label{E:1.1}\\
 x_{1} x_{3}+x_{1}^{2} x_{3}^{2} + x_{2},\label{E:1.2}\\
 x_{1} x_{2} x_{3}.\label{E:1.3}
\end{gather}
```

Rule ■ gather environment

1. Lines are separated with \\. Do not type \\ at the end of the last line!
2. Each line is numbered unless it has a \tag or \notag on the line before the line separator \\.
3. No blank lines are permitted within the environment.

The gather* environment is like gather, except that all lines are unnumbered. They can still be \tag-ged.

It would seem natural to code formulas (1)–(3) with three equation environments:

```
\begin{equation}
   x_{1} x_{2}+x_{1}^{2} x_{2}^{2} + x_{3},\label{E:1.1}
\end{equation}
\begin{equation}
   x_{1} x_{3}+x_{1}^{2} x_{3}^{2} + x_{2},\label{E:1.2}
\end{equation}
\begin{equation}
   x_{1} x_{2} x_{3}.\label{E:1.3}
\end{equation}
```

Note how bad this looks typeset:

$$x_1x_2 + x_1^2x_2^2 + x_3, \tag{1}$$

$$x_1x_3 + x_1^2x_3^2 + x_2, \tag{2}$$

$$x_1x_2x_3. \tag{3}$$

7.3 Splitting long formulas

The `multline` environment is used to split one very long formula into several lines. The first line is set flush left, the last line is set flush right, and the middle lines are centered:

$$\begin{multline} (x_1x_2x_3x_4x_5x_6)^2\\ + (y_1y_2y_3y_4y_5 + y_1y_3y_4y_5y_6 + y_1y_2y_4y_5y_6 + y_1y_2y_3y_5y_6)^2\\ + (z_1z_2z_3z_4z_5 + z_1z_3z_4z_5z_6 + z_1z_2z_4z_5z_6 + z_1z_2z_3z_5z_6)^2\\ + (u_1u_2u_3u_4 + u_1u_2u_3u_5 + u_1u_2u_4u_5 + u_1u_3u_4u_5)^2 \tag{4} \end{multline}$$

This formula is typed as

```
\begin{multline}\label{E:mm2}
   (x_{1} x_{2} x_{3} x_{4} x_{5} x_{6})^{2}\\
   + (y_{1} y_{2} y_{3} y_{4} y_{5}
   + y_{1} y_{3} y_{4} y_{5} y_{6}
   + y_{1} y_{2} y_{4} y_{5} y_{6}
   + y_{1} y_{2} y_{3} y_{5} y_{6})^{2}\\
   + (z_{1} z_{2} z_{3} z_{4} z_{5}
   + z_{1} z_{3} z_{4} z_{5} z_{6}
   + z_{1} z_{2} z_{4} z_{5} z_{6}
   + z_{1} z_{2} z_{3} z_{5} z_{6})^{2}\\
   + (u_{1} u_{2} u_{3} u_{4} + u_{1} u_{2} u_{3} u_{5}
  + u_{1} u_{2} u_{4} u_{5} + u_{1} u_{3} u_{4} u_{5})^{2}
\end{multline}
```

Rule ■ `multline` environment

1. Lines are separated with `\\`. Do not type `\\` at the end of the last line!
2. The formula is numbered *as a whole* unless it is `\tag`-ged or the numbering is suppressed with `\notag`. (Alternatively, use the `multline*` environment.)

3. No blank lines are permitted within the environment.

4. Each line is a subformula (see Section 7.4.2).

If you are very observant, you may have noticed that we failed to type {}+ following the line separators of the formula. In Section 6.1.2, you were told that this omission would result in the second line being typeset as

$$+(y_1 y_2 y_3 y_4 y_5 + y_1 y_3 y_4 y_5 y_6 + y_1 y_2 y_4 y_5 y_6 + y_1 y_2 y_3 y_5 y_6)^2$$

The `multline` environment, however, knows that a long formula is being broken and so typesets + as a binary operation.

A common mistake is to write `multiline` for `multline`, resulting in the message:

```
! LaTeX Error: Environment multiline undefined.
```

In the `multline*` environment, the formula is not numbered but can be `\tag`-ged.

The indentation of the first and last lines is controlled by the `\multlinegap` length command, with a default of 10 points, unless there is a tag on one of those lines. You can adjust the indentation by enclosing the `multline` environment in a `setlength` environment (see Section 14.5.2), as follows:

```
\begin{multline*}
   (x_{1} x_{2} x_{3} x_{4} x_{5} x_{6})^{2}\\
   + (x_{1} x_{2} x_{3} x_{4} x_{5}
   + x_{1} x_{3} x_{4} x_{5} x_{6}
   + x_{1} x_{2} x_{4} x_{5} x_{6}
   + x_{1} x_{2} x_{3} x_{5} x_{6})^{2}\\
   + (x_{1} x_{2} x_{3} x_{4} + x_{1} x_{2} x_{3} x_{5}
   + x_{1} x_{2} x_{4} x_{5} + x_{1} x_{3} x_{4})^{2}
\end{multline*}
\begin{setlength}{\multlinegap}{0pt}
   \begin{multline*}
     (x_{1} x_{2} x_{3} x_{4} x_{5} x_{6})^{2}\\
     + (x_{1} x_{2} x_{3} x_{4} x_{5}
     + x_{1} x_{3} x_{4} x_{5} x_{6}
     + x_{1} x_{2} x_{4} x_{5} x_{6}
     + x_{1} x_{2} x_{3} x_{5} x_{6})^{2}\\
     + (x_{1} x_{2} x_{3} x_{4} + x_{1} x_{2} x_{3} x_{5}
     + x_{1} x_{2} x_{4} x_{5} + x_{1} x_{3} x_{4})^{2}
   \end{multline*}
\end{setlength}
```

which typesets as

$$(x_1 x_2 x_3 x_4 x_5 x_6)^2$$
$$+ (x_1 x_2 x_3 x_4 x_5 + x_1 x_3 x_4 x_5 x_6 + x_1 x_2 x_4 x_5 x_6 + x_1 x_2 x_3 x_5 x_6)^2$$
$$+ (x_1 x_2 x_3 x_4 + x_1 x_2 x_3 x_5 + x_1 x_2 x_4 x_5 + x_1 x_3 x_4)^2$$

$$
\begin{multline*}
(x_{1} x_{2} x_{3} x_{4} x_{5} x_{6})^{2}\\
+ (x_{1} x_{2} x_{3} x_{4} x_{5} + x_{1} x_{3} x_{4} x_{5} x_{6} + x_{1} x_{2} x_{4} x_{5} x_{6} + x_{1} x_{2} x_{3} x_{5} x_{6})^{2}\\
+ (x_{1} x_{2} x_{3} x_{4} + x_{1} x_{2} x_{3} x_{5} + x_{1} x_{2} x_{4} x_{5} + x_{1} x_{3} x_{4})^{2}
\end{multline*}
$$

Notice that the second variant is not indented.

Any line of a `multline` environment can be typeset flush left or right by making it the argument of a `\shoveleft` or `\shoveright` command, respectively (same with `multline*`). For instance, to typeset the second line of formula (4) flush left, as in

$$
\begin{multline*}
(x_{1} x_{2} x_{3} x_{4} x_{5} x_{6})^{2}\\
\shoveleft{+ (x_{1} x_{2} x_{3} x_{4} x_{5} + x_{1} x_{3} x_{4} x_{5} x_{6} + x_{1} x_{2} x_{4} x_{5} x_{6} + x_{1} x_{2} x_{3} x_{5} x_{6})^{2}}\\
+ (x_{1} x_{2} x_{3} x_{4} + x_{1} x_{2} x_{3} x_{5} + x_{1} x_{2} x_{4} x_{5} + x_{1} x_{3} x_{4} x_{5})^{2}
\end{multline*}
$$

type the formula as follows:

```
\begin{multline*}
  (x_{1} x_{2} x_{3} x_{4} x_{5} x_{6})^{2}\\
  \shoveleft{+ (x_{1} x_{2} x_{3} x_{4} x_{5}
    + x_{1} x_{3} x_{4} x_{5} x_{6}
    + x_{1} x_{2} x_{4} x_{5} x_{6}
    + x_{1} x_{2} x_{3} x_{5} x_{6})^{2}}\\
    + (x_{1} x_{2} x_{3} x_{4} + x_{1} x_{2} x_{3} x_{5}
    + x_{1} x_{2} x_{4} x_{5}
                + x_{1} x_{3} x_{4} x_{5})^{2}
\end{multline*}
```

Observe that the entire line is the argument of the `\shoveleft` command, which is followed by `\\` unless it is the last line of the environment.

7.4 Some general rules

7.4.1 General rules

Even though you have only seen a few examples of multiline math environments, I venture to point out now that the multiline math environments and subsidiary math environments share a number of rules.

Rule ■ Multiline math environments

1. Lines are separated with `\\`. Do not type `\\` at the end of the last line!
2. No blank lines are permitted within an environment.
3. No blank line before the environment.

4. If an environment contains more than one formula, then, as a rule, each formula is numbered separately. If you add a `\label` command to a line, then the equation number generated for that line can be cross-referenced.

5. You can suppress the numbering of a line by using a `\notag` command on the line.

6. You can also override numbering with the `\tag` command, which works just as it does for equations (see Section 6.6).

7. `\tag` and `\label` should always precede the line separator `\\` for lines that are regarded as formulas in their own right. For instance, the lines of the `multline` environment cannot be individually numbered or tagged. The `\tag` command works for individual lines, not for the environment as a whole.

8. For cross-referencing, use `\label`, `\ref`, and `\eqref` in the same way you would for an equation (see Section 5.3).

9. Each multiline math environment has a `*`-ed form, which suppresses numbering. Individual formulas can still be `\tag`-ged.

A `\notag` command placed after the environment is ignored, but a `\tag` command gives the message

```
! Package amsmath Error: \tag not allowed here.
```

7.4.2 Subformula rules

A formula in the `multline` environment is split into a number of parts by `\\` commands; for instance, formula (4) is split into three parts:

```
1. (x_{1} x_{2} x_{3} x_{4} x_{5} x_{6})^{2}

2.  + (x_{1} x_{2} x_{3} x_{4} x_{5}
      + x_{1} x_{3} x_{4} x_{5} x_{6}
      + x_{1} x_{2} x_{4} x_{5} x_{6}
      + x_{1} x_{2} x_{3} x_{5} x_{6})^{2}

3.  + (x_{1} x_{2} x_{3} x_{4}+x_{1} x_{2} x_{3} x_{5} +
     x_{1} x_{2} x_{4} x_{5}+x_{1} x_{3} x_{4} x_{5})^{2}
```

Such parts of a formula are called *subformulas.*

The first line of the aligned formula $r^2 = s^2 + t^2$—from the simple alignment example in Section 1.7.3—which is typed as

```
r^{2} &= s^{2} + t^{2}
```

is split into two parts:

```
1.  r^{2}
2.      = s^{2} + t^{2}
```

In general, in a line of an aligned formula, the first part is everything between the beginning of the line and the first & symbol. There can then be a number of parts delimited by two consecutive & symbols. Finally, the last part is from the last & symbol to the end of the line or the line separator \\. These parts are also called *subformulas*.

Here are the last of the general rules.

Rule ■ Subformula

1. Each subformula must be a formula that LaTeX can typeset independently.
2. If a subformula starts with the binary operation + or -, type it as {}+ or {}-.
3. If a subformula ends with the binary operation + or -, type it as +{} or -{}.

Suppose that you want to split the formula

$$x_{1} + y_{1} + \left(\sum_{i < 5} \binom{5}{i} + a^{2} \right)^{2}$$

just before the binomial coefficient. Try

```
\begin{multline}
   x_{1} + y_{1} + \left( \sum_{i < 5}\\
    \binom{5}{i} + a^{2} \right)^{2}
\end{multline}
```

When typesetting this formula, you get the message

```
! Missing \right. inserted.
```

because the first subformula violates the first subformula rule.

```
x_{1} + y_{1} + \left( \sum_{i < 5}
```

cannot be typeset by LaTeX because the `\left(` command must be matched by the `\right` command and some delimiter.

Testing for the first subformula rule is easy. Split the formula into its subformulas, and test each subformula separately by typesetting it.

7.4.3 *Breaking and aligning formulas*

You do not have to know where and how to break inline math formulas because LaTeX does all the work for you.

Unfortunately, multiline formulas are different. LaTeX gives you excellent tools for displaying multiline math formulas, but offers you no advice on deciding where to break a long formula into lines. And that is how it should be. You, the author, are the only judge of where to break a long formula so that the result is mathematically informative and follows the traditions of mathematical typesetting.

A strict set of rules is formulated in *Mathematics into Type* by Ellen Swanson, Arlene Ann O'Sean, and Antoinette Tingley Schleyer [68]. I state only three.

Rule ■ Breaking displayed formulas

1. Try to break a long formula *before* a binary relation or binary operation.
2. If you break a formula before a + or -, start the next line with `{}+` or `{}-`.
3. If you break a formula within a bracket, indent the next line so that it begins *to the right of* the opening bracket.

Formula (4) on page 196 illustrates the first rule. Here is an illustration of the third rule:

$$
\begin{aligned}
f(x, y, z, u) = [(x + y + z) \times (x^2 + y^2 + z^2 - 1) \\
\times (x^3 + y^3 + z^3 - u) \times (x^4 + y^4 + z^4 + u)]^2
\end{aligned}
$$

The rules for aligning columns are similar.

Rule ■ Aligning columns

1. Try to align columns at a binary relation or a binary operation.
2. If you align a column at a binary relation, put the & symbol immediately *to the left* of the binary relation.
3. If you align a column at the binary operation + or -, put the & symbol to the left of the binary operation.

7.4.4 Numbering groups of formulas

With most constructs in this chapter, you have a number of equations typeset together, arranged in some way, aligned or adjusted. Each equation is numbered separately, unless `\tag`-ged or `\notag`-ged. Often, you may want the equations to share a common number, but still be able to reference each equation separately.

You can change the numbering of the equations on page 195 in formulas (1)–(3) to (1), (1a), and (1b) as follows:

```
\begin{gather}
   x_{1} x_{2} + x_{1}^{2} x_{2}^{2} + x_{3},
   \label{E:1}\\
   x_{1} x_{3} + x_{1}^{2} x_{3}^{2} + x_{2},
   \tag{\ref{E:1}a}\\
   x_{1} x_{2} x_{3};\tag{\ref{E:1}b}
\end{gather}
```

produces the desired result:

$$
\begin{aligned}
&(1) & x_1x_2 + x_1^2x_2^2 + x_3,\\
&(1a) & x_1x_3 + x_1^2x_3^2 + x_2,\\
&(1b) & x_1x_2x_3;
\end{aligned}
$$

To obtain $(1')$ or (1') type

```
\tag{\ref{E:1}$'$}
```

or

```
\tag{(\ref{E:1}\textquoteright)}
```

and for (1_{a}), type

```
\tag{\ref{E:1}${}_{\text{a}}$}
```

Alternatively, you may include the gather environment in a subequations environment (see Section 6.6):

$$
\begin{aligned}
&(5a) & x_1x_2 + x_1^2x_2^2 + x_3,\\
&(5b) & x_1x_3 + x_1^2x_3^2 + x_2,\\
&(5c) & x_1x_2x_3,
\end{aligned}
$$

typed as

```
\begin{subequations}\label{E:gp}
   \begin{gather}
      x_{1} x_{2} + x_{1}^{2} x_{2}^{2} + x_{3},
      \label{E:gp1}\\
      x_{1} x_{3} + x_{1}^{2} x_{3}^{2} + x_{2},
      \label{E:gp2}\\
      x_{1} x_{2} x_{3},\label{E:gp3}
   \end{gather}
\end{subequations}
```

Then \eqref{E:gp} references the whole group of equations as (5), while

\eqref{E:gp1}, \eqref{E:gp2}, and \eqref{E:gp3}

reference the individual formulas as (5a), (5b), and (5c).

7.5 Aligned columns

The lines of multiline formulas are naturally divided into columns. In this section, we discuss how to typeset such formulas with *aligned columns*. All of these constructs are implemented with the align math environment and its variants.

In Section 1.7.3, you saw two simple, one-column examples of aligned columns—which we called *simple alignment*—and a special case of aligned columns—which we called *annotated alignment*.

The `align` environment can also create multiple aligned columns. The number of columns is restricted only by the width of the page. In the following example, there are two aligned columns:

$$f(x) = x + yz \qquad g(x) = x + y + z$$
$$h(x) = xy + xz + yz \qquad k(x) = (x + y)(x + z)(y + z) \tag{6}$$

typed as

```
\begin{align}\label{E:mm3}
   f(x) &= x + yz          & g(x) &= x + y + z\\
   h(x) &= xy + xz + yz  & k(x) &= (x + y)(x + z)(y + z)
    \notag
\end{align}
```

Use Figure 7.2 to visualize how the alignment points in the source turn into alignment points in the typeset formula and the role played by the intercolumn space. Remember that the visual layout of the source is for your benefit only.

In a multicolumn `align` environment, the ampersand (&) plays two roles. It is a mark for the *alignment point* and it is also a *column separator*. In the line

```
   f(x) &= x + yz          & g(x) &= x + y + z
```

the two columns are

```
1.    f(x) &= x + yz
2.                              g(x) &= x + y + z
```

In each column, we use a single ampersand to mark the alignment point. Of the three & symbols in the previous example,

- The first & marks the *alignment point* of the first column.
- The second & is a *column separator* that separates the first and second columns.
- The third & marks the *alignment point* of the second column.

I use the convention of typing a space on the left of an alignment point & and no space on the right, and of putting spaces on both sides of & as a column separator.

If the number of columns is three, then there should be five &'s in each line. Even-numbered &'s are column separators and odd-numbered &'s are alignment marks.

Rule ■ Ampersands

If there are n aligned columns, then each line should have at most $2n - 1$ ampersands. Even-numbered &'s are column separators; odd-numbered &'s mark the alignment points.

---

first column second column

```
f(x) &= x + yz         & g(x) &= x + y + z\\
h(x) &= xy + xz + yz   & k(x) &= (x + y)(x + z)(y + z)
```

alignment points of first column

alignment points of second column

start of second column

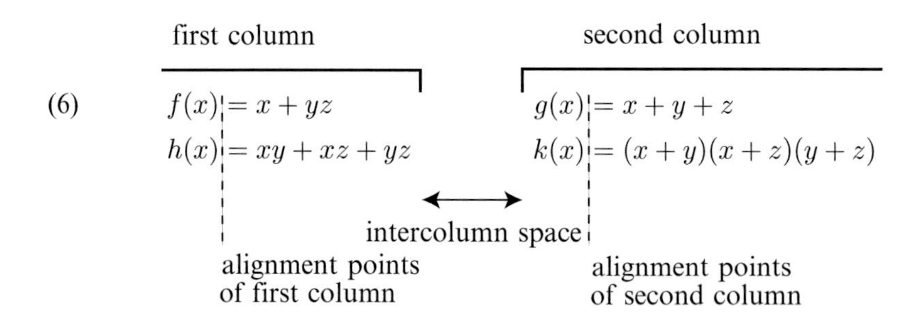

Figure 7.2: Two aligned columns: source and typeset.

So for a single aligned column, you have to place one alignment point for each line. For two aligned columns, you have to place at most three alignment points for each line. The beginning of the line to the second & is the first column, then from the second & to the end of the line is the second column. Each line of each column has an alignment point marked by &.

A column in a line may be empty—a gap is produced—or it may have only a few columns. Both of these are illustrated by

$$\begin{aligned}
& a_1 & & & &c_1\\
& & &b_2 & &c_2\\
& a_3
\end{aligned}$$

typed as

```
\begin{align*}
   & a_1 &  &    &  &c_1\\
   &     &  &b_2 &  &c_2\\
   & a_3
\end{align*}
```

7.5.1 *An* `align` *variant*

A variant of `align` is the flush alignment environment `flalign`, which moves the leftmost column as far left and the rightmost column as far right as space allows, making more room for the formula. Here is formula (6) again, followed by the `flalign` variant:

$$f(x) = x + yz \qquad g(x) = x + y + z \tag{6}$$
$$h(x) = xy + xz + yz \qquad k(x) = (x + y)(x + z)(y + z)$$

$$f(x) = x + yz \qquad g(x) = x + y + z \tag{7}$$
$$h(x) = xy + xz + yz \qquad k(x) = (x + y)(x + z)(y + z)$$

The variant is typed as follows:

```
\begin{flalign}\label{E:mm3fl}
   f(x) &= x + yz         & g(x) &= x + y + z\\
   h(x) &= xy + xz + yz & k(x) &= (x + y)(x + z)(y + z)
    \notag
\end{flalign}
```

7.5.2 `eqnarray`, *the ancestor of* `align`

LaTeX's original aligned math environment is `eqnarray`. Here is an example:

```
\begin{eqnarray}
   x & = & 17y\\
   y & > & a + b + c
\end{eqnarray}
```

which typesets as

$$x \;=\; 17y \tag{8}$$
$$y \;>\; a + b + c \tag{9}$$

You can type the same formulas with `align`:

```
\begin{align}
   x  & =  17y\\
   y  & >  a + b + c
\end{align}
```

which typesets as

$$x = 17y \tag{10}$$
$$y > a + b + c \tag{11}$$

In the `eqnarray` environment the spacing is based on the spacing of the columns rather than on the spacing requirements of the symbols.

I mention `eqnarray` not for historical reasons but for a very practical one. Unfortunately, a large number of journal submissions still use this construct, and have to be recoded in the editorial offices. Be kind to your editor and do not use `eqnarray`.

7.5.3 *The subformula rule revisited*

Suppose that you want to align the formula

$$x_1 + y_1 + \left(\sum_i \binom{5}{i} + a^2 \right)^2$$

with

$$\left(\sum_i \binom{5}{i} + \alpha^2 \right)^2$$

so that the $+\, a^2$ in the first formula aligns with the $+\, \alpha^2$ in the second formula. You might try typing

```
\begin{align*}
   x_{1} + y_{1} + \left( \sum_i \binom{5}{i}
            &+ a^{2} \right)^{2}\\
    \left( \sum_i \binom{5}{i} &+ \alpha^{2} \right)^{2}
\end{align*}
```

But when you typeset this formula, you get the message

```
! Extra }, or forgotten \right.
```

This alignment structure violates the subformula rule because LaTeX cannot typeset

```
x_{1} + y_{1} + \left( \sum_i \binom{5}{i}
```

so it is not a subformula.

As another simple example, try to align the + in $\binom{a+b}{2}$ with the + in $x + y$:

```
\begin{align}
  \binom{a &+ b}{2}\\
    x &+ y
\end{align}
```

When typesetting this formula, you get the message

```
! Missing } inserted.
```

Again, LaTeX cannot typeset the subformula `\binom{a`.

To align the two formulas in the first example, add a `\phantom` command to push the second line to the right:

```
\begin{align*}
   &x_{1} + y_{1} + \left( \sum_{i < 5} \binom{5}{i}
      + a^{2} \right)^{2}\\
   &\phantom{x_{1} + y_{1} + {}}
      \left( \sum_{i < 5} \binom{5}{i} + \alpha^{2}
      \right)^{2}
\end{align*}
```

yielding

$$x_1 + y_1 + \left(\sum_{i<5} \binom{5}{i} + a^2 \right)^2$$

$$\left(\sum_{i<5} \binom{5}{i} + \alpha^2 \right)^2$$

7.5.4 *The* `alignat` *environment*

Another variant of the `align` environment is the `alignat` environment, which is one of the most important alignment environments. While the `align` environment calculates how much space to put between the columns, the `alignat` environment leaves spacing up to the user. It is important to note that the `alignat` environment has a required argument, the number of columns.

Here is formula (6) typed with the `alignat` environment:

```
\begin{alignat}{2}\label{E:mm3A}
   f(x) &= x + yz        & g(x) &= x + y + z\\
   h(x) &= xy + xz + yz & k(x) &= (x + y)(x + z)(y + z)
   \notag
\end{alignat}
```

which typesets as

$$f(x) = x + yz \qquad g(x) = x + y + z \tag{12}$$
$$h(x) = xy + xz + yzk(x) = (x + y)(x + z)(y + z)$$

This attempt did not work very well because `alignat` did not separate the two formulas in the second line. So you must provide the intercolumn spacing. For instance, if you want a `\qquad` space between the columns, as in

$$\begin{alignedat}{2} f(x) &= x + yz \qquad & g(x) &= x + y + z \\ h(x) &= xy + xz + yz \qquad & k(x) &= (x + y)(x + z)(y + z) \end{alignedat} \tag{13}$$

then type the formula as

```
\begin{alignat}{2}\label{E:mm3B}
   f(x) &= x + yz                & g(x) &= x + y + z\\
   h(x) &= xy + xz + yz \qquad & k(x) &= (x+y)(x+z)(y+z)
    \notag
\end{alignat}
```

The `alignat` environment is especially appropriate when annotating formulas where you would normally want a `\quad` between the formula and the text. To obtain

$$\begin{alignedat}{2} x &= x \wedge (y \vee z) &\quad &\text{(by distributivity)} \\ &= (x \wedge y) \vee (x \wedge z) & &\text{(by condition (M))} \\ &= y \vee z \end{alignedat} \tag{14}$$

type

```
\begin{alignat}{2}\label{E:mm4}
   x &= x \wedge (y \vee z) &
   &\quad\text{(by distributivity)}\\
     &= (x \wedge y) \vee (x \wedge z) & &
      \quad\text{(by condition (M))}\notag\\
     &= y \vee z \notag
\end{alignat}
```

The alignat environment is very important for typing systems of equations such as

$$\begin{alignat}{2} (A + BC)x &+{} &C &y = 0, \tag{15}\\ Ex &+{} &(F + G)&y = 23. \tag{16} \end{alignat}$$

typed as follows:

```
\begin{alignat}{2}
   (A + B C)x &+{} &C        &y  = 0,\\
            Ex &+{} &(F + G)&y  = 23.
\end{alignat}
```

Note again +{}. See also the subformula rule in Section 7.4.2.

As a last example, consider

$$\begin{alignat}{4} a_{11}x_1 &+ a_{12}x_2 &&+ a_{13}x_3 && &&= y_1, \tag{17}\\ a_{21}x_1 &+ a_{22}x_2 && &&+ a_{24}x_4 &&= y_2, \tag{18}\\ a_{31}x_1 & &&+ a_{33}x_3 &&+ a_{34}x_4 &&= y_3. \tag{19} \end{alignat}$$

typed as

```
\begin{alignat}{4}
  a_{11}x_1 &+ a_{12}x_2 &&+ a_{13}x_3 &&
    &&= y_1,\\
  a_{21}x_1 &+ a_{22}x_2 &&             &&+ a_{24}x_4
    &&= y_2,\\
  a_{31}x_1 &             &&+ a_{33}x_3 &&+ a_{34}x_4
    &&= y_3.
\end{alignat}
```

Note that the argument of alignat does not have to be precise. If you want two columns, the argument can be 2, or 3, or any larger number. If you want to, you can simply type 10 and just ignore the argument. You may define a new environment (see Section 14.2.1) that does just that.

7.5.5 Inserting text

The \intertext command places one or more lines of text in the middle of an aligned environment. For instance, to obtain

$$h(x) = \int \left(\frac{f(x) + g(x)}{1 + f^{2}(x)} + \frac{1 + f(x)g(x)}{\sqrt{1 - \sin x}} \right) \, dx \tag{20}$$

The reader may find the following form easier to read:

$$= \int \frac{1 + f(x)}{1 + g(x)} \, dx - 2 \arctan(x - 2)$$

you would type

```
\begin{align}\label{E:mm5}
      h(x) &= \int \left(
                         \frac{ f(x) + g(x) }
                              {1 + f^{2}(x)} +
                         \frac{1 + f(x)g(x)}
                              { \sqrt{1 - \sin x} }
                   \right) \, dx\\
   \intertext{The reader may find the following form
         easier to read:}
      &= \int \frac{1 + f(x)}
                   {1 + g(x)}
         \, dx - 2 \arctan(x - 2) \notag
\end{align}
```

Notice how the equal sign in the first formula is aligned with the equal sign in the second formula even though a line of text separates the two.

Here is another example, this one using `align*`:

$$f(x) = x + yz \qquad\qquad g(x) = x + y + z$$

The reader may also find the following polynomials useful:

$$h(x) = xy + xz + yz \qquad\qquad k(x) = (x + y)(x + z)(y + z)$$

is typed as

```
\begin{align*}
   f(x) &= x + yz & \qquad g(x) &= x + y + z\\
   \intertext{The reader may also find the following
    polynomials useful:}
   h(x) &= xy + xz + yz
                  & \qquad k(x) &= (x + y)(x + z)(y + z)
\end{align*}
```

The `\intertext` command must follow a line separator command, `\\` or `\\*` (see Section 7.9). If you violate this rule, you get the message

```
! Misplaced \noalign. \intertext #1->\noalign
                        {\penalty \postdisplaypenalty
                                          \vskip ...
```

The text in `\intertext` can be centered using a `center` environment or with the `\centering` command (see Section 4.3).

7.6 Aligned subsidiary math environments

A *subsidiary math environment* is a math environment that can only be used *inside another math environment.* Think of it as creating a "large math symbol".

In this section, we discuss aligned subsidiary math environments. We discuss adjusted subsidiary math environments, including `cases`, in Section 7.7.

7.6.1 Subsidiary variants

The `align`, `alignat`, and `gather` environments (see Sections 7.5, 7.5.4, and 7.2) have subsidiary versions. They are called `aligned`, `alignedat`, and `gathered`. To obtain

$$\begin{aligned} x &= 3 + \mathbf{p} + \alpha\\ y &= 4 + \mathbf{q}\\ z &= 5 + \mathbf{r}\\ u &=6 + \mathbf{s} \end{aligned} \text{\qquad using\qquad} \begin{gathered} \mathbf{p} = 5 + a + \alpha\\ \mathbf{q} = 12\\ \mathbf{r} = 13\\ \mathbf{s} = 11 + d \end{gathered}$$

type

```
\[
   \begin{aligned}
      x &= 3 + \mathbf{p} + \alpha\\
      y &= 4 + \mathbf{q}\\
      z &= 5 + \mathbf{r}\\
      u &=6 + \mathbf{s}
   \end{aligned}
   \text{\qquad using\qquad}
   \begin{gathered}
      \mathbf{p} = 5 + a + \alpha\\
      \mathbf{q} = 12\\
      \mathbf{r} = 13\\
      \mathbf{s} = 11 + d
   \end{gathered}
\]
```

Note how the list of aligned formulas

$$\begin{aligned} x &= 3 + p + \alpha \\ y &= 4 + \mathbf{q} \\ z &= 5 + \mathbf{r} \\ u &= 6 + \mathbf{s} \end{aligned}$$

and the list of centered formulas

$$\begin{gathered} \mathbf{p} = 5 + a + \alpha \\ \mathbf{q} = 12 \\ \mathbf{r} = 13 \\ \mathbf{s} = 11 + d \end{gathered}$$

are treated as individual large symbols.

The `aligned`, `alignedat`, and `gathered` subsidiary math environments follow the same rules as `align` and `gather`. The `aligned` subsidiary math environment allows any number of columns, but you must specify the intercolumn spacing as in the `alignat` environment.

You can use the `aligned` subsidiary math environment to rewrite formula (4) from Section 1.7.3 so that the formula number is centered between the two lines:

$$\begin{aligned} h(x) &= \int \left(\frac{f(x) + g(x)}{1 + f^{2}(x)} + \frac{1 + f(x)g(x)}{\sqrt{1 - \sin x}} \right) \, dx \\ &= \int \frac{1 + f(x)}{1 + g(x)} \, dx - 2 \arctan (x - 2) \end{aligned} \tag{21}$$

this is typed as

```
\begin{equation}\label{E:mm6}
\begin{aligned}
   h(x) &= \int \left(
                  \frac{ f(x) + g(x) }
                       { 1 + f^{2}(x) } +
                  \frac{ 1 + f(x)g(x) }
                       { \sqrt{1 - \sin x} }
               \right) \, dx\\
        &= \int \frac{ 1 + f(x) }
                     { 1 + g(x) } \, dx - 2 \arctan (x - 2)
\end{aligned}
\end{equation}
```

See Section 7.6.2 for a better way to split a long formula.

Symbols, as a rule, are vertically centrally aligned. This is not normally an issue with math symbols, but it may be important with large symbols created by subsidiary math environments. The subsidiary math environments, `aligned`, `gathered`, and array, take c, t, or b as optional arguments to force vertically centered, top, or bottom alignment, respectively. The default is c (centered). To obtain

$$\begin{aligned} x &= 3 + \mathbf{p} + \alpha\\ y &= 4 + \mathbf{q}\\ z &= 5 + \mathbf{r}\\ u &=6 + \mathbf{s} \end{aligned} \text{\qquad using\qquad} \begin{gathered} \mathbf{p} = 5 + a + \alpha\\ \mathbf{q} = 12\\ \mathbf{r} = 13\\ \mathbf{s} = 11 + d \end{gathered}$$

for example, you would type

```
\[
   \begin{aligned}[b]
      x &= 3 + \mathbf{p} + \alpha\\
      y &= 4 + \mathbf{q}\\

      z &= 5 + \mathbf{r}\\
      u &=6 + \mathbf{s}
   \end{aligned}
   \text{\qquad using\qquad}
   \begin{gathered}[b]
      \mathbf{p} = 5 + a + \alpha\\
      \mathbf{q} = 12\\
      \mathbf{r} = 13\\
      \mathbf{s} = 11 + d
   \end{gathered}
\]
```

There is no numbering or \tag-ing allowed in subsidiary math environments because LaTeX does not number or tag what it considers to be a single symbol.

7.6.2 Split

The split subsidiary math environment is used to split a long formula into aligned parts. There are two major advantages to use split:

1. The math environment that contains it considers the split environment to be a single equation, so it generates only one number for it.
2. If a split environment appears inside an align environment, the alignment point of the split environment is recognized by align and is used in aligning all the formulas in the align environment.

To illustrate the first advantage, consider

$$\begin{multlined} (x_1x_2x_3x_4x_5x_6)^2\\ + (x_1x_2x_3x_4x_5 + x_1x_3x_4x_5x_6 + x_1x_2x_4x_5x_6 + x_1x_2x_3x_5x_6)^2 \end{multlined} \tag{22}$$

typed as

```
\begin{equation}\label{E:mm7}
   \begin{split}
       (x_{1}x_{2}&x_{3}x_{4}x_{5}x_{6})^{2}\\
                   &+ (x_{1}x_{2}x_{3}x_{4}x_{5}
                    + x_{1}x_{3}x_{4}x_{5}x_{6}
                    + x_{1}x_{2}x_{4}x_{5}x_{6}
                    + x_{1}x_{2}x_{3}x_{5}x_{6})^{2}
   \end{split}
\end{equation}
```

See also the two examples of `split` in the `secondarticle.tex` sample article in Section 9.3 and in the `samples` folder (see page 5).

To illustrate the second advantage, here is an example of a `split` subsidiary math environment within an `align` environment:

$$\begin{split} f &= (x_1x_2x_3x_4x_5x_6)^2\\ &= (x_1x_2x_3x_4x_5 + x_1x_3x_4x_5x_6 + x_1x_2x_4x_5x_6 + x_1x_2x_3x_5x_6)^2, \end{split} \tag{23}$$

$$g = y_1y_2y_3. \tag{24}$$

which is typed as

```
\begin{align}\label{E:mm8}
   \begin{split}
      f &= (x_{1} x_{2} x_{3} x_{4} x_{5} x_{6})^{2}\\
        &= (x_{1} x_{2} x_{3} x_{4} x_{5}
          + x_{1} x_{3} x_{4} x_{5} x_{6}
          + x_{1} x_{2} x_{4} x_{5} x_{6}
          + x_{1} x_{2} x_{3} x_{5} x_{6})^{2},
   \end{split}\\
      g &= y_{1} y_{2} y_{3}.\label{E:mm9}
\end{align}
```

Notice the `\\` command following `\end{split}` to separate the lines for `align`.

Rule ■ `split` **subsidiary math environment**

1. `split` can only be used inside another math environment, such as `displaymath`, `equation`, `align`, `gather`, `flalign`, `gathered` and their `*`-ed variants.

2. A `split` formula has only one number, automatically generated by LaTeX, or one tag from a `\tag` command. Use the `\notag` command to suppress numbering.

3. The `\label`, `\tag`, or `\notag` command must precede `\begin{split}` or follow `\end{split}`.

Here is an example of split inside a gather environment:

```
\begin{gather}\label{E:mm10}
   \begin{split}
      f &= (x_{1} x_{2} x_{3} x_{4} x_{5} x_{6})^{2}\\
        &= (x_{1} x_{2} x_{3} x_{4} x_{5}
           + x_{1} x_{3} x_{4} x_{5} x_{6}
           + x_{1} x_{2} x_{4} x_{5} x_{6}
           + x_{1} x_{2} x_{3} x_{5} x_{6})^{2}\\
        &= (x_{1} x_{2} x_{3} x_{4}
           + x_{1} x_{2} x_{3} x_{5}
           + x_{1} x_{2} x_{4} x_{5}
           + x_{1} x_{3} x_{4} x_{5})^{2}
   \end{split}\\
   \begin{aligned}
      g &= y_{1} y_{2} y_{3}\\
      h &= z_{1}^{2} z_{2}^{2} z_{3}^{2} z_{4}^{2}
   \end{aligned}
\end{gather}
```

which produces

$$\begin{split} f &= (x_{1} x_{2} x_{3} x_{4} x_{5} x_{6})^{2}\\ &= (x_{1} x_{2} x_{3} x_{4} x_{5} + x_{1} x_{3} x_{4} x_{5} x_{6} + x_{1} x_{2} x_{4} x_{5} x_{6} + x_{1} x_{2} x_{3} x_{5} x_{6})^{2}\\ &= (x_{1} x_{2} x_{3} x_{4} + x_{1} x_{2} x_{3} x_{5} + x_{1} x_{2} x_{4} x_{5} + x_{1} x_{3} x_{4} x_{5})^{2} \end{split} \tag{25}$$

$$\begin{aligned} g &= y_{1} y_{2} y_{3}\\ h &= z_{1}^{2} z_{2}^{2} z_{3}^{2} z_{4}^{2} \end{aligned}$$

If you try to use split outside a displayed math environment, you get the message

```
! Package amsmath Error: \begin{split} won't work here.
```

You may want to read the discussion of the AMS document classes and amsmath package options in Section 9.5 that modify the placement of equation numbers.

7.7 Adjusted columns

In an *adjusted* multiline math environment, the columns are adjusted so that they are displayed centered, flush left, or flush right, instead of aligned (as in Section 7.5). Since you have no control line by line over the alignment of the columns, & has only one role to play—it is the column separator.

In Sections 7.2 and 7.3, we discussed two adjusted one-column math environments, gather and multline. All the other adjusted constructs are subsidiary math environments. For example, a matrix environment (see Section 7.7.1) produces a multicolumn centered display:

$$\begin{pmatrix} a+b+c & uv & x-y & 27 \\ a+b & u+v & z & 1340 \end{pmatrix} = \begin{pmatrix} 1 & 100 & 115 & 27 \\ 201 & 0 & 1 & 1340 \end{pmatrix}$$

The array environment (see Section 7.7.2) produces a multicolumn adjusted display:

$$\left(\begin{array}{cccr} a+b+c & uv & x-y & 27 \\ a+b & u+v & z & 1340 \end{array} \right) = \left(\begin{array}{rrrr} 1 & 100 & 115 & 27 \\ 201 & 0 & 1 & 1340 \end{array} \right)$$

The columns are centered, flush left, or flush right. In this example, the first matrix has three centered columns and one flush right column, while the second matrix has four flush right columns. A variant, cases (see Sections 1.7.4 and 7.7.3), produces two columns set flush left:

$$f(x) = \begin{cases} -x^2, & \text{if } x < 0; \\ \alpha + x, & \text{if } 0 \leq x \leq 1; \\ x^2, & \text{otherwise.} \end{cases} \tag{26}$$

7.7.1 Matrices

Use the matrix subsidiary math environment to typeset matrices. For example,

```
\begin{equation*}
   \left(
   \begin{matrix}
      a + b + c & uv    & x - y & 27\\
      a + b     & u + v & z     & 1340
   \end{matrix}
   \right) =
   \left(
   \begin{matrix}
      1   & 100 & 115 & 27\\
      201 & 0   & 1   & 1340
   \end{matrix}
   \right)
\end{equation*}
```

produces

$$\begin{pmatrix} a+b+c & uv & x-y & 27 \\ a+b & u+v & z & 1340 \end{pmatrix} = \begin{pmatrix} 1 & 100 & 115 & 27 \\ 201 & 0 & 1 & 1340 \end{pmatrix}$$

If you use matrix on its own, i.e., outside a math environment,

```
\begin{matrix}
   a + b + c & uv    & x - y & 27\\
   a + b     & u + v & z     & 134
\end{matrix}
```

you get the message

```
! Missing $ inserted.
<inserted text>
                $
l.5 \begin{matrix}
```

obliquely reminding you that `matrix` is a subsidiary math environment.

The `matrix` subsidiary math environment provides a matrix of up to 10 centered columns. If you need more columns, you have to ask for them. The following example sets the number of columns to 12:

```
\begin{equation}\label{E:mm12}
   \setcounter{MaxMatrixCols}{12}
   \begin{matrix}
      1 & 2 & 3 & 4 & 5 & 6 & 7 & 8 & 9 & 10 & 11 & 12\\
      1 & 2 & 3 & \hdotsfor{7}                & 11 & 12
   \end{matrix}
\end{equation}
```

produces

$$\begin{matrix} 1 & 2 & 3 & 4 & 5 & 6 & 7 & 8 & 9 & 10 & 11 & 12\\ 1 & 2 & 3 & \hdotsfor{7} & 11 & 12 \end{matrix} \tag{27}$$

We discuss `\setcounter` and other counters further in Section 14.5.1.

You can have dots span any number of columns with the `\hdotsfor` command, as in (27). The argument of this command specifies the number of columns to fill (which is one more than the number of `&`'s the command replaces). The `\hdotsfor` command must either appear at the beginning of a row or immediately following an ampersand (`&`). If you violate this rule, you get the message

```
! Misplaced \omit.
\multispan #1->\omit
                     \mscount #1\relax \loop \ifnum
                                                \mscount ...
l.12 \end{equation}
```

The `\hdotsfor` command also takes an optional argument, a number that multiplies the spacing between the dots. The default is `1`. For instance, if we replace `\hdotsfor{7}` in the previous example by `\hdotsfor[3]{7}`, then we get

$$\begin{matrix} 1 & 2 & 3 & 4 & 5 & 6 & 7 & 8 & 9 & 10 & 11 & 12\\ 1 & 2 & 3 & \hdotsfor[3]{7} & 11 & 12 \end{matrix} \tag{28}$$

We can replace a part of a matrix column with a large symbol.

$$a = \begin{pmatrix} (a_{11}) & \\ \cdots & \text{\huge 0} \\ (a_{n1}) & \end{pmatrix}, \quad (a_{k1}) = \begin{pmatrix} 0\ldots 0 & 1 & 0\ldots 0 \\ & 0 & \\ \text{\huge 0} & \vdots & \text{\huge 0} \\ & 0 & \end{pmatrix}$$

typed as

```
\newcommand{\BigFig}[1]{\parbox{12pt}{\Huge #1}}
\newcommand{\BigZero}{\BigFig{0}}
\[
a=\left( \begin{matrix}
(a_{11})\\
\cdots & \BigZero \\
(a_{n1})\\
\end{matrix}
\right) ,\quad
(a_{k1})=\left(
\begin{matrix}
0\ldots 0 & 1 & 0\ldots 0\\
 & 0\\
\BigZero & \cdots & \BigZero\\
 & 0\\
\end{matrix} \right)
\]
```

Matrix variants

A matrix may be enclosed by delimiters (see Section 5.5.1) in a number of different ways:

$$\begin{matrix} a+b+c & uv \\ a+b & c+d \end{matrix} \qquad \begin{pmatrix} a+b+c & uv \\ a+b & c+d \end{pmatrix} \qquad \begin{bmatrix} a+b+c & uv \\ a+b & c+d \end{bmatrix}$$

$$\begin{vmatrix} a+b+c & uv \\ a+b & c+d \end{vmatrix} \qquad \begin{Vmatrix} a+b+c & uv \\ a+b & c+d \end{Vmatrix} \qquad \begin{Bmatrix} a+b+c & uv \\ a+b & c+d \end{Bmatrix}$$

The first matrix is typed as

```
\begin{matrix}
   a + b + c & uv\\
   a + b     & c + d
\end{matrix}
```

The others are typed in the same way, except that they use the `pmatrix`, `bmatrix`, `vmatrix`, `Vmatrix`, and `Bmatrix` environments, respectively. We can use other delimiters, as in

```
\begin{equation*}
   \left(
   \begin{matrix}
      1      &      0 & \dots  & 0\\
      0      &      1 & \dots  & 0\\
      \vdots & \vdots & \ddots & \vdots\\
```

```
      0       &    0   & \dots   & 1
   \end{matrix}
   \right]
\end{equation*}
```

which produces

$$\begin{bmatrix} 1 & 0 & \dots & 0 \\ 0 & 1 & \dots & 0 \\ \vdots & \vdots & \ddots & \vdots \\ 0 & 0 & \dots & 1 \end{bmatrix}$$

This example also uses *vertical dots* provided by the `\vdots` commands and *diagonal dots* provided by the `\ddots` commands.

Small matrix

If you put a `matrix` in an inline math formula, it may be too large. Instead, use the `smallmatrix` environment. Compare $\begin{pmatrix} a + b + c & uv \\ a + b & c + d \end{pmatrix}$, typed as

```
$\begin{pmatrix}
   a + b + c & uv\\
   a + b     & c + d
\end{pmatrix}$
```

with the small matrix $\left(\begin{smallmatrix} a+b+c & uv \\ a+b & c+d \end{smallmatrix} \right)$, typed as

```
$\left(
\begin{smallmatrix}
   a + b + c & uv\\
   a + b     & c + d
\end{smallmatrix}
\right)$
```

There are no delimited variants of `smallmatrix` similar to those of `matrix`. Instead, use the `\left` and `\right` commands with delimiters to enclose a small matrix. The `\hdotsfor` command does not work in a small matrix.

7.7.2 *Arrays*

The `array` subsidiary math environment is a variant of `matrix`. For `array`, you must specify the alignment of each column and you have more options to customize it.

The first matrix in the introduction to Section 7.7 would be typed as follows using the `array` subsidiary math environment:

```
\begin{equation*}
   \left(
   \begin{array}{cccc}
      a + b + c & uv     & x - y & 27\\
```

```
      a + b       & u + v & z       & 134
   \end{array}
   \right)
\end{equation*}
```

which produces

$$\begin{pmatrix} a+b+c & uv & x-y & 27 \\ a+b & u+v & z & 134 \end{pmatrix}$$

Rule ■ `array` **subsidiary math environment**

1. Adjacent columns are separated by an ampersand (&).
2. The argument of `\begin{array}` is mandatory. The argument is a series of the letters `l`, `r`, or `c`, signifying that the corresponding column in the array should be set flush left, flush right, or centered, respectively.

The matrix

$$\begin{pmatrix} a+b+c & uv & x-y & 27 \\ a+b & u+v & z & 134 \end{pmatrix}$$

could not have been typeset with `matrix` since the last column is set flush right. This is not quite true, of course. In a `matrix` environment, `\hfill 27` would force the number 27 to be set flush right (see Section 3.8.4).

If the argument of `\begin{array}` is missing, as in

```
\begin{equation}
   \begin{array}
      a + b + c & uv      & x - y & 27\\
      a + b     & u + v   & z     & 134
   \end{array}
\end{equation}
```

LaTeX generates the message

```
! Package array Error:  Illegal pream-token (a): 'c' used.
```

If you change the first entry of the matrix to `c + b + a`, then the message is

```
! Extra alignment tab has been changed to \cr.
<recently read> \endtemplate

l.5         c + b + a &
                           uv       & x - y & 27\\
```

Note that the first character in `c + b + a` is not an

```
Illegal character in array arg.
```

because `c` is one possible argument of `\begin{array}`.

If the closing brace of the argument of \begin{array} is missing, as in

```
\begin{equation}
   \begin{array}{cccc
      a + b + c & uv    & x - y & 27\\
      a + b     & u + v & z     & 134
   \end{array}
\end{equation}
```

you get the message

```
Runaway argument?
{cccc a + b + c & uv    & x - y & 27\\ a + b
                        & u + v \ETC.
! Paragraph ended before \@array was complete.
```

In fact, the argument of array can be more complex than stated in the rule. Indeed, the array subsidiary math environment can take any argument that the tabular environment can take (see Section 4.6). For instance, here is a matrix with headers:

$$\begin{array}{r|rrr}
 & a & b & c \\
\hline
1 & 1 & 1 & 1 \\
2 & 1 & -1 & -1 \\
2 & 2 & 1 & 0
\end{array}$$

typed as

```
\[
\begin{array}{r|rrr}
  & a & b & c \\
\hline
1 & 1 &  1 &  1 \\
2 & 1 & -1 & -1 \\
2 & 2 &  1 &  0
\end{array}
\]
```

In Section 7.7.1 we have the matrix example:

$$a = \begin{pmatrix} \begin{matrix} (a_{11}) \\ \cdots \\ (a_{n1}) \end{matrix} & \text{\Huge 0} \end{pmatrix}$$

If rows are spanned, we need to use array instead of matrix:

$$\left[\begin{array}{cc} \begin{matrix} a & b \\ c & d \end{matrix} & \text{\Huge 0} \\ \text{\Huge 0} & \begin{matrix} m & n \\ k & l \end{matrix} \end{array} \right]$$

typed as (the \BigZero command is defined on page 217)

```
\left[ \hspace{-\arraycolsep}
 spacing is automatic with matrix but not with array
\begin{array}{cccc}
a & b &\multicolumn{2}{c}{}\\
c & d &\multicolumn{2}{c}
                    {\raisebox{1.5ex}[0pt]{\BigZero}}\\
\multicolumn{2}{c}{}& m & n \\
\multicolumn{2}{c}
               {\raisebox{1.5ex}[0pt]{\BigZero}}& k & l
\end{array}
\hspace{-\arraycolsep} \right]
\end{equation*}
```

7.7.3 Cases

The cases environment is also a subsidiary math environment. Here is the example from Section 1.7.4 and the introduction to this section:

$$f(x) = \begin{cases} -x^2, & \text{if } x < 0; \\ \alpha + x, & \text{if } 0 \leq x \leq 1; \\ x^2, & \text{otherwise.} \end{cases}$$

It is typed as

```
\begin{equation}
   f(x)=
   \begin{cases}
      -x^{2},          &\text{if $x < 0$;}\\
      \alpha + x,      &\text{if $0 \leq x \leq 1$;}\\
      x^{2},           &\text{otherwise.}
   \end{cases}
\end{equation}
```

It would be easy to code the cases environment as a special case of the array subsidiary math environment:

```
\begin{equation}
   f(x) =
  \left\{
   \begin{array}{ll}
      -x^{2},          &\text{if $x < 0$;}\\
     \alpha + x,       &\text{if $0 \leq x \leq 1$;}\\
      x^{2},           &\text{otherwise.}
   \end{array}
   \right.
\end{equation}
```

or of the `alignedat` subsidiary math environment:

```
\begin{equation*}
   f(x) =
   \left\{
   \begin{alignedat}{2}
     &-x^{2},        &&\quad\text{if $x < 0$;}\\
     &\alpha + x,    &&\quad\text{if $0 \leq x \leq 1$;}\\
     &x^{2},         &&\quad\text{otherwise.}
     \end{alignedat}
   \right.
\end{equation*}
```

7.8 Commutative diagrams

The `amscd` package provides the `CD` subsidiary math environment for typesetting simple commutative diagrams. To use it, make sure that the command

```
\usepackage{amscd}
```

is in the preamble of the document.

For instance, to obtain

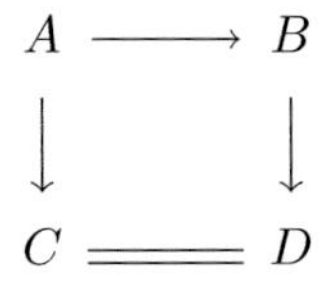

type

```
\[
   \begin{CD}
      A          @>>>          B\\
      @VVV                     @VVV\\
      C          @=            D
   \end{CD}
\]
```

A commutative diagram is a matrix made up of two kinds of rows, *horizontal rows,* that is, rows with horizontal arrows; and *vertical rows,* rows with vertical arrows. For example,

```
A          @>>>        B
```

is a typical horizontal row. It defines two columns and a connecting horizontal arrow `@>>>`. There may also be more than two columns, as in

```
A  @>>>  B  @>>>  C  @=  D  @<<<  E  @<<<  F
```

The connecting pieces can be:

- Stretchable right arrows, `@>>>`
- Stretchable left arrows, `@<<<`
- Stretchable equal signs, `@=`
- Blanks, `@.`

The label above a stretchable arrow should be typed between the first and second > or < symbols, whereas the label below should be typed between the second and third > or < symbols. You can have both.

The following is a typical vertical row containing vertical arrows:

```
@VVV       @VVV       @AAA
```

The vertical pieces could be

- Stretchable down arrows, `@VVV`
- Stretchable up arrows, `@AAA`
- Double vertical lines, `@|` or `@\vert`
- Blanks, `@.`

The vertical arrows are placed starting with the first column.

The label to the left of a stretchable vertical arrow should be typed between the first and second `V` or `A`, whereas the label on the right should be typed between the second and third `V` or `A` symbols. You can have both.

These constructs are illustrated in

$$\begin{CD} \mathbb{C} @>H_{1}>> \mathbb{C} @>H_{2}>> \mathbb{C}\\ @VP_{c,3}VV @VP_{\bar{c},3}VV @VVP_{-c,3}V\\ \mathbb{C} @>H_{1}>> \mathbb{C} @>H_{2}>> \mathbb{C} \end{CD}$$

typed as

```
\[
  \begin{CD}
   \mathbb{C} @>H_{1}>> \mathbb{C} @>H_{2}>>\mathbb{C}\\
   @VP_{c,3}VV  @VP_{\bar{c},3}VV  @VVP_{-c,3}V\\
   \mathbb{C} @>H_{1}>> \mathbb{C} @>H_{2}>> \mathbb{C}
  \end{CD}
\]
```

Here is another example utilizing the `\text` command, followed by its source:

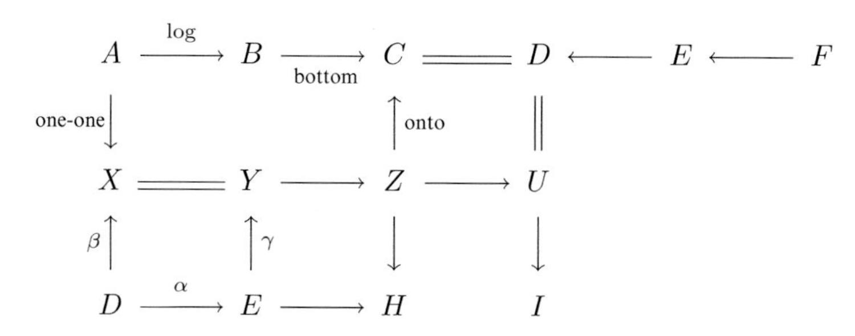


```
\[
  \begin{CD}
    A       @>\log>>     B       @>>\text{bottom}>   C
            @=           D       @<<<                E
            @<<<         F\\
    @V\text{one-one}VV   @.      @AA\text{onto}A    @|\\
    X          @=                Y       @>>>               Z
               @>>>              U\\
    @A\beta AA                   @AA\gamma A     @VVV    @VVV\\
    D          @>\alpha>>        E                @>>>    H
               @.                I\\
  \end{CD}
\]
```

Diagrams requiring more advanced commands should be done with a drawing (or drafting) application or with specialized packages. The AMS recommends Kristoffer H. Rose and Ross Moore's `xy-pic` package (see Section D.1). If you get familiar with the Ti*k*Z package in Chapter 13, then you should utilize the `tikzcd` package of Florêncio Neves to draw commutative diagrams, see Section 13.4.

7.9 Adjusting the display

By default, the math environments described in this chapter do not allow page breaks. While a page break in a `cases` environment is obviously not desirable, it may be acceptable in an `align` or `gather` environment. You can allow page breaks by using the

```
\allowdisplaybreaks
```

command. It allows page breaks in a multiline math environment within its scope. For instance,

```
{\allowdisplaybreaks
\begin{align}\label{E:mm13}
  a &= b + c,\\
  d &= e + f,\\
```

```
    x &= y + z,\\
    u &= v + w.
\end{align}
}% end of \allowdisplaybreaks
```

allows a page break after any one of the first three lines.

Within the scope of an `\allowdisplaybreaks` command, use the `\\*` command to prohibit a break after that line. The line separators `\\` and `\\*` can use an optional argument to add some additional interline space (see Section 3.7.2).

Just before the line separator command (`\\`), include a `\displaybreak` command to force a break, or a

```
\displaybreak[0]
```

command to allow one. `\displaybreak[n]`, where *n* is 1, 2, or 3, specifies the intermediate steps between allowing and forcing a break. `\displaybreak[4]` is the same as `\displaybreak`. You can easily visualize these rules:

allow display break =
`\displaybreak[0]` `\displaybreak[1]` ... `\displaybreak[4]`
= `\displaybreak`
= force display break

Note the similarity between the `displaybreak` sequence and the `pagebreak` sequence in Section 3.7.3.

If you want to allow page breaks in all multiline math environments in your document, place the `\allowdisplaybreaks[1]` command in the preamble of your document. The optional argument can be varied from `1` to `4`, in order of increasing permissiveness.

Note that none of the subsidiary math environments are affected by any variant of the `\displaybreak` or the `\allowdisplaybreaks` commands.

PART III

Document Structure

CHAPTER 8

Documents

In this chapter, we take up the organization of shorter documents. Longer documents and books are discussed in Part VI.

If you are writing a *simple article,* start with a template (see Section 9.4), then you can safely ignore much of the material discussed in this chapter. In more complicated articles you may need the material discussed in this chapter.

Section 8.1 discusses document structure in general, the preamble is presented in Section 8.2. Section 8.3 discusses the top matter, in particular, the `abstract` environment. Section 8.4 presents the main matter, including sectioning, cross-referencing, tables, and figures. Section 8.5 covers the back matter, including the bibliography and index.

In Section 8.1–8.5 we discuss the logical design of a LaTeX document. The visual design is largely left to the document class. In Section 8.6, however, we briefly discuss one frequently adjusted aspect of visual design, the page style.

G. Grätzer, *More Math Into* LaTeX, DOI 10.1007/978-3-319-23796-1_8

8.1 The structure of a document

The source file of a LaTeX document is divided into two main parts: the preamble and the body (see Figure 8.1).

Preamble This is the portion of the source file before the

```
\begin{document}
```

command. It contains definitions and instructions that affect the entire document.

Body This is the content of the document environment. It contains all the material to be typeset.

These statements oversimplify the situation somewhat. For instance, you can define a command in the preamble to typeset some text that will appear wherever the command is used in the body, but the text is actually typed in the preamble. Nevertheless, I hope the division between the preamble and the body is clear.

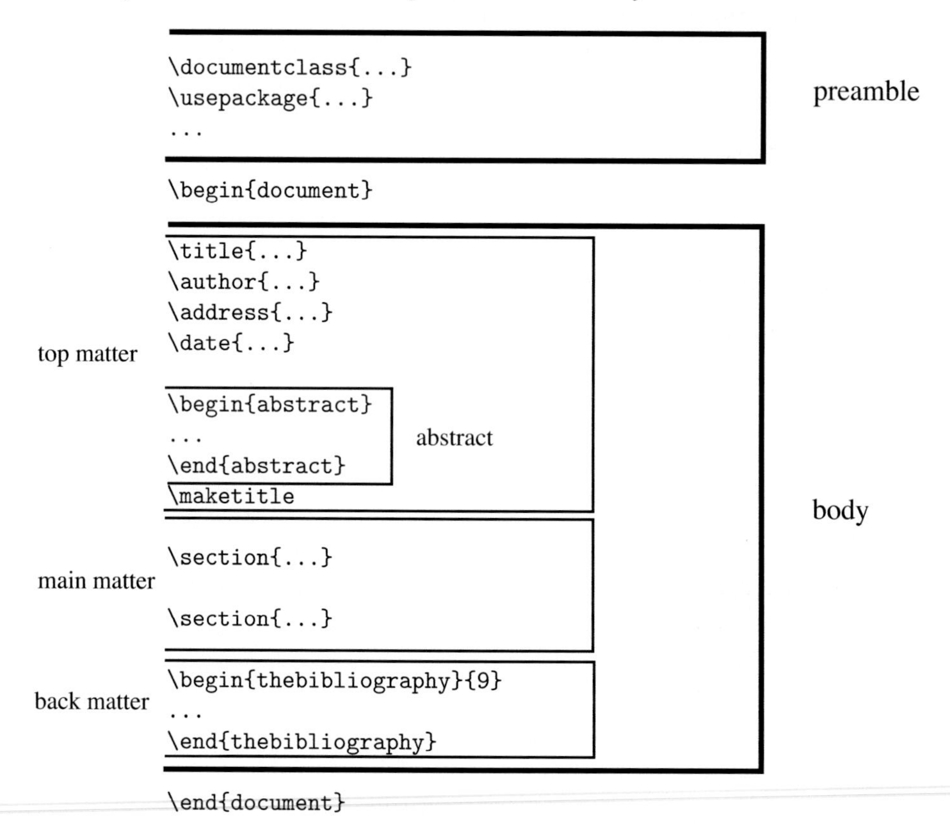

Figure 8.1: The structure of a LaTeX document.

The body is divided into three parts:

Top matter This is the first part of the body. It is concluded with the `\maketitle` command. Traditionally it included only the `\title`, the `\author`, and the `\date` commands. The top matter is derived from these commands and from it the title page of an article was designed. This evolved to include a lot more information about the author(s), for instance, their e-mail addresses, academic affiliations, home pages, and about the article, for instance, research support, subject classification. The typeset top matter now is split into several locations, the top and bottom of the first page and the bottom of the last page. See page 4 and pages 272–275 for two examples and Section 17.1.2 for more components that can be used in longer documents and books.

Main matter This is the main part of the document, including any appendices.

Back matter This is the material that is typeset at the end of the document. For a typical shorter document, the back matter is just the bibliography. See Section 17.1.2 for more information about additional components—such as the index—that are often used in longer documents and books.

8.2 The preamble

You were introduced to the preamble of a document in Section 1.8. Recall that the preamble contains the crucial `\documentclass` line, specifying the document class and the options that modify its behavior. For instance,

```
\documentclass[draft,reqno]{amsart}
```

loads the document class `amsart` with the `draft` option, which paints a slug in the margin indicating lines that are too wide (see Section 3.7.1), and the `reqno` option, which places the equation numbers on the right (see Section 9.5).

`article` is the most popular legacy document class (see Section 10.1). The command

```
\documentclass[titlepage,twoside]{article}
```

loads the document class `article` with the `titlepage` option, which creates a separate title page and places the abstract on a separate page, and the `twoside` option, which formats the typeset article for printing on both sides of the paper.

The `\documentclass` command is usually followed by the `\usepackage` commands, which load LaTeX enhancements called *packages*. For instance,

```
\usepackage{latexsym}
```

loads a package that defines some additional LaTeX symbol names (see Section 10.3), whereas

```
\usepackage[demo]{graphicx}
```

loads the graphicx package (see Section 8.4.3) with the demo option that inserts rectangles in place of the illustrations. Document class options are also passed on to the packages as possible options, so

```
\documentclass[demo]{amsart}
\usepackage{graphicx}
```

would also load the graphicx package with the demo option unless it is invoked with

```
\usepackage[final]{graphicx}
```

Any document class options that are not relevant for a package are ignored.

\usepackage commands can also be combined:

```
\usepackage{amssymb,latexsym}
```

is the same as

```
\usepackage{amssymb}
\usepackage{latexsym}
```

Document class files have a cls extension, whereas package files are designated by the sty extension. The document class amsart is defined in the amsart.cls file, the graphicx package is defined in the graphicx.sty file. You may define your own packages, such as the newlattice package described in Section 14.3.

The preamble normally contains any custom commands (see Chapter 14) and the proclamation definitions (see Section 4.4). Some commands can only be in the preamble. \DeclareMathOperator is such a command (see Section 14.1.6) and so is \numberwithin (see Section 5.3). If you put such a command in the body, for example, \DeclareMathOperator, you get a message:

```
! LaTeX Error: Can be used only in preamble.
l.103 \DeclareMathOperator
```

There is one command that may only be placed *before* the

```
\documentclass{...}
```

line:

```
\NeedsTeXFormat{LaTeX2e}[2005/12/01]
```

This command checks the version of LaTeX being used to typeset the document and issues a warning if it is older than December 1, 2005 or whatever date you specified. Use this optional date argument if your document contains a feature that was introduced on or after the date specified or if an earlier version had a bug that would materially affect the typesetting of your document.

For instance, if you use the \textsubscript command, introduced in the December 1, 2005 release (see page 293), then you may use the \NeedsTeXFormat line shown above. LaTeX now hardly changes from year to year, so this command is rarely used except in document class files or package files. See, however, the discussion on page 293.

8.3 Top matter

The top matter of an article is part of the article body and, as a rule, it contains the material used to create the "title page" and, optionally, an abstract.

Discussion of the top matter should take place in the context of a particular document class. We discuss the top matter of the amsart document class in Section 1.8, and we continue discussing it in much more detail in Section 9.2. The top matter of the article document class is covered in Section 10.1.1.

Long documents, such as books, have rather complicated top matter such as tables of contents (see Chapter 17). In this section, we only discuss the abstract.

8.3.1 Abstract

Most standard document classes, except those for letters and books, make provision for an abstract, typed in an abstract environment.

The document class formats the heading as ABSTRACT, or some variant, and, as a rule, typesets the text of the abstract in smaller type with wider margins.

The amsart document class requires that you place the abstract environment *before* the \maketitle command (see Figure 8.1). See the abstract in the sample article firstarticle.tex on page 4. If you forget to place it there, you get the warning

```
Class amsart Warning:
              Abstract should precede \maketitle in AMS
documentclasses; reported on input line 21.
```

and the abstract is typeset wherever the abstract environment happens to be placed.

In the article document class you place the abstract *after* the \maketitle command. If you place the abstract before the \maketitle command, the abstract is placed on page 1, and the article starts on page 2.

If the abstract and the "footnotes" from the top matter fill the first page, the second page has no running head. To fix this, follow the \maketitle command with the \clearpage command (see Section 3.7.3).

8.4 Main matter

The main matter contains most of the essential parts of the document, including the appendices.

We discuss now how to structure the main matter. We describe sectioning in Section 8.4.1, cross-referencing in Section 8.4.2, and tables and figures in Section 8.4.3.

8.4.1 Sectioning

The main matter of a typical shorter document is divided into *sections*. We discuss sectioning of longer documents in Section 17.1.1.

Sections

LaTeX is instructed to start a section with the `\section` command, which takes the title of the section as its argument. This argument may also be used for the running head and it is also placed in the table of contents (see Section 17.2), which means that you need to protect fragile commands with the `\protect` command (see Section 3.3.3). LaTeX automatically assigns a section number and typesets the section number followed by the section title.

Any `\section` command may be followed by a `\label` command, so that you can refer to the section number generated by LaTeX, as in

```
\section{Introduction}\label{S:intro}
```

The command `\ref{S:intro}` refers to the number of the section and the command `\pageref{S:intro}` refers to the number of the typeset page where the section title appears.

You save a lot of work if in the source file you type in the cross-reference:

```
\section{Introduction}\label{S:intro}
%Section~\ref{S:intro}
```

Other sectioning commands

A section may be subdivided into *subsections*, which may themselves be divided into *subsubsections*, *paragraphs*, and *subparagraphs*. Subsections are numbered within a section (in Section 1, they are numbered 1.1, 1.2, and so on). Here is the whole hierarchy:

```
\section
   \subsection
      \subsubsection
         \paragraph
            \subparagraph
```

It is important to understand that the five levels of sectioning are not just five different styles for typesetting section headers but they form a hierarchy. You should never have a subsection outside a section, a subsubsection outside a subsection, and so on. For instance, if the first sectioning command in your document is `\subsection`, the subsections are numbered 0.1, 0.2, ... Or if in the first section of your document the first sectioning command is `\subsubsection`, the subsubsections are numbered 1.0.1, 1.0.2, ... Both are clearly undesirable.

There are two additional sectioning commands provided by the `report` and by the book document classes (`book` and `amsbook`): `\chapter` and `\part` (discussed in Section 17.1.1).

Any sectioning command may be followed by a `\label` command so that you can refer to the number (if any) generated by LaTeX and the page on which it appears (see Section 8.4.2).

There is also the seldom used top level `\specialsection` command. Articles do not have parts and chapters, but sometimes a long article may require further division using the `\specialsection` command.

The form of sectioning commands

All sectioning commands take one of the following three forms, illustrated below with the `\section` command:

Form 1 The simplest form is

```
\section{title}
```

where *title* is the section title, of course. You need to protect any fragile commands in *title* with the `\protect` command (see Section 3.3.3).

Form 2 The sectioning command may have an optional argument

```
\section[short_title]{title}
```

The optional *short_title* argument is used in the running head. See Section 17.2 on what goes into the table of contents. Protect any fragile commands in *short_title* with the `\protect` command (see Section 3.3.3).

Form 3 Finally, we consider the *-ed version

```
\section*{title}
```

There are no section numbers printed and the *title* is not included in the running head. Remember that if you * a section, all subsections, and so on, must also be *-ed to avoid having strange section numbers.

Sectioning commands typeset

Consider the following text:

```
\section{Introduction}\label{S:Intro}
We shall discuss the main contributors of this era.
\subsection{Birkhoff's contributions}\label{SS:contrib}
\subsubsection{The years 1935--1945}\label{SSS:1935}
Going to Oxford was a major step.
\paragraph{The first paper}
What should be the definition of a universal algebra?
\subparagraph{The idea}
One should read Whitehead very carefully.
```

This is how it looks typeset in the amsart document class:

1 Introduction

We shall discuss the main contributors of this era.

1.1 Birkhoff's contributions

1.1.1 The years 1935–1945

Going to Oxford was a major step.

The first paper What should be the definition of a universal algebra?

The idea One should read Whitehead very carefully.

Notice that paragraphs and subparagraphs are not displayed prominently by the AMS.

By contrast, look at the same text typeset in the legacy article document class:

1. INTRODUCTION

We shall discuss the main contributors of this era.

1.1. **Birkhoff's contributions.**

1.1.1. *The years 1935–1945.* Going to Oxford was a major step.
The first paper. What should be the definition of a universal algebra?
The idea. One should read Whitehead very carefully.

This illustrates vividly one huge difference between the two document classes, the visual handling of sectioning.

Section 14.5.1 discusses how you can change the format of the section numbers, and how to specify which sectioning levels are to be numbered.

Section 2.2 of *The LaTeX Companion,* 2nd edition [56] explains how to change the layout of section headings, especially useful for document class designers.

Appendix

In the main matter, if the article contains appendices, mark the beginning of the appendices with the `\appendix` command. After the `\appendix` command, the `\section` command starts the appendices (for books, see Section 17.1.2):

```
\appendix
\section{A proof of the Main Theorem}\label{S:geom}
```

This produces `Appendix A` with the given title, typeset just like a section.

Note that appendices may be labeled and cross-referenced like any other section. In an appendix, subsections are numbered A.1, A.2, and so on, subsubsections within A.1 are numbered A.1.1, A.1.2, and so on.

Let me repeat, `\appendix` is not like `\section`. It is not a command with an argument. Appendices are named by arguments of the `\section`— commands (for books, by the `\chapter`—commands) *placed after* the `\appendix` command.

8.4.2 Cross-referencing

There are three types of cross-referencing available in LaTeX:

1. Symbolic referencing with `\ref` and `\eqref` for equations
2. Page referencing with `\pageref`
3. Bibliographic referencing with `\cite`

In this section, we discuss the first two, while bibliographies are discussed in Section 8.5.1 and in Chapter 15.

Symbolic referencing

Wherever LaTeX can automatically generate a number in your document, you can place a `\label` command

```
\label{symbol}
```

Then, at any place in your document, you can use the `\ref` command

```
\ref{symbol}
```

to place that number in the document. We call *`symbol`* the *label.* You can use labels for sectioning units, equations, figures, tables, items in an enumerated list environment (see Section 4.2.1), as well as for theorems and other proclamations.

If the equation labeled `E:int` is the fifth equation in an article, then LaTeX stores the number 5 for the label `E:int`, so `\ref{E:int}` produces the number 5. If equations are numbered within sections (see Section 5.3), and an equation is the third equation in Section 2, then LaTeX stores the number `2.3` for the label `E:int`, so the reference `\ref{E:int}` produces the number 2.3.

Example 1 The present section starts with the command

```
\section{Main matter}\label{S:MainMatter}
```

So `\ref{S:MainMatter}` produces the number 8.4 and we get the number of the typeset page where the section title appears with `\pageref{S:MainMatter}`, which is 234.

Tip Type

```
\section{Main matter}\label{S:MainMatter}
%Section~\ref{S:MainMatter}
```

to make cross-referencing quicker.

Example 2

```
\begin{equation}\label{E:int}
   \int_{0}^{\pi} \sin x \, dx = 2.
\end{equation}
```

In this case, `\ref{E:int}` produces the number of the equation, `\eqref{E:int}` produces the number of the equation in parentheses.

Tip Type

```
\begin{equation}\label{E:int}%\eqref{E:int}
   \int_{0}^{\pi} \sin x \, dx = 2.
\end{equation}
```

to make cross-referencing quicker.

Tip If you have to reference an equation in the statement of a theorem, always use `\eqref`. Do not use `\eqref` to reference anything but proclamations. (See the `\itemref` command introduced in Section 14.1.2.)

Example 3

```
\begin{theorem}\label{T:fund}
   Statement of theorem.
\end{theorem}
```

The reference `\ref{T:fund}` produces the number of the theorem.

Tip Type

```
\begin{theorem}\label{T:fund}%Theorem~\ref{T:fund}
   Statement of theorem.
\end{theorem}
```

to make cross-referencing quicker.

Tip Typeset a document twice to see a change in a cross-reference.

See Section C.2.4 for a discussion of how LaTeX stores these numbers and why you have to typeset twice. If you typeset only once, and LaTeX suspects that the cross-references have not been updated, you get a warning:

```
 LaTeX Warning: Label(s) may have changed.
Rerun to get cross-references right.
```

Rule 1 ■ `\label` **command**

The argument of the `\label` command is a string of letters, punctuation marks, and digits. It is case sensitive, so `S:intro` is different from `S:Intro`.

Rule 2 ■ `\label` **command**

Place a `\label` command immediately after the command that generates the number.

The following is not compulsory but advisable.

Tip When referencing:

```
see Section~\ref{S:Intro} proved in Theorem~\ref{T:main}
```

or

```
see Sections~\ref{S:Intro} and~\ref{S:main}
```

use ties (~).

It is difficult to overemphasize how useful automatic cross-referencing can be when writing a document.

Tip Make your labels meaningful to yourself, so they are easy to remember. Systematize your labels. For example, start the label for a section with `S:`, theorem with `T:`, lemma with `L:`, and so on.

When you are cross-referencing, even if you follow these tips, it may not be easy to remember a label. David Carlisle's `showkeys` package may help you out. It is part of the tools distribution (see Section 10.3.1 and Section D.1). Include the line

```
\usepackage{showkeys}
```

in the preamble of your document. The `showkeys` package shows all symbolic references in the margin of the typeset document. With the `notcite` option, my preference,

```
\usepackage[notcite]{showkeys}
```

`showkeys` does not show the labels for bibliographic references. When the document is ready for final typesetting, then comment out this line.

Section 2.4 of *The LaTeX Companion,* 2nd edition [56] describes `varioref`, a package which extends the power of `\ref`, and `xr`, a package for referencing external documents.

Absolute referencing

There are two forms of absolute referencing.

1. Equations can be *tagged.* The `\tag{name}` command attaches a name to the formula. The tag replaces the equation number.

2. Items in an `itemize` environment can be tagged with the `\item[name]` construct. The tag replaces the item number.

Our first example is the equation

$$\int_0^\pi \sin x \, dx = 2 \tag{Int}$$

is typed as

```
\begin{equation}
  \int_{0}^{\pi} \sin x \, dx = 2 \tag{Int}
\end{equation}
```

Our second example is the numbered list:

This space has the following properties:
(a) Grade 2 Cantor;
(b) Half-smooth Hausdorff;
(c) Metrizably smooth.

typed as

```
\noindent This space has the following properties:
\begin{enumerate}
   \item[(a)] Grade 2 Cantor\label{Cantor};
   \item[(b)] Half-smooth Hausdorff\label{Hausdorff};
   \item[(c)] Metrizably smooth\label{smooth}.
\end{enumerate}
```

Tags are *absolute*. This equation is *always* referred to as (Int). Equation numbers, on the other hand, are *relative*, they may change when the file is edited.

Tip Do not label absolute references. It may lead to problems that are hard to explain.

Page referencing

The command

\pageref{*symbol*}

produces the number of the typeset page corresponding to the location of the command \label{*symbol*}. For example, if the following text is typeset on page 5,

```
There may be three types of problems with the
construction of such lattices.\label{problem}
```

and you type

```
Because of the problems associated with
the construction (see page~\pageref{problem})
```

anywhere in the document, LaTeX produces

> Because of the problems associated with the construction (see page 5)

Because of the way LaTeX typesets a page, page references may be off by one. See the discussion in Section 17.6 on how to guarantee that the page number is correct.

8.4.3 *Floating tables and illustrations*

Many documents contain tables and illustrations. These must be treated in a special way since they cannot be broken across pages. If necessary, LaTeX moves—floats—a table or an illustration to the top or bottom of the current or the next page if possible and further away if not.

LaTeX provides the `table` and the `figure` environments for typesetting floats. The two are essentially identical except that the `figure` environments are named Figure 1, Figure 2, and so on, whereas the `table` environments are numbered as Table 1, Table 2, and so on.

Tables

A `table` environment is set up as follows:

```
\begin{table}
    Place the table here
  \caption{title}\label{Ta:xxx}
\end{table}
```

The `\caption` command is optional and may also precede the table. The optional `\label` command must be placed between the command `\caption` and the command `\end{table}`. The label is used to reference the table's number. A `table` environment can have more than one table, each with its own caption.

Tip Type

```
\begin{table}
  Place the table here
  \caption{title}\label{Ta:xxx}%Table~\ref{Ta:xxx}
\end{table}
```

to make cross-referencing quicker.

The `table` environment is primarily used for tables made with the `tabular` or similar environments (see Section 4.6). There are many examples of tables in this book, for instance, Section 3.4 has four.

If your document uses the `twocolumn` document class option, the `table` environment produces tables that span only one column and the `table*` environment produces tables that span both columns. Such tables can be placed only at the top of a page.

Figures

Illustrations, also called *graphics* or *figures*, include drawings, scanned images, digitized photos, and so on. These can be inserted with a `figure` environment:

```
\begin{figure}
    Place the graphics here
   \caption{title}\label{Fi:xxx}
\end{figure}
```

The above discussion of captions and labels for tables also applies to figures. Like the `table` environment, if your document uses the `twocolumn` document class option, the `figure` environment produces figures that span only one column, but the `figure*` environment produces figures that span both columns. However, these figures can be placed only at the top of a page.

Tip Type

```
\begin{figure}
   Place the table here
   \caption{title}\label{Fi:xxx}%Table~\ref{Fi:xxx}
\end{figure}
```

to make cross-referencing quicker.

The standard way of including a graphics file is with the commands provided by the `graphicx` package by David Carlisle and Sebastian Rahtz, which is part of the LaTeX distribution (see Section 10.3). Save your graphics in PDF (Portable Document Format) format—as a rule.

Your graphics can also be made within a `picture` environment, an approach that is neither encouraged nor discussed in this book. To draw within LaTeX, use TikZ, see Chapter 13.

Using the `graphicx` package, a typical `figure` is specified as follows:

```
\begin{figure}
   \centering\includegraphics{file}
   \caption{title}\label{Fi:xxx}
\end{figure}
```

The illustration `circle.pdf` is included with the command

```
\includegraphics{circle}
```

without the extension! LaTeX and the `graphicx` package assumes the `pdf` extension.

If you have to scale the graphics image, say to 68% of its original size, use the command

```
\includegraphics[scale=.68]{file}
```

For instance, the figure on page 32 is included with the commands

```
\begin{figure}
   \centering\includegraphics[scale=.8]{StrucLaT}
   \caption{The structure of \protect\la.}
   \label{Fi:StrucLaT}
\end{figure}
```

For another use of the `graphicx` package, see Section 6.2.1.

Float control

The `table` and `figure` environments may have an optional argument, with which you can influence LaTeX's placement of the typeset table. The optional argument consists of one to four letters:

- `b`, the bottom of the page
- `h`, here (where the environment appears in the text)
- `t`, the top of the page
- `p`, a separate page

For instance,

```
\begin{table}[ht]
```

requests LaTeX to place the table "here" or at the "top" of a page. The default is `[tbp]` and the order of the optional arguments is immaterial, for example, `[th]` is the same as `[ht]`. If `h` is specified, it takes precedence, followed by `t` and `b`.

LaTeX has more than a dozen internal parameters that control a complicated algorithm to determine the placement of tables and figures. If you want to override these parameters *for one table or figure only,* add an exclamation mark (`!`) to the optional argument. For instance, `[!h]` requests that this table or figure be placed where it is in the source file even if this placement violates the algorithm. For a detailed discussion of the float mechanism, see Chapter 6 of *The LaTeX Companion,* 2nd edition [56].

The `\suppressfloats` command stops LaTeX from placing any more tables or figures on the page it appears on. An optional argument `t` or `b` (but not both) prohibits placement of floats at the top or bottom of the current page. The table or figure that is *suppressed* appears on the next page or later in the document, if necessary.

Your demands and LaTeX's float mechanism may conflict with one another with the result that LaTeX may not place material where you want it. The default values of the float placement parameters are good only for documents with a small number of floating objects. Combining two tables or illustrations into one sometimes helps. The `\clearpage` command not only starts a new page with the `\newpage` command, but also forces LaTeX to print all the tables and figures it has accumulated but not yet placed in the typeset document. See also some related commands discussed in Section 3.7.3.

For more information on graphics, see Chapter 10 of *The LaTeX Companion,* 2nd edition [56] and Chapter 2.4 of *The LaTeX Graphics Companion* [17]. See also the documentation for the `graphicx` package in the LaTeX distribution (see Section 10.3).

8.5 Back matter

The back matter of an article is very simple, as a rule. It is either empty or consists of only a bibliography. A long document, such as a book, may have more complicated back matter (see Chapter 17). In this section, we discuss only the *bibliography* and a very simple *index.*

8.5.1 Bibliographies in articles

The simplest way to typeset a bibliography is to type it directly into the article. For an example, see the bibliography in the `secondarticle.tex` article (on page 275).

The following bibliography contains two examples, one short and one long, of each of the seven most frequently used kinds of items.

You type the text of a bibliography in a `thebibliography` environment, as shown in the following examples.

```
\begin{thebibliography}{99}
\bibitem{hA70}
   Henry~H. Albert,
   \emph{Free torsoids},
   Current trends in lattice theory.
   D.~Van Nostrand, 1970.
\bibitem{hA70a}
```

```
   Henry~H. Albert,
   \emph{Free torsoids},
   Current trends in lattice theory
   (G.\,H. Birnbaum, ed.).
   vol.~7, D.~Van Nostrand, Princeton, January, 1970,
   no translation available, pp.~173--215 (German).
\bibitem{sF90}
   Soo-Key Foo,
   \emph{Lattice Constructions},
   Ph.D. thesis, University of Winnebago, 1990.
\bibitem{sF90a}
   Soo-Key Foo,
   \emph{Lattice Constructions},
   Ph.D. thesis, University of Winnebago, Winnebago, MN,
   December 1990, final revision not yet available.
\bibitem{gF86}
   Grant~H. Foster,
   \emph{Computational complexity in lattice theory},
   tech. report, Carnegie Mellon University, 1986.
\bibitem{gF86a}
   Grant~H. Foster,
   \emph{Computational complexity in lattice theory},
   Research Note 128A, Carnegie Mellon University,
   Pittsburgh, PA, December, 1986,
   research article in preparation.
\bibitem{pK69}
   Peter Konig,
   \emph{Composition of functions}.
   Proceedings of the Conference on Universal Algebra
   (Kingston, 1969).
\bibitem{pK69a}
   Peter Konig,
   \emph{Composition of functions}.
   Proceedings of the Conference on Universal Algebra
   (G.~H. Birnbaum, ed.).
   vol.~7, Canadian Mathematical Society,
   Queen's Univ., Kingston, ON,
   available from the Montreal office,
   pp.~1--106 (English).
\bibitem{wL75}
   William~A. Landau,
   \emph{Representations of complete lattices},
```

```
    Abstract: Notices Amer. Math. Soc. \textbf{18}, 937.
\bibitem{wL75a}
    William~A. Landau,
    \emph{Representations of complete lattices},
    Abstract: Notices Amer. Math. Soc. \textbf{18}, 937,
    December, 1975.
\bibitem{gM68}
    George~A. Menuhin,
    \emph{Universal algebra}.
    D.~Van Nostrand, Princeton, 1968.
\bibitem{gM68a}
    George~A. Menuhin,
    \emph{Universal algebra}. 2nd ed.,
    University Series in Higher Mathematics, vol.~58,
    D.~Van Nostrand, Princeton,
    March, 1968 (English), no Russian translation.
\bibitem{eM57}
    Ernest~T. Moynahan,
    \emph{On a problem of M. Stone},
    Acta Math. Acad. Sci. Hungar.
    \textbf{8}~(1957), 455--460.
\bibitem{eM57a}
    Ernest~T. Moynahan,
    \emph{On a problem of M. Stone},
    Acta Math. Acad. Sci. Hungar.
    \textbf{8}~(1957), 455--460
    (English), Russian translation available.
\end{thebibliography}
```

Figure 8.2 shows a typeset version of this bibliography in the `amsart` document class. By contrast, look at the same bibliography typeset in the legacy `article` document class in Figure 8.3.

You can find these entries in the document `inbibl.tpl` in the `samples` folder (see page 5).

I use the convention that the label for a `\bibitem` consists of the initials of the author and the year of publication. The first cited publication by Andrew B. Reich in 1987 would have the label `aR87` and the second, `aR87a`. Of course, you can use any label you choose, but such conventions make the items easier to reuse.

The `thebibliography` environment takes an argument—in the previous example, this argument is 99—telling LaTeX that the widest reference number it must generate is two digits wide. For fewer than 10 items, use 9 and for 100 or more items, use 999.

If the argument of `\begin{thebibliography}` is missing, you get the message

```
! LaTeX Error: Something's wrong--perhaps
                                   a missing \item.
```

Each bibliographic item is introduced with `\bibitem`, which is used the same as the `\label` command. In your text, use `\cite`, in a similar way to `\eqref`—it provides the number enclosed in brackets. So if the 13th bibliographic item is introduced with

```
\bibitem{eM57}
```

then

```
\cite{eM57}
```

refers to that item and typesets it as [13]. The bibliography of the article itself is automatically numbered by LaTeX. It is up to the author to make sure that the listing of the bibliographic items is in the proper order.

REFERENCES

[1] Henry H. Albert, *Free torsoids*, Current trends in lattice theory. D. Van Nostrand, 1970.
[2] Henry H. Albert, *Free torsoids*, Current trends in lattice theory (G. H. Birnbaum, ed.). vol. 7, D. Van Nostrand, Princeton, January, 1970, no translation available, pp. 173–215 (German).
[3] Soo-Key Foo, *Lattice Constructions*, Ph.D. thesis, University of Winnebago, 1990.
[4] Soo-Key Foo, *Lattice Constructions*, Ph.D. thesis, University of Winnebago, Winnebago, MN, December 1990, final revision not yet available.
[5] Grant H. Foster, *Computational complexity in lattice theory*, tech. report, Carnegie Mellon University, 1986.
[6] Grant H. Foster, *Computational complexity in lattice theory*, Research Note 128A, Carnegie Mellon University, Pittsburgh, PA, December, 1986, research article in preparation.
[7] Peter Konig, *Composition of functions*. Proceedings of the Conference on Universal Algebra (Kingston, 1969).
[8] Peter Konig, *Composition of functions*. Proceedings of the Conference on Universal Algebra (G. H. Birnbaum, ed.). vol. 7, Canadian Mathematical Society, Queen's Univ., Kingston, ON, available from the Montreal office, pp. 1–106 (English).
[9] William A. Landau, *Representations of complete lattices*, Abstract: Notices Amer. Math. Soc. **18**, 937.
[10] William A. Landau, *Representations of complete lattices*, Abstract: Notices Amer. Math. Soc. **18**, 937, December, 1975.
[11] George A. Menuhin, *Universal algebra*. D. Van Nostrand, Princeton, 1968.
[12] George A. Menuhin, *Universal algebra*. 2nd ed., University Series in Higher Mathematics, vol. 58, D. Van Nostrand, Princeton, March, 1968 (English), no Russian translation.
[13] Ernest T. Moynahan, *On a problem of M. Stone*, Acta Math. Acad. Sci. Hungar. **8** (1957), 455–460.
[14] Ernest T. Moynahan, *On a problem of M. Stone*, Acta Math. Acad. Sci. Hungar. **8** (1957), 455–460 (English), Russian translation available.

Figure 8.2: The most important bibliographic entry types.

References

[1] Henry H. Albert, *Free torsoids*, Current trends in lattice theory, D. Van Nostrand, 1970.

[2] Henry H. Albert, *Free torsoids*, Current trends in lattice theory (G. H. Birnbaum, ed.), vol. 7, D. Van Nostrand, Princeton, January, 1970, no translation available, pp. 173–215 (German).

[3] Soo-Key Foo, *Lattice Constructions*, Ph.D. thesis, University of Winnebago, 1990.

[4] Soo-Key Foo, *Lattice Constructions*, Ph.D. thesis, University of Winnebago, Winnebago, MN, December 1990, final revision not yet available.

[5] Grant H. Foster, *Computational complexity in lattice theory*, tech. report, Carnegie Mellon University, 1986.

[6] Grant H. Foster, *Computational complexity in lattice theory*, Research Note 128A, Carnegie Mellon University, Pittsburgh, PA, December, 1986, research article in preparation.

[7] Peter Konig, *Composition of functions*. Proceedings of the Conference on Universal Algebra (Kingston, 1969).

[8] Peter Konig, *Composition of functions*. Proceedings of the Conference on Universal Algebra (G. H. Birnbaum, ed.), vol. 7, Canadian Mathematical Society, Queen's Univ., Kingston, ON, available from the Montreal office, pp. 1–106 (English).

[9] William A. Landau, *Representations of complete lattices*, Abstract: Notices Amer. Math. Soc., **18**, 937.

[10] William A. Landau, *Representations of complete lattices*, Abstract: Notices Amer. Math. Soc. **18**, 937, December, 1975.

[11] George A. Menuhin, *Universal algebra*. D. van Nostrand, Princeton, 1968.

[12] George A. Menuhin, *Universal algebra*. Second ed., University Series in Higher Mathematics, vol. 58, D. van Nostrand, Princeton, March, 1968 (English), no Russian translation.

[13] Ernest T. Moynahan, *On a problem of M. Stone*, Acta Math. Acad. Sci. Hungar. **8** (1957), 455–460.

[14] Ernest T. Moynahan, *On a problem of M. Stone*, Acta Math. Acad. Sci. Hungar. **8** (1957), 455–460 (English), Russian translation available.

Figure 8.3: Bibliography in the `article` document class.

Tip Do not leave spaces in a `\cite` command. For example, `\cite{eM57␣}` produces [?] indicating an unknown reference.

You can use `\cite` to cite two or more items in the form

```
\cite{hA70,eM57}
```

which typesets as [1, 13]. There is also an optional argument for `\cite` to specify additional information. For example,

```
\cite[pages~2--15]{eM57}
```

typesets as [13, pages 2–15].

If you wish to use labels rather than numbers to identify bibliographic items, then you can specify those labels with an optional argument of the `\bibitem` command:

[EM57] Ernest T. Moynahan, *On a problem of M. Stone*, Acta Math. Acad. Sci. Hungar. **8** (1957), 455–460.

typed as

```
\bibitem[EM57]{eM57}
  Ernest~T. Moynahan, \emph{On a problem of M. Stone},
   Acta Math. Acad. Sci. Hungar.
   \textbf{8} (1957), 455--460.
```

If this optional argument of `\bibitem` is used, then the `\cite` command produces [EM57]. The argument of `\begin{thebibliography}` must be set wide enough to allow for such labels.

Rule ■ Label for a bibliographic item
A label cannot contain a comma or a space.

The examples I have used follow the formatting rules set by the AMS. Only titles are italicized, and only volume numbers of journals are set in boldface. You also have to watch the order in which the items are given, the punctuation, and the capitalization.

If an author appears repeatedly, use the `\bysame` command, which replaces the author's name with a long dash followed by a thin space. For example,

```
\bibitem{gF86}
   Grant~H. Foster,
   \emph{Computational complexity in lattice theory},
```

```
   tech. report, Carnegie Mellon University, 1986.
\bibitem{gF86a}
   \bysame,
   \emph{Computational complexity in lattice theory},
   Research Note 128A, Carnegie Mellon University,
   Pittsburgh PA, December 1986,
   research article in preparation.
```

See the third page of `secondarticle.pdf` on page 275 for a typeset example.

Tip If you want a different title for your bibliography, say Bibliography, place the command

```
\renewcommand{\refname}{Bibliography}
```

anywhere before the `thebibliography` environment (see Section 14.1.7). If you use a legacy document class or `amsbook.cls`, use the line

```
\renewcommand{\bibname}{Bibliography}
```

Tip You may have more than one `thebibliography` environment in a document. Because each bibliography would number the entries from 1, you should provide labels as optional parameters of the `\bibitem` commands for cross-referencing.

8.5.2 Simple indexes

Using the `\label` and `\pageref` commands (see Section 8.4.2), it is quite simple to produce a small index in a `theindex` environment. At each point in the text that you want to reference in the index, place a `\label` command. The corresponding entry in the index typesets the page number with the `\pageref` command.

The `\item`, `\subitem`, and `\subsubitem` commands create an entry, subentry, and subsubentry, respectively. If you need additional vertical spacing when the first letter changes, for instance, between the "h" entries and the "i" entries, you can use the `\indexspace` command. Here are some examples of index entries:

```
\begin{theindex}
\item Lakser, H., \pageref{Lakser}
\item Lattice, \pageref{Lattice_intro},
               \textbf{\pageref{Lattice}}
   \subitem distributive, \pageref{Lattice_distributive}
   \subitem modular, \pageref{Lattice_distributive},
```

```
      \textbf{\pageref{Lattice_distributive2}}
\item Linear subspace, \pageref{Linear_subspace}
\end{theindex}
```

And here is the typeset index:

INDEX

For a larger index, you should use the *MakeIndex* application (see Chapter 16).

8.6 Visual design

In this chapter, we have discussed the logical design of a LaTeX document. The visual design is largely left to the document class. But there is one small aspect of the visual design we have to discuss, the page style.

To get a visual representation of the page style of your document, use `layout` package of Kent McPherson. Load the package with

```
\usepackage{layout}
```

and place the `\layout` command somewhere in the body of your article. LaTeX produces a graphical representation of the page layout. Figure 8.4 shows the page layout for odd pages for the `amsart` document class with no options.

A typeset page has three parts, the *running head* or *header*, the *body*, and the *footer*. As a rule, the document class takes care of the contents and formatting of all three parts. For the running head and footer, however, you can override the page design of the document class with the command

```
\pagestyle{style}
```

where the argument *`style`* is one of the following:

`plain` The running head is empty and the footer contains only the page number

`empty` Both the running head and the footer are blank

`headings` The running head contains the information provided by the document class and the footer is empty

`myheadings` The running head contains the information provided by the commands `\markboth` and `\markright`, the footer is empty

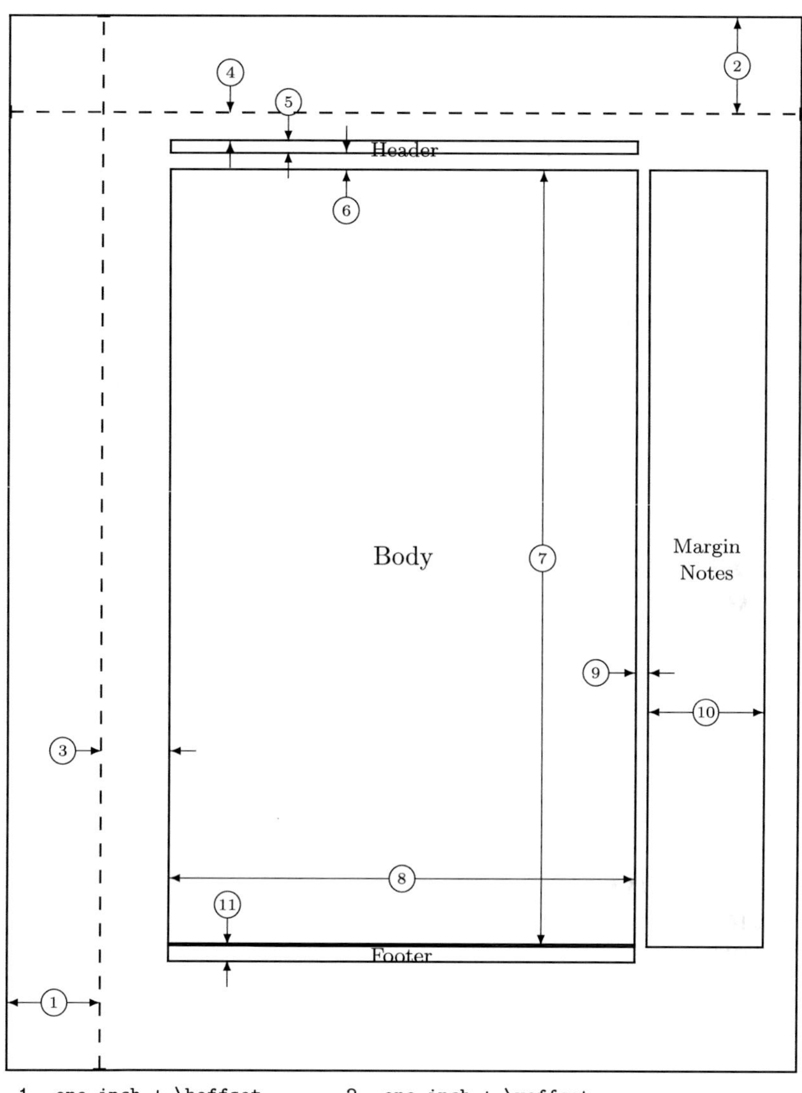

1	one inch + \hoffset	2	one inch + \voffset
3	\oddsidemargin = 54pt	4	\topmargin = 22pt
5	\headheight = 8pt	6	\headsep = 14pt
7	\textheight = 584pt	8	\textwidth = 360pt
9	\marginparsep = 11pt	10	\marginparwidth = 90pt
11	\footskip = 12pt		\marginparpush = 5pt (not shown)
	\hoffset = 0pt		\voffset = 0pt
	\paperwidth = 614pt		\paperheight = 794pt

Figure 8.4: Page layout for the amsart document class.

The \markright command takes only one argument. The last \markright on a page provides the running head information for that page. The \markboth command has two arguments. The first provides the running head information for a left-hand page, the second provides the running head information for a right-hand page. The AMS document classes also have a \markleft command for the running head information for a left-hand page.

The \thispagestyle command is the same as \pagestyle except that it affects only the current page.

For instance, if the current page is a full-page graphic, you might want to issue the command

```
\thispagestyle{empty}
```

The \maketitle command automatically issues a

```
\thispagestyle{plain}
```

command, so if you want to suppress the page number on the first page of a document, you have to put

```
\thispagestyle{empty}
```

immediately after the \maketitle command.

The commands listed in Figure 8.4 are length commands (see Section 14.5.2) and can be changed with the commands introduced in that section. As a rule, you do not have to worry about these settings, they are chosen by the document class for you. Sometimes, however, you have a job that requires such changes. I once had to submit a research plan on a form with a 7.5 inch by 5 inch box. To be able to cut and paste the typeset report, I had to produce the text with a \textwidth of 7 inches. If I simply set

```
\setlength{\textwidth}{7in}
```

the text would overflow the printed page and the last few characters of each line would be missing. So I had to change the margins by starting the document with

```
\documentclass[12pt]{report}
\setlength{\textwidth}{7in}
\setlength{\oddsidemargin}{0pt}
```

All of Chapter 4 of *The LaTeX Companion,* 2nd edition [56] deals with page layouts. There you can find a description of the geometry package of Hideo Umeki, which computes all the parameters from the ones you supply. Also you find there a discussion of Piet van Oostrum's excellent package, fancyhdr, which allows you to create your own page style (see also [24]).

However, if you submit an article to a journal, do not change the type size, page dimensions, headers. Use the document class and the article templates the journal provides (if any). This will make your submission easier for you and the journal. See also Section 2.6.

CHAPTER

The AMS article document class

In this chapter, we discuss amsart, the main AMS document class for journal articles. The AMS book document class is discussed in Chapter 17.

In Section 9.1, I argue that there are good reasons why you should write your articles for publication in amsart. Section 9.2 introduces the rules governing the top matter in the amsart document class. The amsart sample article secondarticle.tex is presented in Section 9.3. In Section 9.4, you are guided through the process for creating detailed templates.

A document class is finely tuned by its options. In Section 9.5, we discuss the options of amsart. Section 9.6 briefly describes the various packages in the AMS distribution and their interdependencies.

9.1 Why amsart?

9.1.1 Submitting an article to the AMS

You want to submit an article written with the amsart document class to the Proceedings of the American Mathematical Society.

G. Grätzer, *More Math Into* LaTeX, DOI 10.1007/978-3-319-23796-1_9

For general information on the AMS journals, go to the AMS Web site

```
http://www.ams.org/
```

and start discovering the wealth of relevant information for `Author`, in the `Author Resource Center`.

To find the class file for the Proceedings at the `http://www.ams.org/` site, type in the search field: `proc_amslatex`. Click on `Proceedings of the AMS`, choose the `TeX package AMS-LaTeX`, and finally, choose `proc-l.cls` to download it.

Now, in the preamble of your article, replace the line

```
\documentclass{amsart}
```

with

```
\documentclass{proc-l}
```

Typeset the article and you are done. Your article is formatted as it will appear in the Proceedings.

9.1.2 Submitting an article to Algebra Universalis

There are many journals whose document classes are based on `amsart`. For instance,

```
http://www.algebrauniversalis.com
```

takes you to the home page of the journal *Algebra Universalis*. To find the document class, click on `Instructions for Authors` and in Section D, click on `au.cls`. Now in your article make the replacement

```
\documentclass{au}
```

and your article typesets in the format appropriate for this journal.

9.1.3 Submitting to other journals

A large number of journals use document classes based on `amsart`. Not all are as friendly as *Algebra Universalis*, but as a rule a small number of changes in the article suffice.

All of them share the attribute that the top matter is given as the arguments of several commands. In the introductory sample article, `firstarticle.tex`, on page 4, there were only four, but in the sample article `secondarticle.tex` in Section 9.3, there are nine—there could be more. Contrast this with the legacy `article` class (see Section 10.1.1). As a result, this document class is able to shape the top matter as the journal requires. Even if the names of some of these commands are different (e.g., `affiliation` for `address`), the principles you learn from the `amsart` document class apply.

Many journals insist that you use their own document classes. For these, you may have to add the AMS packages (see Section 9.6) to continue using the enhancements of the AMS.

A shrinking number of journals use document classes incompatible with the AMS packages. If you can, avoid these journals.

9.1.4 Submitting to conference proceedings

The AMS also has a document class for articles for book-form proceedings of meetings. The differences in the rules for the `amsart` and `amsproc` document classes are minor, for instance, `amsproc` does not access `\date`.

9.2 The top matter

See the typeset top matter of the `secondarticle.tex` article on pages 272 and 275 for a fairly representative example. As you may recall from Section 1.8, part of the author information is moved to the end of the typeset article—see page 275.

Title page information is provided as arguments of several commands. For your convenience, I divide them into three groups: information about the article, information about the author, and AMS related information.

There is only one general rule.

Rule ■ Top matter commands
All top matter commands are *short*.

This means that there can be no blank line (or `\par` command) in the argument of any of these commands (see Section 3.3.3).

9.2.1 Article information

You have to supply five pieces of information about the article.

Rule ■ Title

- Command: `\title`
- Separate lines with `\\`
- Optional argument: Short title for running head

- Do not put a period at the end of a title
- Do not use custom commands in the title

The typeset title is placed on the front page of the typeset article.

Many titles are too long to be typeset on a single line. If the way LaTeX breaks the title is not satisfactory, you can indicate where the title should be broken with the \\ command. Alternatively, you may nudge LaTeX in the right direction with ~ (see Section 3.4.3). For instance, the title:

```
The \texttt{amsart} document class
```

is broken by LaTeX between `document` and `class`. So either add \\:

```
The \texttt{amsart}\\ document class
```

or replace `document class` with `document~class`:

```
The \texttt{amsart} document~class
```

The *running head* (see Section 8.6) is the title on odd-numbered pages, set in capital letters. If the title is more than a few words long, use an optional argument to specify a short title for the running head. Do not use \\ in the short title.

Example of a title:

```
\title{A construction of distributive lattices}
```

A title with a short title:

```
\title[Complete-simple distributive lattices]
{A construction of\\ complete-simple
distributive lattices}
```

Note the AMS rules about short titles and the table of contents in Section 17.2.1.

Rule ■ Translator

- Command: `\translator`
- Do not put a period at the end of the argument.

The typeset `\translator` is placed on the last page of the typeset article, before the address(es). There can be more than one translator. Each should be given as the argument of a separate `\translator` command.

Example:

```
\translator{Harry~M. Goldstein}
```

Rule ■ Dedication

- Command: `\dedicatory`
- Separate lines with `\\`

The typeset dedication is placed under the author(s).
Example:

```
\dedicatory{To the memory of my esteemed
   friend and teacher,\\ Harry~M. Goldstein}
```

Rule ■ Date

- Command: `\date`

Example:

```
\date{January 22, 2015}
```

The typeset `\date` is placed on the front page of the typeset article as a footnote. Do not use this when you submit an article; specify the submission date.

To suppress the date, use `\date{}` or omit the `\date` command entirely.

9.2.2 Author information

There are seven pieces of information about yourself.

Rule ■ Author

- Command: `\author`
- Optional argument: Short form of the name for the running head

The typeset author is placed on the front page of the typeset article.
Examples:

```
\author{George~A. Menuhin}
```

With a short form of the name for the running head:

```
\author[G.\,A. Menuhin]{George~A. Menuhin}
```

Section 9.2.4 discusses how to specify multiple authors.

Rule ■ Contributor

- Command: `\contrib`
- Optional argument: Describing the contribution

The typeset contributor's name is placed on the front page of the typeset article.
Examples:
A contributor authoring an appendix:

```
\contrib[with an appendix by]{John Blaise}
```

If this appendix has two authors:

```
\contrib[with an appendix by]{J. Blaise}
\contrib[]{W. Brock}
```

This typesets (with author G. A. Menuhin) the author line as

G. A. MENUHIN, WITH AN APPENDIX BY J. BLAISE AND W. Brock

Contributors can have addresses, current addresses, etc., just like authors.

Rule ■ Address

- Command: `\address`
- Separate lines with `\\`
- Optional argument: Name of author

The typeset address is placed at the end of the typeset article.
Example:

DEPARTMENT OF APPLIED MATHEMATICS, UNIVERSITY OF WINNEBAGO, WINNEBAGO, MN 53714

which is typed as

```
\address{Department of Applied Mathematics\\
         University of Winnebago\\
         Winnebago, MN 53714}
```

Notice that LaTeX replaces the `\\` line separators with commas.

If there are several authors, you can use the author's name as an optional argument of \address to avoid ambiguity. See Example 4 in Section 9.2.5 (page 267) for a complete example.

Rule ■ Current address

- Command: \curraddr
- Separate lines with \\
- Optional argument: name of author

The typeset current address is placed at the end of the typeset article.

Example:

Current address: Department of Mathematics, University of York, Heslington, York, England

is typed as

```
\curraddr{Department of Mathematics\\
          University of York\\
          Heslington, York, England}
```

If there are several authors, you can use the author's name as an optional argument of \curraddr to avoid ambiguity; for some examples, see Section 9.2.5.

Rule ■ E-mail address

- Command: \email
- Optional argument: Name of author

The typeset e-mail address is placed at the end of the typeset article.

Example:

```
\email{gmen@ccw.uwinnebago.edu}
```

Tip Some e-mail addresses contain the special underscore character (_). Recall (see Section 3.4.4) that you have to type _ to get _.

Example:

```
\email{George\_Gratzer@umanitoba.ca}
```

Tip Some older e-mail addresses contain the percent symbol (%); recall that you have to type \% to get % (see Section 3.4.4).

Example:

```
\email{h1175moy\%ella@relay.eu.net}
```

Rule ■ Web (home) page (URL)

- Command: \urladdr
- Optional argument: Name of author

The typeset Web (home) page is placed at the end of the typeset article.
Example:

```
\urladdr{http://www.maths.umanitoba.ca/homepages/gratzer/}
```

Tip Many Internet addresses contain the tilde (~), indicating the home directory of the user. Type ~ to get ~ and not \~, as recommended in Section 3.4.4. $\sim$ is also unacceptable.

Example:

```
\urladdr{http://kahuna.math.hawaii.edu/~ralph/}
```

Rule ■ Research support or other acknowledgments

- Command: \thanks
- Do not specify linebreaks.
- Terminate the sentence with a period.

The typeset research support or other acknowledgments is placed on the front page of the typeset article as an unmarked footnote.
Example:

```
\thanks{Supported in part by NSF grant PAL-90-2466.}
```

A `\thanks{}` command is ignored in typesetting.

9.2.3 *AMS information*

The AMS requires that you supply two more pieces of information about the article.

The following are collected at the bottom of the first page as unmarked footnotes along with the arguments of the `\thanks` and `\date` commands.

Rule ■ AMS subject classifications

- Command: `\subjclass`
- Optional argument: `2010`—the default is 1991.
- `amsart` supplies the phrase 1991 *Mathematics Subject Classification* and a period at the end of the subject classification—with the optional argument `2010`, the phrase is 2010 *Mathematics Subject Classification*
- The argument should be either a five-character code or the phrase `Primary:` followed by a five-character code, a semicolon, the phrase `Secondary:` and one or more additional five-character codes.

The typeset AMS subject classifications is placed at the bottom of the front page of the typeset article as a footnote.
Examples:

```
\subjclass[2010]{06B10}
\subjclass[2010]{Primary: 06B10; Secondary: 06D05}
```

The current subject classification scheme for mathematics was adopted in 2010, making the 1991 classification scheme obsolete. Thus, `2010` should be considered as a *compulsory* optional argument—maybe the only one in all of LaTeX.

The current subject classification scheme, MSC 2010, is available from the AMS Web site

```
http://www.ams.org/
```

Search for `MSC`. Or in the `Author Resource Center` click on `MSC`.

Rule ■ Keywords

- Command: `\keywords`
- Do not indicate line breaks.
- `amsart` supplies the phrase *Key words and phrases.* and a period at the end of the list of keywords.

The typeset keywords are placed on the front page of the typeset article as a footnote. *Example:*

```
\keywords{Complete lattice, distributive lattice,
complete congruence, congruence lattice}
```

Keywords are optional for many journals.

Further footnotes An additional `\thanks` command creates an unmarked footnote. *Examples:*

```
\thanks{This is a preliminary version of this article,
        prepared for the Second Annual Meeting of the
        Statistical Association of Winnebago.}

\thanks{This article is in final form, and no version
                of it will be submitted elsewhere.}
```

9.2.4 Multiple authors

If an article has several authors, repeat the author information commands for each one. Take care that the e-mail address follows the address.

If two authors share the same address, omit the `\address` command for the second author, who can still have a different e-mail address and Web home page. An additional `\thanks` command for the first author should precede any `\thanks` commands for the second author. Since the footnotes are not marked, the argument of the `\thanks` command for research support should contain a reference to the author:

```
\thanks{The research of the first author was supported
        in part by NSF grant PAL-90-2466.}

\thanks{The research of the second author was supported by
        the Hungarian National Foundation for Scientific
        Research, under Grant No.~9901.}
```

Finally, if an article has more than two authors, supply the author information for each author as usual, but explicitly specify the running heads with the `\markleft` command:

```
\markleft{FIRST AUTHOR ET AL.}
```

where *FIRST AUTHOR* must be all capitals.

If there are multiple authors, sometimes it may not be clear whose address, current address, e-mail address, or Web home page is being given. In such cases, give the name of the authors as optional arguments for these commands. For example,

Email address, Ernest T. Moynahan: emoy@ccw.uwinnebago.edu.

is typed as

```
\email[Ernest~T. Moynahan]{emoy@ccw.uwinnebago.edu}
```

See also Example 4 in Section 9.2.5.

9.2.5 *Examples*

The following examples show typical top matter commands and can be found in the `topmat.tpl` file in the `samples` folder (see page 5).

Example 1 One author.

```
%Article information
\title[Complete-simple distributive lattices]
      {A construction of complete-simple\\
       distributive lattices}
\date{\today}

%Author information
\author{George~A. Menuhin}
\address{Computer Science Department\\
         University of Winnebago\\
         Winnebago, MN 53714}
\email{gmen@ccw.uwinnebago.edu}
\urladdr{http://math.uwinnebago.edu/homepages/menuhin/}
\thanks{This research was supported by
        the NSF under grant number 23466.}

%AMS information
\keywords{Complete lattice, distributive lattice,
          complete congruence, congruence lattice}
\subjclass[2010]{Primary: 06B10; Secondary: 06D05}
```

In the \title command, supplying the optional argument for the running head is the rule, not the exception. The only required item is \title. If it is missing, you get the strange message:

```
! Undefined control sequence.
<argument> \shorttitle

l.49 \maketitle
```

Example 2 Two authors but only the first has a Web home page. I only show the author information section here. The other commands are the same as in Example 1.

```
%Author information
\author{George~A. Menuhin}
\address{Computer Science Department\\
         University of Winnebago\\
         Winnebago, MN 53714}
\email{gmen@ccw.uwinnebago.edu}
\urladdr{http://math.uwinnebago.edu/homepages/menuhin/}
\thanks{The research of the first author was
        supported by the NSF under grant number 23466.}
\author{Ernest~T. Moynahan}
\address{Mathematical Research Institute
         of the Hungarian Academy of Sciences\\
         Budapest, P.O.B. 127, H-1364\\
         Hungary}
\email{h1175moy\%ella@relay.eu.net}
\thanks{The research of the second author
        was supported by the Hungarian
        National Foundation for Scientific Research,
        under Grant No. 9901.}
```

Example 3 Two authors, same department. I only show the author information section here. The other commands are identical to those in Example 1.

```
%Author information
\author{George~A. Menuhin}
\address{Computer Science Department\\
         University of Winnebago\\
         Winnebago, MN 53714}
\email[George~A. Menuhin]{gmen@ccw.uwinnebago.edu}
\urladdr[George~A. Menuhin]%
        {http://math.uwinnebago.edu/homepages/menuhin/}
```

```
\thanks{The research of the first author was
        supported by the NSF under grant number~23466.}
\author{Ernest~T. Moynahan}
\email[Ernest~T. Moynahan]{emoy@ccw.uwinnebago.edu}
\thanks{The research of the second author was supported
        by the Hungarian National Foundation for
        Scientific Research, under Grant No. 9901.}
```

Note that the second author has no `\address`.

Example 4 Three authors, the first two from the same department, the second and third with e-mail addresses and research support. I only show the author information section. The other commands are unchanged. There are various ways of handling this situation. This example shows one solution.

```
%Author information
\author{George~A. Menuhin}
\address[George~A. Menuhin and Ernest~T. Moynahan]
   {Computer Science Department\\
    University of Winnebago\\
    Winnebago, MN 53714}
\email[George~A. Menuhin]{gmen@ccw.uwinnebago.edu}
\urladdr[George~A. Menuhin]%
        {http://math.uwinnebago.edu/homepages/menuhin/}
\thanks{The research of the first author was
        supported by the NSF under grant number 23466.}
\author{Ernest~T. Moynahan}
\email[Ernest~T. Moynahan]{emoy@ccw.uwinnebago.edu}
\thanks{The research of the second author was supported
        by the Hungarian National Foundation for
        Scientific Research, under Grant No. 9901.}
\author{Ferenc~R. Richardson}
\address[Ferenc~R. Richardson]
   {Department of Mathematics\\
    California United Colleges\\
    Frasco, CA 23714}
\email[Ferenc~R. Richardson]{frich@ccu.frasco.edu}
\thanks{The research of the third author was
        supported by the NSF under grant number 23466.}
```

The most common mistake in the top matter is the misspelling of a command name; for instance, `\adress`. LaTeX sends the error message

```
! Undefined control sequence.
l.37 \adress
            {Computer Science Department\\
```

which tells you exactly what you mistyped. Similarly, if you drop a closing brace, as in

```
\email{menuhin@ccw.uwinnebago.edu
```

you are told clearly what went wrong. Because the top matter commands are short (see Section 3.3.3), LaTeX gives the message

```
Runaway argument?
{menuhin@ccw.uwinnebago.edu \thanks
      {The research of th\ETC.
!File ended while scanning use of \\email.
```

If you drop an opening brace,

```
\author George~A. Menuhin}
```

you get the message

```
! Too many }'s.
l.43 \author George~A. Menuhin}
```

If you enclose an optional argument in braces instead of brackets,

```
\title{Complete-simple distributive lattices}%
      {A construction of complete-simple\\
       distributive lattices}
```

LaTeX uses the short title as the title and the real title is typeset before the title of the typeset article.

9.2.6 *Abstract*

As we discussed in Section 8.3.1, you type the abstract in an `abstract` environment, which you place as the last item before the `\maketitle` command. The abstract should be self-contained; do not include cross-references and do not cite from the bibliography. Avoid custom commands.

If you place the abstract *after* the `\maketitle` command, LaTeX typesets it wherever it happens to be and sends a warning.

9.3 *The sample article*

`secondarticle.tex` is the source file for our more advanced sample article (in the `samples` folder, see page 5) using the `amsart` document class. A simpler article, `firstarticle.tex`, is presented in Part I (see Section 1.8).

Look up the full text of `secondarticle.tex` in the `samples` folder. On the next few pages, we present some important parts—from a LaTeX point of view—of the source file and the full typeset file.

```
% Sample file: secondarticle.tex

\documentclass{amsart}
\usepackage{amssymb,latexsym}

\theoremstyle{plain}
\newtheorem{theorem}{Theorem}
\newtheorem{corollary}{Corollary}
\newtheorem*{main}{Main~Theorem}
\newtheorem{lemma}{Lemma}
\newtheorem{proposition}{Proposition}

\theoremstyle{definition}
\newtheorem{definition}{Definition}
\theoremstyle{remark}
\newtheorem*{notation}{Notation}
\numberwithin{equation}{section}

\begin{document}
\title[Complete-simple distributive lattices]
      {A construction of complete-simple\\
       distributive lattices}
\author{George~A. Menuhin}
\address{Computer Science Department\\
         University of Winnebago\\
         Winnebago, MN 53714}
\email{menuhin@ccw.uwinnebago.edu}
\urladdr{http://math.uwinnebago.edu/menuhin/}
\thanks{Research supported by the NSF under grant number
23466.}
\keywords{Complete lattice, distributive lattice,
   complete congruence, congruence lattice}
\subjclass[2010]{Primary: 06B10; Secondary: 06D05}
\date{March 15, 2015}
\begin{abstract}
   In this note we prove that there exist
   \emph{complete-simple distributive lattices,}
   that is, complete distributive lattices, in which
   there are only two complete congruences.
\end{abstract}

\maketitle
```

```
\section{Introduction}\label{S:intro}
In this note we prove the following result:

\begin{main}
   There exists an infinite complete distributive lattice~$K$
   with only the two trivial complete congruence relations.
\end{main}

\section{The $D^{\langle 2 \rangle}$ construction}\label{S:Ds}
For the basic notation in lattice theory and universal algebra,
see Ferenc~R. Richardson~\cite{fR82} and
George~A. Menuhin~\cite{gM68}.
We start with some definitions:

\begin{definition}\label{D:prime}
   Let $V$ be a complete lattice, and let $\mathfrak{p} = [u, v]$
...
\end{definition}

Now we prove the following result:

\begin{lemma}\label{L:ds}
   Let $D$ be a complete distributive lattice satisfying
   conditions \eqref{m-i} and~\eqref{j-i}.  Then
...
\end{lemma}

\begin{proof}
   By conditions~\eqref{m-i} and \eqref{j-i},
...

\end{proof}

\begin{corollary}\label{C:prime}
   If $D$ is complete-prime, then so is $D^{\langle 2 \rangle}$.
\end{corollary}

The motivation for the following result comes
from Soo-Key Foo~\cite{sF90}.

\begin{lemma}\label{L:ccr}
```

```
   Let $\Theta$ be a complete congruence relation of
...
\end{lemma}

\begin{proof}
   Let $\Theta$ be a complete congruence relation of
   $D^{\langle 2 \rangle}$ satisfying \eqref{E:rigid}.
   Then $\Theta = \iota$.
\end{proof}

\section{The $\Pi^{*}$ construction}\label{S:P*}
The following construction is crucial to our proof
of the Main Theorem:

\begin{definition}\label{D:P*}
   Let $D_{i}$, for $i \in I$, be complete distributive lattices
   satisfying condition~\eqref{j-i}.
...
\end{definition}

\begin{notation}
   If $i \in I$ and $d \in D_{i}^{-}$, then
...
\end{notation}

See also Ernest~T. Moynahan \cite{eM57a}.  Next we verify:

\begin{theorem}\label{T:P*}
   Let $D_{i}$, for $i \in I$, be complete distributive lattices
   satisfying condition~\eqref{j-i}.  Let $\Theta$ be a complete
...
\end{theorem}

\begin{proof}
...
\end{proof}

\begin{theorem}\label{T:P*a}
   Let $D_{i}$ for $i \in I$ be complete distributive lattices
...
\end{theorem}
```

```
\begin{proof}
   Let $\Theta$ be a complete congruence on
   $\Pi^{*} ( D_{i} \mid i \in I )$. Let $i \in I$.
   \end{proof}

The Main Theorem follows easily from \ref{T:P*} and \ref{T:P*a}.

\begin{thebibliography}{9}
   \bibitem{sF90}
      Soo-Key Foo, \emph{Lattice Constructions}, Ph.D. thesis,
      University of Winnebago, Winnebago, MN, December, 1990.
   \bibitem{gM68}
      George~A. Menuhin, \emph{Universal algebra},
      D.~van Nostrand, Princeton, 1968.
   \bibitem{eM57}
      Ernest~T. Moynahan, \emph{On a problem of M. Stone},
      Acta Math. Acad. Sci. Hungar. \textbf{8} (1957), 455--460.
   \bibitem{eM57a}
      \bysame, \emph{Ideals and congruence relations in lattices}.
      II, Magyar Tud. Akad. Mat. Fiz. Oszt. K\"{o}zl.
      \textbf{9} (1957), 417--434  (Hungarian).
   \bibitem{fR82}
      Ferenc~R. Richardson, \emph{General lattice theory},
      Mir, Moscow, expanded and revised ed., 1982 (Russian).

\end{thebibliography}
\end{document}
```

A CONSTRUCTION OF COMPLETE-SIMPLE DISTRIBUTIVE LATTICES

GEORGE A. MENUHIN

ABSTRACT. In this note we prove that there exist *complete-simple distributive lattices,* that is, complete distributive lattices in which there are only two complete congruences.

1. INTRODUCTION

In this note we prove the following result:

Main Theorem. *There exists an infinite complete distributive lattice K with only the two trivial complete congruence relations.*

2. THE $D^{\langle 2\rangle}$ CONSTRUCTION

For the basic notation in lattice theory and universal algebra, see Ferenc R. Richardson [5] and George A. Menuhin [2]. We start with some definitions:

Definition 1. Let V be a complete lattice, and let $\mathfrak{p} = [u, v]$ be an interval of V. Then $\mathfrak{p}$ is called *complete-prime* if the following three conditions are satisfied:

(1) u is meet-irreducible but u is *not* completely meet-irreducible;
(2) v is join-irreducible but v is *not* completely join-irreducible;
(3) $[u, v]$ is a complete-simple lattice.

Now we prove the following result:

Lemma 1. *Let D be a complete distributive lattice satisfying conditions* (1) *and* (2). *Then $D^{\langle 2\rangle}$ is a sublattice of D^2; hence $D^{\langle 2\rangle}$ is a lattice, and $D^{\langle 2\rangle}$ is a complete distributive lattice satisfying conditions* (1) *and* (2).

Proof. By conditions (1) and (2), $D^{\langle 2\rangle}$ is a sublattice of D^2. Hence, $D^{\langle 2\rangle}$ is a lattice.

Since $D^{\langle 2\rangle}$ is a sublattice of a distributive lattice, $D^{\langle 2\rangle}$ is a distributive lattice. Using the characterization of standard ideals in Ernest T. Moynahan [3], $D^{\langle 2\rangle}$ has a zero and a unit element, namely, $\langle 0, 0\rangle$ and $\langle 1, 1\rangle$. To show that $D^{\langle 2\rangle}$ is complete, let $\varnothing \neq A \subseteq D^{\langle 2\rangle}$, and let $a = \bigvee A$ in D^2. If $a \in D^{\langle 2\rangle}$, then $a = \bigvee A$ in $D^{\langle 2\rangle}$; otherwise, a is of the form $\langle b, 1\rangle$ for some $b \in D$ with $b < 1$. Now $\bigvee A = \langle 1, 1\rangle$ in D^2 and the dual argument shows that $\bigwedge A$ also exists in D^2. Hence D is complete. Conditions (1) and (2) are obvious for $D^{\langle 2\rangle}$. □

Corollary 1. *If D is complete-prime, then so is $D^{\langle 2\rangle}$.*

Date: March 15, 2015.
2010 *Mathematics Subject Classification.* Primary: 06B10; Secondary: 06D05.
Key words and phrases. Complete lattice, distributive lattice, complete congruence, congruence lattice.
Research supported by the NSF under grant number 23466.

The motivation for the following result comes from Soo-Key Foo [1].

Lemma 2. *Let Θ be a complete congruence relation of $D^{\langle 2 \rangle}$ such that*

$$\langle 1, d \rangle \equiv \langle 1, 1 \rangle \pmod{\Theta}, \tag{2.1}$$

for some $d \in D$ with $d < 1$. Then $\Theta = \iota$.

Proof. Let Θ be a complete congruence relation of $D^{\langle 2 \rangle}$ satisfying (2.1). Then $\Theta = \iota$. □

3. The Π^* construction

The following construction is crucial to our proof of the Main Theorem:

Definition 2. Let D_i, for $i \in I$, be complete distributive lattices satisfying condition (2). Their Π^* product is defined as follows:

$$\Pi^*(D_i \mid i \in I) = \Pi(D_i^- \mid i \in I) + 1;$$

that is, $\Pi^*(D_i \mid i \in I)$ is $\Pi(D_i^- \mid i \in I)$ with a new unit element.

Notation. If $i \in I$ and $d \in D_i^-$, then

$$\langle \dots, 0, \dots, \overset{i}{d}, \dots, 0, \dots \rangle$$

is the element of $\Pi^*(D_i \mid i \in I)$ whose i-th component is d and all the other components are 0.

See also Ernest T. Moynahan [4]. Next we verify:

Theorem 1. *Let D_i, for $i \in I$, be complete distributive lattices satisfying condition (2). Let Θ be a complete congruence relation on $\Pi^*(D_i \mid i \in I)$. If there exist $i \in I$ and $d \in D_i$ with $d < 1_i$ such that for all $d \leq c < 1_i$,*

$$\langle \dots, 0, \dots, \overset{i}{d}, \dots, 0, \dots \rangle \equiv \langle \dots, 0, \dots, \overset{i}{c}, \dots, 0, \dots \rangle \pmod{\Theta}, \tag{3.1}$$

then $\Theta = \iota$.

Proof. Since

$$\langle \dots, 0, \dots, \overset{i}{d}, \dots, 0, \dots \rangle \equiv \langle \dots, 0, \dots, \overset{i}{c}, \dots, 0, \dots \rangle \pmod{\Theta}, \tag{3.2}$$

and Θ is a complete congruence relation, it follows from condition (3) that

$$\begin{aligned} &\langle \dots, \overset{i}{d}, \dots, 0, \dots \rangle \\ &\quad \equiv \bigvee (\langle \dots, 0, \dots, \overset{i}{c}, \dots, 0, \dots \rangle \mid d \leq c < 1) \equiv 1 \pmod{\Theta}. \end{aligned} \tag{3.3}$$

Let $j \in I$ for $j \neq i$, and let $a \in D_j^-$. Meeting both sides of the congruence (3.2) with $\langle \dots, 0, \dots, \overset{j}{a}, \dots, 0, \dots \rangle$, we obtain

$$\begin{aligned} 0 &= \langle \dots, 0, \dots, \overset{i}{d}, \dots, 0, \dots \rangle \wedge \langle \dots, 0, \dots, \overset{j}{a}, \dots, 0, \dots \rangle \\ &\equiv \langle \dots, 0, \dots, \overset{j}{a}, \dots, 0, \dots \rangle \pmod{\Theta}. \end{aligned} \tag{3.4}$$

Using the completeness of Θ and (3.4), we get:

$$0 \equiv \bigvee (\langle \dots, 0, \dots, \overset{j}{a}, \dots, 0, \dots \rangle \mid a \in D_j^-) = 1 \pmod{\Theta},$$

hence $\Theta = \iota$. □

Theorem 2. *Let* D_i *for* $i \in I$ *be complete distributive lattices satisfying conditions* (2) *and* (3). *Then* $\Pi^*(D_i \mid i \in I)$ *also satisfies conditions* (2) *and* (3).

Proof. Let Θ be a complete congruence on $\Pi^*(D_i \mid i \in I)$. Let $i \in I$. Define

$$\widehat{D}_i = \{\langle \dots, 0, \dots, \overset{i}{d}, \dots, 0, \dots \rangle \mid d \in D_i^- \} \cup \{1\}.$$

Then $\widehat{D}_i$ is a complete sublattice of $\Pi^*(D_i \mid i \in I)$, and $\widehat{D}_i$ is isomorphic to D_i. Let Θ_i be the restriction of Θ to $\widehat{D}_i$.

Since D_i is complete-simple, so is $\widehat{D}_i$, and hence Θ_i is ω or ι. If $\Theta_i = \rho$ for all $i \in I$, then $\Theta = \omega$. If there is an $i \in I$, such that $\Theta_i = \iota$, then $0 \equiv 1 \pmod{\Theta}$, hence $\Theta = \iota$. □

The Main Theorem follows easily from Theorems 1 and 2.

REFERENCES

[1] Soo-Key Foo, *Lattice Constructions*, Ph.D. thesis, University of Winnebago, Winnebago, MN, December, 1990.
[2] George A. Menuhin, *Universal algebra*. D. Van Nostrand, Princeton, 1968.
[3] Ernest T. Moynahan, *On a problem of M. Stone*, Acta Math. Acad. Sci. Hungar. **8** (1957), 455–460.
[4] ———, *Ideals and congruence relations in lattices*. II, Magyar Tud. Akad. Mat. Fiz. Oszt. Közl. **9** (1957), 417–434 (Hungarian).
[5] Ferenc R. Richardson, *General lattice theory*. Mir, Moscow, expanded and revised ed., 1982 (Russian).

COMPUTER SCIENCE DEPARTMENT, UNIVERSITY OF WINNEBAGO, WINNEBAGO, MN 53714
E-mail address: menuhin@ccw.uwinnebago.edu
URL: http://math.uwinnebago.edu/homepages/menuhin/

9.4 Article templates

In this section, we create a template to be used for amsart articles. Open it with a text editor and save it under a different name. You can then start to write your new article using the new file, without having to remember the details governing the preamble and the top matter.

Create the template, which contains a customized preamble and top matter with sample bibliographic items, in several steps.

Step 1 In your text editor, open the amsart.tpl document from the samples folder (see page 5) and save it in your work subfolder as myams.tpl. Alternatively, type in the lines as shown in this section.

The first few lines of the file are

```
% Sample file: amsart.tpl

%Preamble
\documentclass{amsart}
\usepackage{amssymb,latexsym}
```

Notice the use of commented out lines (lines that start with %) that have been added as comments about the file.

Edit line 1 to read

```
% Template file: myams.tpl
```

The lines

```
\documentclass{amsart}
\usepackage{amssymb,latexsym}
```

specify the amsart document class and the use of the amssymb and latexsym packages to gain access, by name, to all the symbols listed in Appendices A and B.

Step 2 After the \usepackage command, there are sets of proclamation definitions corresponding to the examples in Section 4.4.2. Choose Option 5 for myams.tpl by deleting all the lines related to the other options. You are left with the lines

```
%Theorems, corollaries, lemmas, and propositions, in the
%most emphatic (plain) style. All are numbered separately.
%There is a Main Theorem in the most emphatic (plain)
%style, unnumbered. There are definitions, in the less
%emphatic(definition) style. There are notations, in the
%least emphatic (remark) style, unnumbered.


\theoremstyle{plain}
\newtheorem{theorem}{Theorem}
\newtheorem{corollary}{Corollary}
\newtheorem*{main}{Main Theorem}
\newtheorem{lemma}{Lemma}
\newtheorem{proposition}{Proposition}

\theoremstyle{definition}
\newtheorem{definition}{Definition}

\theoremstyle{remark}
\newtheorem*{notation}{Notation}
```

Step 3 Two more choices are presented. You can have either one or two authors—for more complex situations, see Section 9.2.4. For the myams.tpl template, choose one author by deleting everything between

```
%Two authors
```

and

```
%End Two authors
```

You are left with

```
\begin{document}
%One author
\title[shorttitle]{titleline1\\
                   titleline2}
\author{name}
\address{line1\\
         line2\\
         line3}
\email{name@address}
\urladdr{http://homepage}
\thanks{thanks}
%End one author

\keywords{keywords}
\subjclass[2010]{Primary: subject; Secondary: subject}
\date{date}

\begin{abstract}
   abstract
\end{abstract}
\maketitle

\begin{thebibliography}{99}

\end{thebibliography}
\end{document}
```

In the top matter, fill in your own personal information. For instance, I edited

```
\author{name}
```

to read

```
\author{George~Gr\"{a}tzer}
```

I also edited `\address`, `\email`, `\urladdr`, and `\thanks`. After the editing, I had the following:

```
%top matter
\title[shorttitle]{titleline1\\
                   titleline2}
```

```
\author{George~Gr\"{a}tzer}
\address{University of Manitoba\\
         Department of Mathematics\\
         Winnipeg, MB R3T 2N2\\
         Canada}
\email{gratzer@ms.umanitoba.ca}
\urladdr{http://server.maths.umanitoba.ca/homepages/gratzer/}
\thanks{Research supported by the NSERC of Canada.}

\keywords{keywords}
\subjclass[2010]{Primary: subject; Secondary: subject}
\date{date}

\begin{abstract}
   abstract
\end{abstract}
\maketitle

\begin{thebibliography}{99}

\end{thebibliography}
\end{document}
```

Since this template is meant to be used for all my future articles, I do not edit the lines that change from article to article (`\title`, `\keywords`, and so on).

Remember that the short title is for running heads, the title shown at the top of every odd-numbered page other than the title page. If the title of your article is only one line long, delete the separation mark `\\` and the second line, except for the closing brace. If the full title of your article is short, delete `[shorttitle]`.

Now save `myams.tpl`. I saved my template under the name `ggamsart.tpl` (in the `samples` folder, see page 5). You can also make an additional template with two authors to be used as a template for joint articles. Note that at the end of the template, just before the line `\end{document}`, there are two lines:

```
\begin{thebibliography}{99}

\end{thebibliography}
```

The argument of `\begin{thebibliography}` should be 9 if there are fewer than 10 references, 99 with 10–99 references, and so forth. We discuss how to format bibliographic items in 8.5.1. The templates for bibliographic items are listed after the `\end{document}` line.

To make sure that you do not overwrite your template, I recommend that you make it read-only. How you do this depends on your computer's operating system.

You should modify the template you create in this section to the template of the journal you submit your article to. In the samples folder, you find the AMS template for the Proceedings of the AMS, called amsproc.tpl.

9.5 Options

The amsart document class supports a number of options, affecting many attributes. For each attribute there is a *default value* that is used if a value is not specified.

Font size

Options:	9pt	
	10pt	*default*
	11pt	
	12pt	

This option declares the default font size. You may want to use the 12pt option for proofreading:

```
\documentclass[12pt]{amsart}
```

Remember, however, that changing the font size changes the line breaks, so changing the 12pt option back to 10pt may require that you make some adjustments in the text (see Section 1.4).

Paper size

Options:	letterpaper	(8.5 inches by 11 inches)	*default*
	legalpaper	(8.5 inches by 14 inches)	
	a4paper	(210 mm by 297 mm)	

Equations and equation numbers

A number of options deal with the placement of equations and equation numbers.

Options:	leqno	*default*
	reqno	

By default, equation numbers are placed on the left, the default leqno option. The reqno option places the equation numbers on the right.

Option: fleqn

This option positions equations a fixed distance from the left margin rather than centering them. The fleqn option is typically used in conjunction with the reqno option. Here is how an equation looks with the fleqn and reqno options:

$$\int_{0}^{\pi} \sin x \, dx = 2 \tag{1}$$

typed as

```
\begin{equation}\label{E:firstInt}
  \int_{0}^{\pi} \sin x \, dx = 2
\end{equation}
```

Options: `tbtags`
`centertags` *default*

The `tbtags` option uses *top-or-bottom tags* for a `split` environment, that is, it places the equation number level with the last line if numbers are on the right, or level with the first line if the numbers are on the left:

$$\begin{aligned} (1) \qquad f &= (x_1x_2x_3x_4x_5x_6)^2 \\ &= (x_1x_2x_3x_4x_5 + x_1x_3x_4x_5x_6 + x_1x_2x_4x_5x_6 + x_1x_2x_3x_5x_6)^2 \\ &= (x_1x_2x_3x_4 + x_1x_2x_3x_5 + x_1x_2x_4x_5 + x_1x_3x_4x_5)^2 \end{aligned}$$

tbtags

The `centertags` option (the default) vertically centers the equation number in a `split` subsidiary math environment.

Limits

Options: `intlimits`
`nointlimits` *default*

The `intlimits` option places the subscripts and superscripts of integral symbols above and below the integral symbol rather than on the side in a displayed math formula—with this option you can use the `\nolimit` command to disable the option for one integral. The `nointlimits` option positions the subscripts and superscripts of integral symbols on the side.

Options: `sumlimits` *default*
`nosumlimits`

The `sumlimits` option places the subscripts and superscripts of large operators, such as $\sum$, $\prod$, $\coprod$, $\bigotimes$, $\bigoplus$, above and below the large operator in a displayed math formula. `nosumlimits` positions them on the side (see Table 5.7 and Section A.7.1).

Options: `namelimits` *default*
`nonamelimits`

The `namelimits` option places the subscripts and superscripts of operators with limits such as det, inf, lim, max, min, and so on, above and below the operator in a displayed math formula. `nonamelimits` positions them on the side (see Tables 5.4, 5.5, and Section A.7).

Two-sided printing

Options: `twoside` *default*
`oneside`

The `twoside` option formats the output for printing on both sides of the paper. The alternative is the `oneside` option. This option influences running heads, the placement of page numbers, and so on.

Two-column printing

Options: `twocolumn`
`onecolumn` *default*

The `twocolumn` option typesets the document in two columns.

Title page

Options: `titlepage`
`notitlepage` *default*

The `titlepage` option creates a separate title page including the abstract.
The `notitlepage` option splits the top matter between the first and last pages of the typeset article.

Draft

Options: `draft`
`final` *default*

The `draft` option prints a slug in the margin next to each line that is too wide. The `final` option does not. Note that this option is passed on to some packages, such as `graphicx`.

Fonts

Option: `noamsfonts`

With this option, the document class does not load the packages necessary for the use of the AMSFonts font set.

Option: `psamsfonts`

The `psamsfonts` option tells LaTeX to use the PostScript version of the AMSFonts set.

No math

Option: nomath

By default, amsart loads the amsmath package (which, in turn, loads three more math packages). If you want to use the title page and related features without the math features, you can use the nomath option.

9.6 The AMS packages

If you follow the recommendation of this book and begin each article with

```
\documentclass{amsart}
\usepackage{amssymb,latexsym}
```

then you can safely ignore most of the information in this section. There are two minor exceptions, the packages amsxtra and upref.

However, if you use a document class that does not load the same packages that amsart loads, then you have to load the packages needed for your work. Typically, you have

```
\usepackage{amsmath,amsfonts,amsthm}
\usepackage{amssymb,latexsym}
```

as a minimum.

The AMS distribution contains many packages that can be loaded together or by themselves.

Math enhancements

amsmath The primary math enhancement package, which loads the four packages, amsgen, amsbsy, amsopn, and amstext.

amsbsy Provides two commands for the use of bold math symbols, \boldsymbol and \pmb (see Section 6.4.3).

amscd Commands for creating simple commutative diagrams (see Section 7.8).

amsgen An auxiliary package that is never invoked directly. It is loaded by all the AMS math packages (except for upref).

amsopn Provides operator names and also the \DeclareMathOperator command for defining new ones (see Section 5.6).

`amstext` Defines the `\text` command and redefines commands such as `\textrm` and `\textbf` to behave like the `\text` command (see Section 5.4.6).

`amsxtra` Provides the "sp" math accents (see Sections 5.7 and A.8) and loads the `amsmath` package.

`upref` Ensures that the `\ref` command always produces upright numbers.

AMSFonts

`amsfonts` Contains the basic commands needed to utilize the AMSFonts. It also defines the `\mathfrak` command which makes the Euler Fraktur math alphabet available (see Section 6.4.2). If you use the PostScript AMSFonts font set, you should load this package with the option

```
\usepackage[psamsfonts]{amsfonts}
```

In addition, if you want to use the `12pt` document class option, then you must also load the `exscale` package (see Section 10.3):

```
\usepackage{exscale}
```

`amssymb` Defines the symbol names for `amsfonts`. It loads `amsfonts`.

`eucal` Replaces the calligraphic math alphabet with the Euler Script math alphabet (see Section 6.4.2). If you load it with the option `mathscr`, as in

```
\usepackage[mathscr]{eucal}
```

then both the `\mathscr` and the `\mathcal` commands are available, so you can have both $\mathcal{C}\mathcal{E}$ and $\mathscr{C}\mathscr{E}$, typed as

```
$\mathcal{C}\mathcal{E}$ and $\mathscr{C}\mathscr{E}$
```

`eufrak` Defines the Euler Fraktur math alphabet (see Section 6.4.2).

Loading packages

`amsart` contains code to provide more flexible formatting of proclamations and the `proof` environment (see Sections 4.4.2 and 4.5). By loading the `amsthm` package you can add this functionality to a non-AMS document class. The `amsthm` package loads the `amsgen` package.

`amsart` loads four packages from the math enhancements group, the `amsmath`, `amsbsy`, `amstext`, `amsopn`, and `amsgen` packages, and the `amsfonts` package from the AMSFonts group.

A typical article using the legacy article document class (see Section 10.1) and the AMS enhancements would normally have

```
\documentclass{article}
\usepackage{amsmath}% math enhancements
\usepackage{amssymb,latexsym}% AMSFonts and LaTeX symbol names
\usepackage{amsthm}% proclamations with style
```

and perhaps the following:

```
\usepackage{eucal}% Euler Script
```

Note that it is not critical for you to remember which packages load others. No harm is done if you type

```
\usepackage{amsmath}
\usepackage{amsbsy}
```

The amsbsy package is loaded by the amsmath package, and the

```
\usepackage{amsbsy}
```

line is ignored by LaTeX.

All the math related options of amsart (see Section 9.5) are also options of the amsmath package. So, for instance, if you want the equation numbers on the right, load amsmath with the reqno option:

```
\usepackage[reqno]{amsmath}
```

Multiple indices

The AMS distribution also contains the package amsmidx for creating multiple indices for amsbook. This package is discussed in Section 16.5.

CHAPTER

Legacy documents

Even though the AMS spent a few decades refining the amsart document class, some of the legacy document classes of LaTeX are still around. If you want to whip up a quick report or write up a research note, the legacy article or report document classes may serve you well.

In this chapter, we discuss some of the legacy LaTeX document classes. We take up the book document class in Chapter 17. We do not discuss the slides document class for preparing slides. It is now considered obsolete. Use instead the beamer class which we discuss in Chapter 12.

We conclude this chapter with a description of the components of the standard LaTeX distribution.

10.1 Articles and reports

The article and report document classes are very similar. There are two substantive differences to remember:

1. The report document class provides a separate page for the abstract by default, the article document class does not.

G. Grätzer, *More Math Into* LaTeX, DOI 10.1007/978-3-319-23796-1_10

2. The `report` document class has two additional sectioning commands, `\chapter` and `\part`. We discuss these commands in Section 17.1.1.

In the `samples` folder (see page 5) is the document `legacy-article.tex`, a variant of the introductory sample article, `firstarticle.tex` of Chapter 1. The first page of this article is shown typeset on the following page.

10.1.1 Top matter

For a detailed discussion of the top matter, refer to Chapter 8, in particular Sections 8.1 and 8.3—see also Figure 8.1. Here is the top matter of the legacy article:

```
\title{A construction of complete-simple\\
       distributive lattices}
\author{George~A. Menuhin\thanks{Research supported
   by the NSF under grant number 23466.}\\
   Computer Science Department\\
   Winnebago, MN 23714\\
   menuhin@cc.uwinnebago.edu}
\date{March 15, 2010}
\maketitle
```

There are four commands for the top matter: `\title`, `\author`, `\thanks`, and `\date`.

Rule ■ Top matter for the `article` document class

1. If necessary, break the title into separate lines with `\\`. Do not put a `\\` at the end of the last line.
2. `\thanks` places a footnote at the bottom of the first page. If it is not needed, omit it.
3. Separate the lines of the address with `\\`. Do not put a `\\` at the end of the last line.
4. Multiple authors are separated by `\and`.
5. There is only one `\author` command, and it contains *all the information*—name, address, support—about *all the authors.* There is no `\\` command before the `\and` command.
6. If there is no `\date` command, LaTeX will insert the date on which you typeset the file (`\date{\today}` will produce the same result). If you do not want *any* date to appear, type `\date{}`. For a specific date, such as February 21, 2007, type `\date{February 21, 2007}`.
7. The `\title` command is the only required command. The others are optional.

A construction of complete-simple distributive lattices

George A. Menuhin*
Computer Science Department
Winnebago, MN 23714
menuhin@cc.uwinnebago.edu

March 15, 2006

Abstract

In this note, we prove that there exist *complete-simple distributive lattices,* that is, complete distributive lattices in which there are only two complete congruences.

1 Introduction

In this note, we prove the following result:

Theorem 1 *There exists an infinite complete distributive lattice K with only the two trivial complete congruence relations.*

2 The Π^* construction

The following construction is crucial in the proof of our Theorem:

Definition 1 *Let D_i, for $i \in I$, be complete distributive lattices satisfying condition* (J). *Their Π^* product is defined as follows:*

$$\Pi^*(D_i \mid i \in I) = \Pi(D_i^- \mid i \in I) + 1;$$

that is, $\Pi^*(D_i \mid i \in I)$ is $\Pi(D_i^- \mid i \in I)$ with a new unit element.

Notation 1 *If $i \in I$ and $d \in D_i^-$, then*

$$\langle \dots, 0, \dots, d, \dots, 0, \dots \rangle$$

is the element of $\Pi^*(D_i \mid i \in I)$ whose i-th component is d and all the other components are 0.

*Research supported by the NSF under grant number 23466.

Figure 10.1: The first page of the legacy article typeset

As you see, the rules for the \date command here differ slightly from the rules for the \date command in the amsart document class. However, the rules for the command \author here are *very different* from the rules for the \author command in the amsart document class.

For two authors use the following template:

```
\author{name1\thanks{support1}\\
   address1line1\\
   address1line2\\
   address1line3
   \and
   name2\thanks{support2}\\
   address2line1\\
   address2line2\\
   address2line3}
```

Note the use of the \and command, which separates the two authors.

One more difference to keep in mind. Place the abstract *after* the \maketitle command.

10.1.2 Options

The article and report document classes have a similar range of options. These are listed below.

Font size

Options: 10pt *default*
11pt
12pt

Each option declares the specified size to be the default font size.

Paper size

Options:	letterpaper	(8.5 inches by 11 inches)	*default*
	legalpaper	(8.5 inches by 14 inches)	
	executivepaper	(7.25 inches by 10.5 inches)	
	a4paper	(210 mm by 297 mm)	
	a5paper	(148 mm by 210 mm)	
	b5paper	(176 mm by 250 mm)	

Draft

Options: draft
final *default*

The `draft` option places a slug in the margin next to each line that is too wide (see Section 1.4). The `final` option does not. Note that this option is passed on to some packages, such as `graphicx`. To prevent this, invoke `graphicx` with the `final` option.

Landscape printing

Option: `landscape`

The `landscape` option typesets the document in landscape format, swapping the width and height of the paper.

Two-sided printing

Options: `twoside`
`oneside` *default*

The `twoside` option formats the output for printing on both sides of the paper.

Two-column printing

Options: `twocolumn`
`onecolumn` *default*

The `twocolumn` option typesets the document in two-column format. This option has many problems. It is better to use the `multicol` package (see Section 10.3.1).

Title page

Options: `titlepage` *default* for `report`
`notitlepage` *default* for `article`

The `titlepage` option creates a separate title page and places the abstract on a separate page. The `notitlepage` option places the title and the abstract together on the first page.

Equations and equation numbers

Options: `leqno`
`reqno` *default*

The `leqno` option places any equation number in the document on the left side and `reqno` places them on the right.

Option: `fleqn`

The `fleqn` option sets displayed formulas flush left. This option is typically used in conjunction with the `reqno` option.

[1] Soo-Key Foo.
Lattice Constructions.
PhD thesis, University of Winnebago, Winnebago, MN, December 1990.

[2] George A. Menuhin.
Universal Algebra.
D. Van Nostrand, Princeton, 1968.

[3] Ernest T. Moynahan.
Ideals and congruence relations in lattices. II.
Magyar Tud. Akad. Mat. Fiz. Oszt. Közl., 7:417–434, 1957.

[4] Ernest T. Moynahan.
On a problem of M. Stone.
Acta Math. Acad. Sci. Hungar., 8:455–460, 1957.

[5] Ferenc R. Richardson.
General Lattice Theory.
Mir, Moscow, expanded and revised edition, 1982.

Figure 10.2: The `openbib` option.

Bibliography

Option: `openbib`

The `openbib` option typesets the bibliography in a spread out "open" format (see Figure 10.2).

Combinations

Of course, these options can be combined with each other and are also used by most legacy document classes. For instance,

```
\documentclass[12pt,a4paper,twoside,twocolumn]{report}
```

produces a double-columned, two-sided report on A4 paper, the European standard, at the 12-point font size.

10.2 Letters

The `letter` document class was developed for writing letters. One document can contain any number of letters, each in its own `letter` environment. In the following example (`letter.tex` in the `samples` folder) there is only a single letter:

```
% Sample file: letter.tex
\documentclass{letter}

\begin{document}
```

```
\address{George Gr\"{a}tzer\\
        Department of Mathematics\\
        University of Manitoba\\
        Winnipeg, MB, R3T 2N2\\
        Canada}
\signature{George Gr\"{a}tzer}
\date{}

\begin{letter}{Prof.~John Hurtig\\
              Computer Science Department\\
              University of Winnebago\\
              Winnebago, Minnesota 23714}
\opening{Dear John,}
Enclosed you will find the first draft of the
five-year plan.
\closing{Friendly greetings,}
\cc{Carla May\\
    Barry Bold}
\encl{Five-year plan}
\ps{P.S. Remember our lunch meeting tomorrow! G.}
\end{letter}

\end{document}
```

Figure 10.3 shows the typeset letter.

The argument of the `letter` environment is the name and address of the recipient. It is a required argument and if it is omitted, you get a message such as

```
! Incomplete \iffalse; all text was ignored
                                       after line 21.
<inserted text>
               \fi
l.21 \end{letter}
```

As with all multiline arguments, the lines are separated by `\\`.

The arguments of some commands may apply to all the letter environments in the document. Such commands should be placed before the first `letter` environment. In the example, `\signature` and `\address` are so placed.

If the `\date` command is absent, today's date is typeset. If you want no date, use an empty argument `\date{}`, as in the example. If you want all the letters in the same document to have the same date, the `\date` command should precede the first `letter` environment.

George Grätzer
Department of Mathematics
University of Manitoba
Winnipeg, MB, R3T 2N2
Canada

Prof. John Hurtig
Computer Science Department
University of Winnebago
Winnebago, Minnesota 23714

Dear John,

Enclosed you will find the first draft of the five-year plan.

Friendly greetings,

George Grätzer

cc: Carla May
Barry Bold

encl: Five-year plan

P.S. Remember our lunch meeting tomorrow! G.

Figure 10.3: A sample letter.

Many of the options listed in Section 10.1.2 can also be invoked for the `letter` document class.

10.3 The LaTeX distribution

The LaTeX distribution contains a number of document classes and packages, most of which you have probably received with your TeX software. If you find that you are missing some files, see Section D.1 on how to get them.

The files of the LaTeX distribution on CTAN are grouped in the directory

`/pub/tex/macros/latex`

into four subdirectories.

`base` contains all the files necessary to install the system. As a rule, for every package, say, `exscale`, it contains two files, `exscale.ins` and `exscale.dtx`. Typesetting the first gives you `exscale.sty` and typesetting the second produces the user guide and the commented source code. Since most LaTeX implementations install the content of the `unpacked` directory, this directory is not for the average user.

`doc` contains LaTeX documentation in PDF files and also the LaTeX News.

`required` contains the directories `amslatex`, `babel`, `cyrillic`, `graphics`, `psnfss`, `tools`.

`unpacked` contains the unpacked LaTeX distribution. Since most LaTeX implementations install this, you may never need it.

Of the packages and `tex` files included in the `unpacked` folder, the following should be of special interest to readers of this book.

`latexsym` Some symbol definitions (see the tables in Appendix A).

`alltt` The `alltt` environment, which is like the `verbatim` environment except that
`\ { }`
retain their usual meanings.

`exscale` Scaled versions of the math extension font.

`makeidx` Commands for producing indexes (see Chapter 16).

`showidx` A package to allow you to typeset the index entries in the margin of your typeset document (see Section 16.1).

`nfssfont.tex` Generates font tables for use with the `\symbol` command (see Section 3.4.4).

There is also the file `fixltx2e.sty` in the `unpacked` directory (and the corresponding `fixltx2e.dtx` and `fixltx2e.ins` in the `base` directory). This file contains fixes to `latex.ltx`, the main LaTeX file, and also some new commands that did not make it into the current release. For instance, `fixltx2e.sty` of Dec. 2005 contains two important additions. It complements the `\textsuperscript` command (see Section B.4) with a `\textsubscript` command.

More importantly, it introduces a very useful new command `\TextOrMath`. This command has two arguments and it typesets the first in text and the second in math. For instance, if in `newlattice.sty` (see 14.3) you define

```
\newcommand{\ga}{\TextOrMath{$\alpha$\xspace}{\alpha}}
```

then you can type \ga in both test and math. Indeed.

```
this is \ga in text, and this is $\ga - x^2$ in math
```

typesets as

this is α in text, and this is $\alpha - x^2$ in math

If you want to use these commands, include the line

```
\usepackage{fixltx2e}
```

in the preamble. The last `fixltx2e.sty` is dated Sept. 29, 2014. The recent files contain only small bug fixes.

In the `required` folder there are some major software distributions related to LaTeX.

`amslatex` Discussed in detail in this book, this directory contains the AMS math packages and document classes, while the font-related AMS files are in the directory

```
/tex-archive/fonts/amsfonts/latex/
```

`babel` For typesetting languages other than American English.

`cyrillic` For typesetting Cyrillic characters.

`graphicx` For the inclusion and transformation of graphics and for typesetting in color (see Section 8.4.3). This package requires that you have a suitable printer driver.

`psnfss` For typesetting with a wide range of PostScript fonts (see Section E.1).

`tools` A range of tools for managing document production discussed in the next section.

Each of these packages comes with its own documentation. They are also described in *The* LaTeX *Companion,* 2nd edition [56].

10.3.1 Tools

Some of these packages are so important that they could well have been incorporated into LaTeX proper. Here is a brief listing.

`afterpage` Implements the `\afterpage` command. The commands specified in its argument are expanded after the current page is output.

`array` Contains extended versions of the `array` and `tabular` environments with many extra features.

`bm` Gives access to bold math symbols.

`calc` Allows algebraic manipulation of lengths and counter values when specifying lengths and counters.

`dcolumn` Provides alignment on decimal points in tabular entries. It requires the `array` package.

`delarray` Adds "large delimiters" around arrays. It requires the `array` package.

`enumerate` Provides customized `enumerate` environments (see Sections 4.2.4 and also Sections 14.2.1, 14.3, 14.4).

`fileerr` Helps with missing files.

`fontsmpl` Produces a test file for displaying "font samples".

`ftnright` Places all footnotes in the right-hand column of documents typeset with the `twocolumn` document class option.

`hhline` Provides control over horizontal lines in tables.

`indentfirst` Indents the first paragraph of each section.

`layout` Shows the page layout defined by a document class (see Section 8.6).

`longtable` Helps to create multipage tables. It does not require the `array` package, but it uses `array`'s extended features if both packages are loaded.

`multicol` Provides multicolumn typesetting with some advanced features.

`rawfonts` Preloads fonts using the old font names of LaTeX 2.09.

`showkeys` Selectively prints the labels used by `\label`, `\ref`, `\cite`, and so forth, in the margin (see Section 8.4.2).

`somedefs` Elective handling of package options. It is used by the `rawfonts` package.

`tabularx` Defines a variant of the `tabular` environment where all the columns are the same width. It requires the `array` package.

`theorem` Allows the definition of proclamations in flexible formats. The AMS variant, the `amsthm` package, is discussed in Section 4.4.2.

`trace` Tracing help for macro writers.

`varioref` Provides smart as well as multilingual handling of page references.

`verbatim` Extends the `verbatim` environment and provides the `comment` environment (see Sections 3.5.1 and 4.8).

`xr` Creates cross-references among documents.

`xspace` Provides a "smart space" command that helps you avoid the common mistake of missing space after commands. It is mainly used in commands that expand to some text (see Section 14.1.1).

All of these packages are discussed in *The* LaTeX *Companion,* 2nd edition [56].

PART IV

PDF Documents

CHAPTER

11

The PDF file format

11.1 PostScript and PDF

11.1.1 PostScript

PostScript is the preeminent platform and device independent page-description and programming language, introduced by Adobe Systems Inc. in 1982. It describes the placement and shapes of all the elements in the document, including the fonts. Documents placed on the Web in PostScript format can be downloaded to any computer and print identically on all PostScript printers. Until the appearance of PDF, PostScript was the format of choice for sharing LaTeX articles with diagrams or complex forms.

There are a number of disadvantages to using PostScript files on the Web:

- The files tend to be very large.
- They cannot be viewed until the whole file has been downloaded.
- If a PostScript file does not include a particular font used in the document and you do not have that font installed on your computer, then another font—usually Courier—is substituted causing graphically unacceptable rendering.

G. Grätzer, *More Math Into* LaTeX, DOI 10.1007/978-3-319-23796-1_11

11.1.2 PDF

All of these concerns have been addressed by Adobe's Portable Document Format (PDF). See Adobe Systems' *PDF Reference, Version 1.7,* 1st edition [2] for a complete description of this file format. PDF is based on the PostScript language, with some important differences:

- PDF is much more concise than PostScript. A PDF file is normally about 10 percent of the size of the corresponding PostScript file.
- Missing fonts are usually substituted by fonts with the same metrics, so that the size of the substituted text is the same as that of the original. In particular, there are no incorrect line breaks caused by the substitution.
- PDF files allow *partial inclusion of fonts.* As a result, it is much easier to obtain permission to include proprietary fonts in PDF documents.
- PDF files can be downloaded and viewed in a Web browser one page at a time, without having to wait for the whole file to download first.
- Most LaTeX implementations produce PDF files.

PDF files and Adobe Acrobat Professional offer many nice features, including:

- Efficient navigational tools
- Searching and indexing capability for documents and even for collections of documents
- Bookmarks
- Thumbnails of pages
- Limited editing
- Annotations (notes, text, and voice) and markups
- Hyperlinks to the same document or to another document or Web site (see Section 11.1.3)
- The inclusion of programs, particularly JavaScript
- The creation of interactive features
- The inclusion of multimedia objects such as video and sound files

PDF files can also be used to make legacy documents available on the Internet. For instance, if you go to my home page,

`http://www.maths.umanitoba.ca/homepages/gratzer/`

and click on `Mathematical articles`, then `1980-89`, in the bottom frame you will find entry `102`, which links to a PDF file. I created that PDF file by scanning the pages of the original article, converting them to PDF files, and finally stringing them together into a single document. The scanned pages totalled 32 MB, the PDF file is 320 KB. The printed version of the PDF file is somewhat lower in quality than the original, but it is still quite satisfactory. See my article [33] on some practical pointers about scanning and PDF files.

11.1.3 Hyperlinks

With Adobe Acrobat Professional you can place *hyperlinks* in PDF documents. Clicking on a hyperlink, you jump to another location in the same document, to an electronic document, or to a Web site. For instance, in the table of contents, you can put a hyperlink to Chapter 3, so that clicking on it takes you to Chapter 3. Adobe Acrobat's help system has ample information on how to set up links.

11.2 Hyperlinks for LaTeX

It is tedious to set hyperlinks one at a time in your PDF file. Would it not be nice if hyperlinks corresponding to cross-references were set automatically? For instance, clicking on Lemma 6 in

> This follows from Lemma 6 and the relevant definitions.

would cause the display to jump to the page containing Lemma 6.

Sebastian Rahtz's `hyperref` package (maintained now by Heiko Oberdiek) does just that.

11.2.1 *Using* `hyperref`

You invoke the `hyperref` package with the command

```
\usepackage{hyperref}
```

as the *last* `\usepackage` line in the preamble of your LaTeX document. If this does not do the job, try this format, specifying the printer driver:

```
\usepackage[driver]{hyperref}
```

The *`driver`* is one of `hypertex`, `dvips`, `dvipsone`, `ps2pdf`, `tex4ht`, `pdftex`, `dvipdf`, `dvipdfm`, `dvipdfmx`, `dviwindo`, `vtex`. If none of these work, you are out of luck.

Figure 11.1 shows a page fragment from a mathematical article with hyperlinks to some sections, theorems, and citations automatically created by `hypertex`.

The construction of the uniquely complemented lattice representing a given monoid is introduced in Section 4. It is based on V. Koubek and J. Sichler [12]. Section 5 proves that this construct has many simple sublattices. Finally, in Section 6, we put all these pieces together to construct the lattice L for the Main Theorem.

To prove Theorem 1, we need a different construction, which is presented in Section 7.

Figure 11.1: The `hyperref` package with the `\autoref` command.

See Section 11.2.4 for the `autoref` command. To see how `hypertex` works, look up the `secondarticle-ref.tex` article in the `samples` folder. It is the sample article `secondarticle.tex` enhanced with the `hyperref` package. The article uses the `hyperref` options

```
pagebackref,colorlinks,bookmarks=true
```

See Section 11.2.2 for the `pagebackref` and `colorlinks` options. The third option, `bookmarks=true`, is discussed in Section 11.2.3.

Copy `secondarticle-ref.tex` into the `work` folder and typeset it twice. The PDF file created for you has some of the `hyperref` features, but not all. Open the `pdf` file with Adobe Reader. Look at the left pane. `Bookmarks` is a table of contents of the article, with links to the named sections. `Pages` is a thumbnail sketches of the pages, with links to them.

11.2.2 `backref` and `colorlinks`

A useful addition to `hyperref` is David Carlisle's `backref` package. It is invoked as an option of `hyperref`:

```
\usepackage[backref]{hyperref}
```

The items in your bibliography will be followed by a list of sections in which the bibliographic reference is cited. Each number printed after the cited reference becomes a hyperlink to the relevant section. Alternatively, you can use the `pagebackref` option, which produces a list of page numbers. Figure 11.2 shows a page fragment from a bibliography displaying lists of section numbers. `backref` can be used to check if all items in the bibliography have actually been referenced in the article. Any reference that has not been cited does not have a page listed.

Another popular option is `colorlinks`, which colors the text of the links instead of underlining them.

[10] G. Grätzer and J. Sichler, *On the endomorphism semigroup (and category) of bounded lattices*, Pacific J. Math. **35** (1970), 639–647. 1, 1

[11] ———, *On the endomorphism monoid of complemented lattices*, AMS Abstract 97T-06-98. 1

[12] V. Koubek and J. Sichler, *Universality of small lattice varieties*, Proc. Amer. Math. Soc. **91** (1984), 19–24. 1, 2, 4, 4.2, 4.3, 4.3, 4.3, 7

[13] H. Lakser, *Simple sublattices of free products of lattices*, Abstract, Notices Amer. Math. Soc. **19** (1972), A 509. 1, 3, 3

Figure 11.2: The `hyperref` package with the `backref` option.

Tip Make sure that in the bibliography any two items are separated by a blank line and the last item is separated from `\end{bibliography}` by a blank line. Otherwise, you may get very confusing error messages with `\backref`.

If your bibliography is created by BIBTeX, see Chapter 15, it has these blank. Don't edit them out!

11.2.3 Bookmarks

An important navigational feature of Acrobat is the ability to set and use bookmarks. If you choose `View>Navigation Tabs>Bookmarks` in Adobe Reader, the navigation pane opens up showing the bookmarks.

The `hyperref` package option `bookmarks=true` makes bookmarks from the sectioning commands of the LaTeX document, thereby producing a table of contents even if the document had none. You can invoke all these options together:

```
\usepackage[backref,colorlinks,bookmarks=true]{hyperref}
```

Typesetting your LaTeX document with the `bookmarks=true` option produces an `out` file, which contains entries such as

```
\BOOKMARK [1][-]{section.1}{1. Introduction}{}
\BOOKMARK [1][-]{section*.2}{References}{}
```

Once you have produced the final version of your document, you should edit this file to make sure that it contains no LaTeX code. Math formulas in titles create havoc. So do accented characters. `hyperref` does its best to convert internal encodings for accented characters to the encoding used by Acrobat Reader, but it is still best to avoid them. Once this file has been edited, add the line

```
\let\WriteBookmarks\relax
```

at the start of the file to prevent it from being overwritten.

11.2.4 Additional commands

The hyperref package has dozens of commands and parameters, but we will discuss only four more commands.

Preventing links

If you do not want a \ref or \pageref command to appear as a link, you can use their *-ed forms, \ref* and \pageref*.

Long links

An often heard complaint is that in the link Theorem 6, only the 6 can be clicked to activate the link, and it is too short. hyperref provides the \autoref command to help out. Instead of

```
Theorem~\ref{T:new}
```

you can simply type

```
\autoref{T:new}
```

and hyperref will provide the word Theorem so that the link becomes Theorem 6. The names supported by the \autoref command are listed in Table 11.1.

For my own use, I redefine:

```
\renewcommand{\chaptername}{Chapter}
\renewcommand{\sectionname}{Section}
\renewcommand{\subsectionname}{Section}
\renewcommand{\subsubsectionname}{Section}
```

External links

External links can be links to websites or other files that are located on the Internet. Use the

```
\href{address}{text}
```

command to typeset *text* and make it into a link to the Web address (URL).

For instance, in your references, you may have

```
Robert Miner and Jeff Schaefer,
 \emph{Gentle intoduction to MathML.}
 \href{http://www.webeq.com/mathml/gitmml/}
{http://www.webeq.com/mathml/gitmml/}
```

Then the last line of the address becomes a link and clicking on it takes you to the Web site. As an even fancier example, note the top matter command \urladdr (see Section 9.2.2) in secondarticle-ref.tex:

Command	Meaning
\figurename	Figure
\tablename	Table
\partname	Part
\appendixname	Appendix
\equationname	Equation
\Itemname	item
\chaptername	chapter
\sectionname	section
\subsectionname	subsection
\subsubsectionname	subsubsection
\paragraphname	paragraph
\Hfootnotename	footnote
\AMSname	Equation
\theoremname	Theorem

Table 11.1: Redefinable names supported by \autoref.

```
\urladdr{\href{http://math.uwinnebago.edu/menuhin/}
http://math.uwinnebago.edu/homepages/menuhin/}
```

Then, as part of Menuhin's address, you will find

```
http://math.uwinnebago.edu/menuhin/
```

Now clicking on the Web address will link to his Web page.

hyperref, of course, offers a lot more than I have presented here. For more detail, see the user manual and *The* LaTeX *Web Companion* [18].

CHAPTER

12

Presentations

In Section 1.11, we describe how a *presentation* is a PDF file that you open with Adobe Reader. You can put it in full screen mode[1] (View>Full screen), and then project the presentation one page at a time by pressing the space bar or the arrow keys.

Remember overhead transparencies? If we want to see half of what is on the transparency, we cover up the bottom part so that only the top part is projected. This way we have control over what the audience sees and when. We sometimes used overlays: placing another transparency on top of the projected one to modify it by adding text or graphics.

In this chapter, we discuss Till Tantau's beamer package in more detail to help you prepare presentations with overlays and with stunning visual effects. beamer relies on other packages such as the hyperref package (see Section 11.2) to establish links, Till Tantau's Portable Graphics Format package for creating graphics, Uwe Kern's xcolor package for coloring, the AMS packages for formatting math formulas and defining declarations, and some others.

The documentation for these packages runs to about a thousand pages. The good news is that you can use beamer "out of the box". You only have to learn about **20**

[1] If you have Adobe Acrobat Pro, open File>Document Properties and check mark Full Screen Mode. Then the PDF document automatically opens in full screen mode.

G. Grätzer, *More Math Into* LaTeX, DOI 10.1007/978-3-319-23796-1_12

commands—this is more than the **four** new command we had to learn in Section 1.11 but still an easily manageable task.

So we set ourselves in this chapter a modest goal, using beamer "out of the box". It is amazing how much you can achieve with a small investment of your time.

Chances are beamer is already installed for you. If not, consult Section D.1 on how to get it.

12.1 *Quick and dirty* beamer

We convert the article firstarticle.tex (in the samples folder) to a beamer presentation. We will remove some commands that are appropriate for an article but not for a presentation and add some commands that are specific to presentations. This will not produce a very good presentation. Nevertheless, the conversion is a really quick introduction to some basic beamer concepts.

12.1.1 *First changes*

Open firstarticle.tex, save it as quickbeamer.tex in the work folder. The converted tex version and the presentation quickbeamer.pdf are both in the samples folder.

Make the following changes in the preamble and top matter:

1. Change the first line to
   `% Introductory beamer presentation: quickbeamer.tex`
2. Change the documentclass to beamer.
3. Delete the six \usepackage and \newtheorem lines—beamer loads the necessary packages and defines these declarations.
4. Change the \address to \institute—this is the beamer command for address.
5. Delete the abstract environment—this is not needed for the presentation.

Here is the new version of the preamble and top matter.

```
%Introductory beamer presentation: quickbeamer.tex
\documentclass{beamer}
\begin{document}
\title{A construction of complete-simple\\
       distributive lattices}
\author{George~A. Menuhin}
\institute{Computer Science Department\\
          University of Winnebago\\
          Winnebago, MN 53714}
```

```
\date{March 15, 2015}

\maketitle
```

12.1.2 Changes in the body

1. Delete the `notation` and `proof` environments, but not the contents, that is, delete the four lines

   ```
   \begin{notation}
   \end{notation}
   \begin{proof}
   \end{proof}
   ```

   Both environments could theoretically stay, but the `notation` environment is not needed since in the next step we put `Notation` in the frame title. The `proof` environment (see Section 3.3.2) is not suitable for presentations because an environment can only be used within a frame, and proofs are typically longer.

2. Cut the presentation into *frames* (pages, transparencies, foils) with the `frame` environments. After each `\begin{frame}` we put a `\frametitle` command. The argument of the command is the "title" for the frame, displayed prominently at the top of the display.

   It would be tedious to give you precise instructions on how to do this, instead refer to the `quickbeamer.tex` document (in the `samples` folder) for all the `frame` environments and `\frametitle` commands we added.

3. Cut out the `figure` environment, except for the line

   ```
   \centering\includegraphics{products}
   ```

   which should be moved to follow the

   ```
   \frametitle{Illustrating the construction}
   ```

   line and accordingly delete `(see Figure~\ref{Fi:products})`.

Now copy over the illustration `products` from the `samples` folder to the `work` folder and typeset. That's it, enjoy your first presentation.

12.1.3 Making things prettier

Now you make some small changes to `quickbeamer.tex` to utilize `beamer`'s power for wonderful effects. Changes 1 and 3 are quite dramatic.

Save `quickbeamer.tex` with the name `quickbeamer1.tex` in the `work` folder. The edited version,`quickbeamer1.tex`, is in the `samples` folder along with the presentation `quickbeamer1.pdf`.

1. Add `\usetheme{Berkeley}` after the documentclass line.
2. Change `\maketitle` to

   ```
   \begin{frame}
     \titlepage
   \end{frame}
   ```

   Make sure that the last (sub)section is followed by a frame, otherwise it will be missing from the table of contents.
3. Add this frame after the titlepage frame:

   ```
   \begin{frame}
   \frametitle{Outline}
   \tableofcontents[pausesections]
   \end{frame}
   ```

   This creates a table of contents frame, with the section titles appearing one at a time.
4. Replace all instances of `{equation}` by `{equation*}`. In a presentation a reference to another frame is not recommended so equations should not be numbered. You might as well delete all the `\label` commands since these are not needed either.
5. In the second to last frame there are two references to equation numbers. Replace the text `the congruence \eqref{E:cong2}` with `the congruence`, and also replace `\eqref{E:comp}` with `the penultimate equation`, or similar.
6. Change the bibliographic reference to
   `See also Ernest T. Moynahan, 1957.`

Turning `quickbeamer1.tex` into a PDF file will get you a much prettier presentation. The first four pages of the new presentation are displayed in Figures 12.1 and 12.2—unfortunately, without the pretty colors.

12.1.4 Adjusting the navigation

Looking at Figures 12.1 and 12.2, we see that the `Berkeley` theme turns the sidebar into a navigation device and the section titles produce the table of contents (the Outline frame). Remember to typeset a few times! But a number of problems come to light.

1. The title of the presentation is too long for the sidebar, so is the title of the second section.
2. There is no need to repeat the author's name in the sidebar.
3. It was natural for the article `firstarticle.tex` to have only two sections. But sections play a different role in a `beamer` presentation, they are navigation tools.

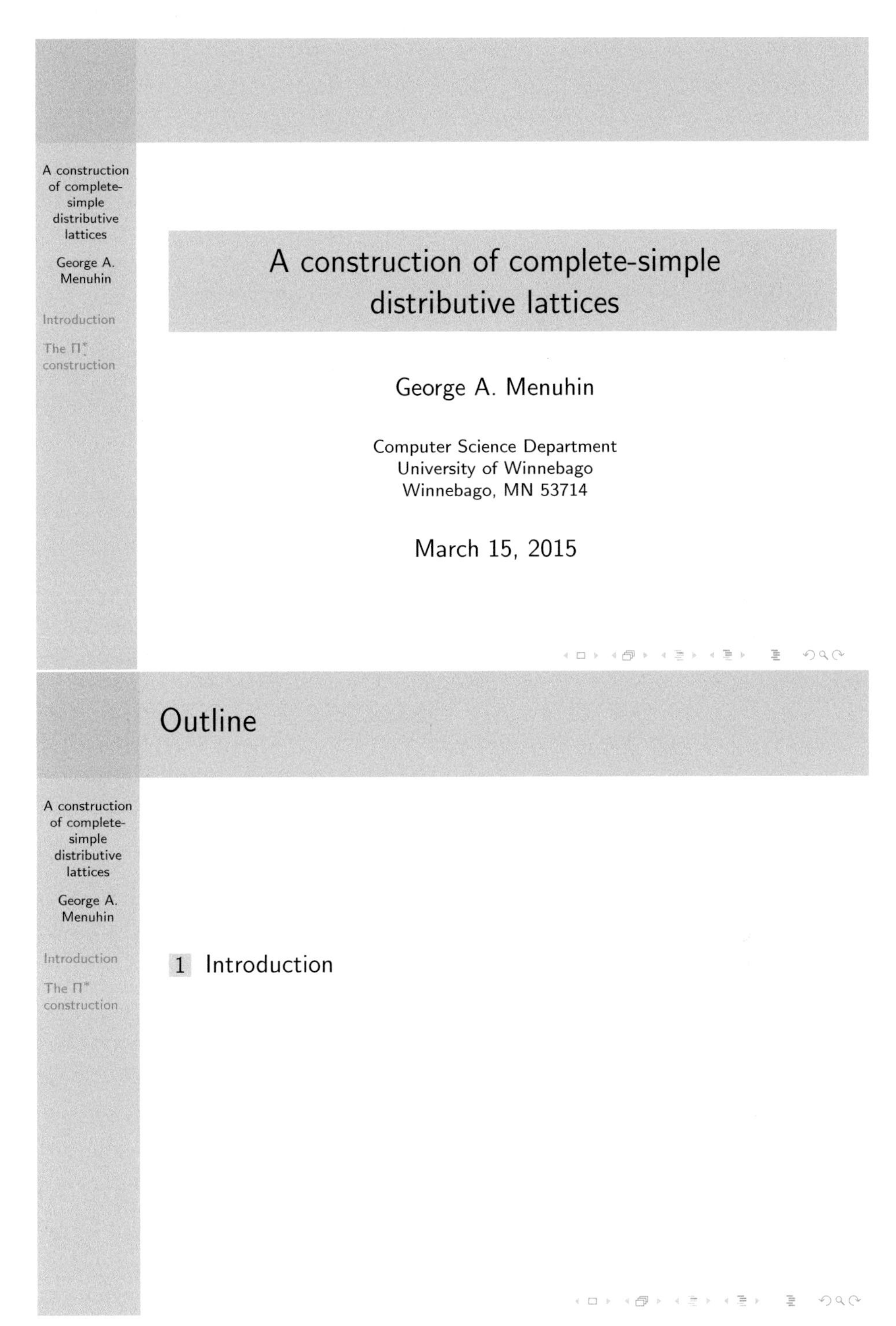

Figure 12.1: quickbeamer1 presentation, pages 1 and 2.

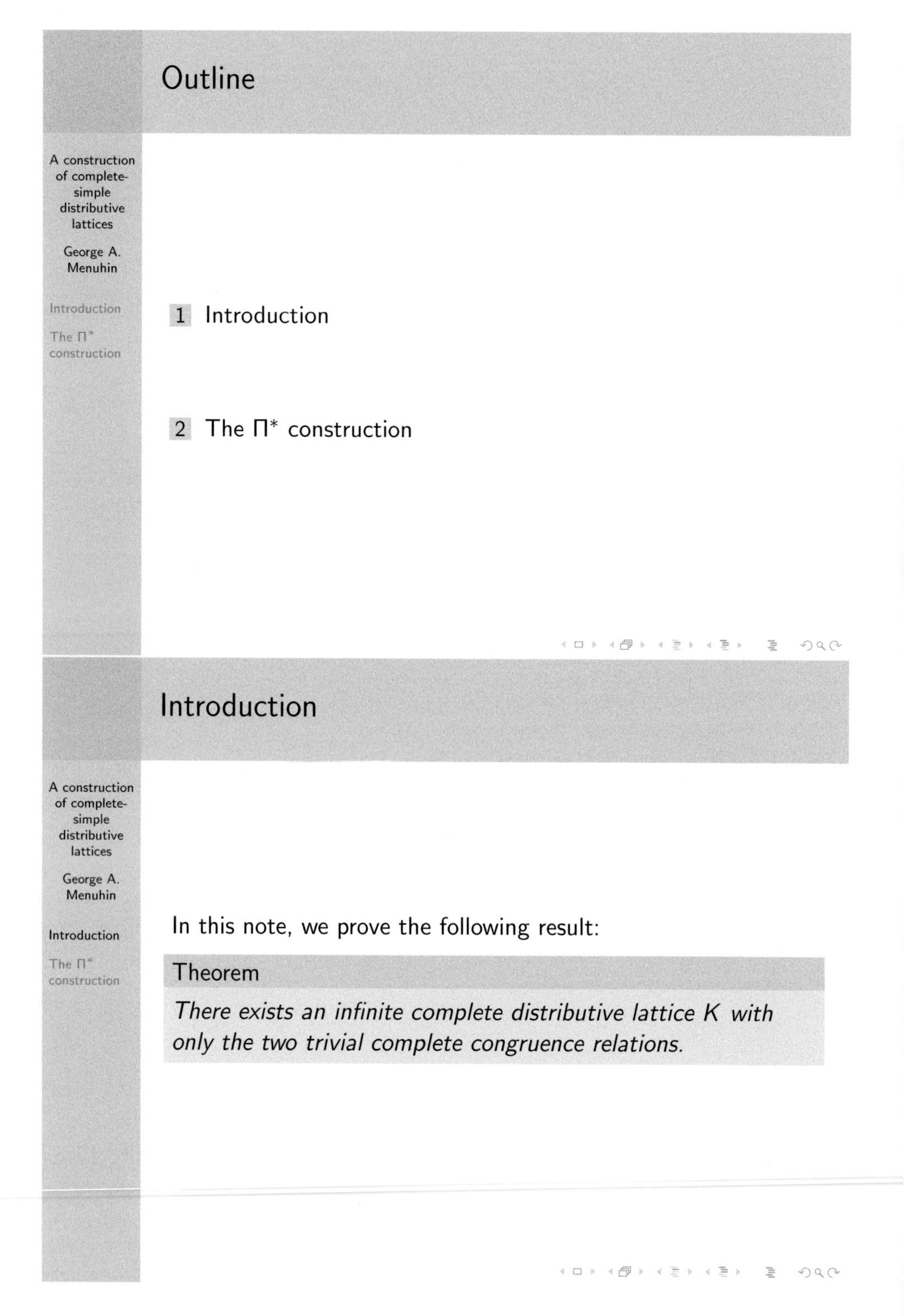

Figure 12.2: `quickbeamer1` presentation, pages 3 and 4.

The sidebar lists all the sections. It also highlights the section we are in. Moreover, by clicking on the name of a section, the presentation jumps there.

To correct these deficiencies, save the file `quickbeamer1.tex` as `quickbeamer2.tex` in the `work` folder. The edited version is in the `samples` folder along with the PDF file.

1. Change the `\title` command to

```
\title[Complete-simple distributive lattices]%
{A construction of complete-simple\\
distributive lattices}
```

   and the second `\section` command to

```
\section[Construction]{The $\Pi^{*}$ construction}
```

   The bracketed parts are the short versions used in the sidebar.

2. Change the `\author` command to

```
\author[]{George~A. Menuhin}
```

   The short version of the `author` command is blank, so the author's name will not be displayed in the sidebar.

3. Add the command

```
\section[Second result]{The second result}
```

   before the frame of the same title and

```
\section{Proof}
```

   before the proof. We even add

```
\section{References}
```

   before the frame of the same name.

Figure 12.3 shows page 7 of the `quickbeamer2` presentation—this corresponds to page 4 of the `quickbeamer1` presentation; the Outline accounts for the difference. Note how all the deficiencies listed above have been corrected. Compare page 7 of this presentation with the `Berkeley` theme in Figure 12.3 and with the `Warsaw` theme in Figure 12.4. Themes are discussed in Section 12.5.

12.2 *Baby* beamers

In the previous discussions you may have noticed two interesting features. First, the Outline frame (table of contents) created *two* pages in the `quickbeamer1` presentation and *five* pages in the `quickbeamer2` presentation. We discuss this in some detail now with the `babybeamer` presentations. You can find all the `babybeamer` presentations as `tex` and PDF files in the `samples` folder. Second, the sidebar shows some links. More about this in Section 12.2.7.

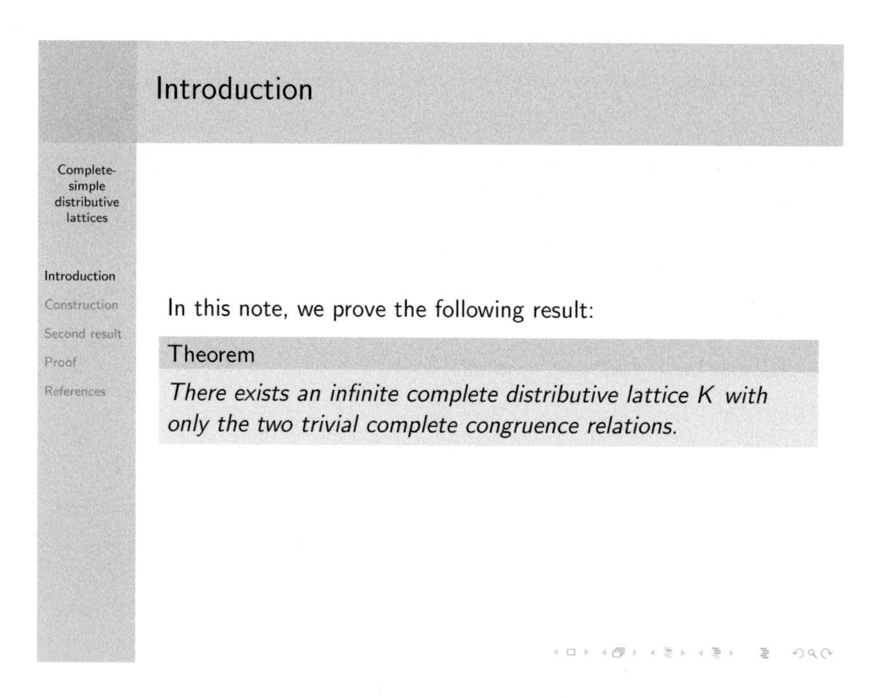

Figure 12.3: `quickbeamer2` presentation, page 7 with `Berkeley` theme.

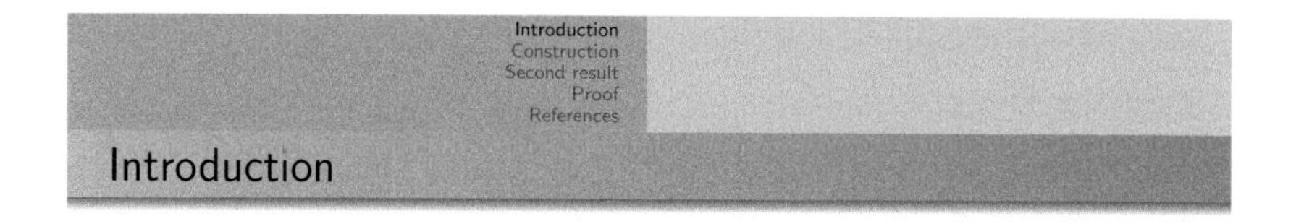

In this note, we prove the following result:

Theorem

There exists an infinite complete distributive lattice K with only the two trivial complete congruence relations.

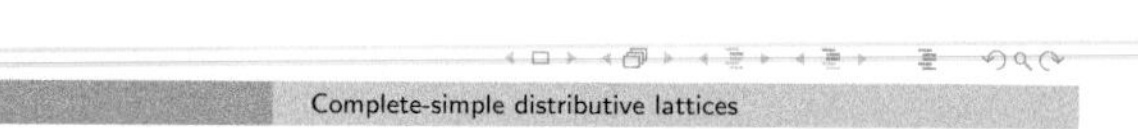

Figure 12.4: `quickbeamer2` presentation, page 7 with `Warsaw` theme.

12.2.1 Overlays

The outline frame of the quickbeamer2 presentation created five pages in the PDF file. Observe how each page, from the second on, completely overlaps the previous one, making it appear that the previous one stayed put and an additional line is displayed "on top of it". In beamer terminology these pages are *overlays* or *slides*. The five overlays will be referenced as overlay 1, ..., overlay 5. A single frame may create one or many overlays. The subsequent sections discuss many more variants.

beamer has many commands creating overlays. We start with some examples of \pause, then \only, and \onslide.

We introduce overlays with some presentations. The first, babybeamer1, introduces the \pause command to create overlays.

```
%babybeamer1 presentation
\documentclass{beamer}
\begin{document}

\begin{frame}
\frametitle{Some background}

We start our discussion with some concepts.
\pause

The first concept we introduce originates with Erd\H os.
\end{frame}
\end{document}
```

produces the presentation of Figure 12.5.

Rule ■ The \pause command

1. A frame may have many \pause commands.

2. The \pause command cannot be given in an AMS multiline math environment.

You move past a \pause command the same way as you get to the next frame, by pressing the space bar or the forward arrow key.

Using the \pause commands you can create many overlays, each containing a little more material on the overlays. If this is all you need, skip to Section 12.2.7, you do not need the more detailed discussion of overlays in the next few pages.

We could have coded the same presentation with the \only command:

```
%babybeamer2 presentation
\documentclass{beamer}
```

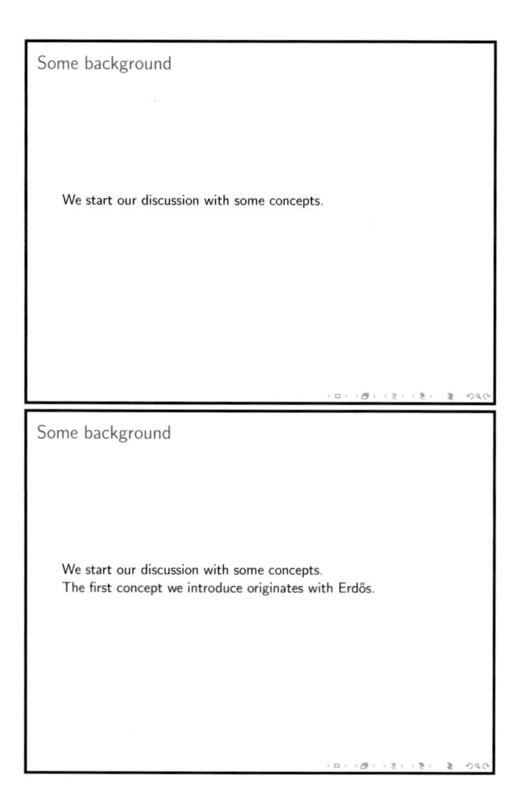

Figure 12.5: babybeamer1 presentation.

```
\begin{document}

\begin{frame}
\frametitle{Some background}

\only<1,2>{We start our discussion with some concepts.}

\only<2>{The first concept we introduce originates
        with Erd\H os.}
\end{frame}
\end{document}
```

This presentation is slightly different from `babybeamer1`. Overlay 1 ignores the second `\only` command and displays the line as appropriate to display one line. Overlay 2 displays the two lines as appropriate to display two lines. As a result, the first line moves slightly up when passing from overlay 1 to overlay 2. The argument of the `\only` command is typeset only on the overlays specified. On the other overlays, it is ignored.

If instead of the `\only` command you use the `\onslide` command (on slide, get it?), as in

```
\onslide<1,2>{We start our discussion with some concepts.}

\onslide<2>{The first concept we introduce originates
         with Erd\H os.}
```

then the first line of overlay 2 completely overlaps the first line of overlay 1, so the first line seems to stay put. The argument of the `\onslide` command is typeset on the overlays specified and on the other overlays it is typeset but invisible. This is the behavior you would want most often, but you may find that sometimes you prefer `\only`.

12.2.2 Understanding overlays

We introduced overlays in Section 12.2.1—probably the most important new concept for presentations. LaTeX typesets the content of a frame and the typeset material

- appears on all overlays for the parts of the source (maybe all) not modified by any command with an overlay specification;
- appears only on the overlays specified and is ignored on the other overlays for the arguments of the `\only` commands;
- appears on the overlays specified and is typeset but made invisible on the other overlays for the arguments of the `\onslide` commands.

More on overlay specifications at the end of this section. Here are some illustrations.

Example 1

```
This is a very \only<1>{very, very} important concept.
\only<1,2>{To start the definition \dots}
```

will typeset overlay 1 as

This is a very very, very important concept. To start the definition ...

and will typeset overlay 2 as

This is a very important concept. To start the definition ...

Example 2

```
What is $2+2$? It is \onslide<2>{$4$}.
\only<1>{Can you figure it out?}

\onslide<2>{I hope you all got it right.}
```

will typeset overlay 1 as

What is $2 + 2$? It is . Can you figure it out?

and will typeset overlay 2 as

What is $2 + 2$? It is 4.
I hope you all got it right.

Note that there is room in overlay 1 for the number 4.

Example 3

```
What is $2+2$?

\onslide<2>{It is {$4$}.}

Can you figure it out?
```

will typeset overlay 1 as

What is $2 + 2$?

Can you figure it out?

and will typeset overlay 2 as

What is $2 + 2$?
It is 4.
Can you figure it out?

Note that there is room in overlay 1 for the “missing” second line.

Overlay specifications

The angle brackets contain the *overlay specification*. Here are some more examples:

`<1-2,4->` means all overlays from 1 to 2, and all overlays from 4 onwards.

`<-3>` means all overlays up to 3.

`<2,4,6>` means overlays 2, 4, and 6.

We have two overlay specifications in the presentation `babybeamer2`: `<1,2>` and `<2>`. Maybe, `<1->` and `<2->` would be better, so that if you add a third overlay you do not have to change these.

The command `\pause` can only take the simplest overlay specification, a number. `\pause<3>` takes effect from overlay 3 on.

Note that overlay specifications are attached to commands but the overlays created are overlays of the frame in which the commands appear.

12.2.3 *More on the* `\only` *and* `\onslide` *commands*

The `\only` and `\onslide` commands can accomplish everything the `\pause` command can and a lot more.

The basic syntax

The syntax of `\only` is

```
\only<overlay spec>{source}
```

where *overlay spec* is the overlay specification and *source* is the code typeset by LaTeX.

A (partial) syntax of `\onslide` is

```
\onslide<overlay spec>{source}
```

With the same syntax you can give overlay specifications to many commands, including `\textbf`, `\textit`, `\alert`—`beamer`'s alternative to the `\emph` command—and then the command is in effect only on the overlays specified.

```
%babybeamer3 presentation
\documentclass{beamer}
\begin{document}

\begin{frame}
\frametitle{Some background}

\textbf<1>{We start our discussion with some concepts.}

\textbf<2>{The first concept we introduce originates
          with Erd\H os.}
```

```
\end{frame}
\end{document}
```

So the `babybeamer3` presentation (see Figure 12.6) has two overlays, each with two lines of text. On overlay 1 the first line is bold, on overlay 2 the second line is bold.

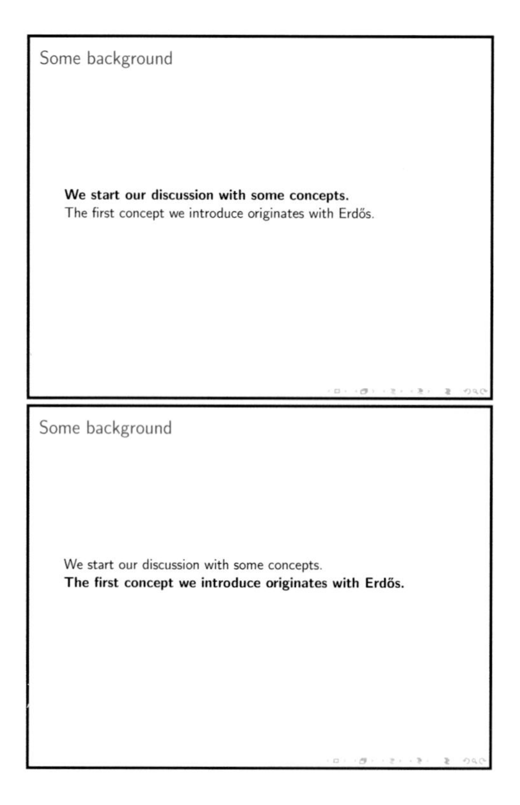

Figure 12.6: `babybeamer3` presentation.

A different syntax

The command `\only` has an alternate syntax:

`\only{`*`source`*`}<`*`overlay spec`*`>`

So

```
\only<1>{Can you figure it out?}
```

and

```
\only{Can you figure it out?}<1>
```

accomplish the same.

With this syntax, you can define your own commands that allow overlay specifications. For instance, using the command \color{blue} defined in Section 12.2.9, you can define the command

```
\newcommand{\myblue}{\only{\color{blue}}}
```

Then

```
\myblue<2>{Some more text}
```

will color the text blue on overlay 2 only.

12.2.4 Lists as overlays

Lists may be presented one item at a time, for example the babybeamer4 presentation in Figure 12.7 (in the samples folder) shows the four overlays of a list. R. Padmanabhan appears on the first, R. Padmanabhan and Brian Davey appear on the second, and so on. This is accomplished simply by adding the overlay specification <1-> to the item for R. Padmanabhan, the overlay specification <2-> to the item for Brian Davey, and so on.

```
%babybeamer4 presentation
\documentclass{beamer}
\begin{document}

\begin{frame}
\frametitle{Overlaying lists}

We introduce our guests:
\begin{itemize}
\item<1-> R. Padmanabhan
\item<2-> Brian Davey
\item<3-> Harry Lakser
\item<4-> Dick Koch
\end{itemize}
\end{frame}
\end{document}
```

Such an overlay structure is used so often that beamer has a shorthand for it, [<+->]. Here it is in babybeamer5.

```
%babybeamer5 presentation
\documentclass{beamer}
\begin{document}
```

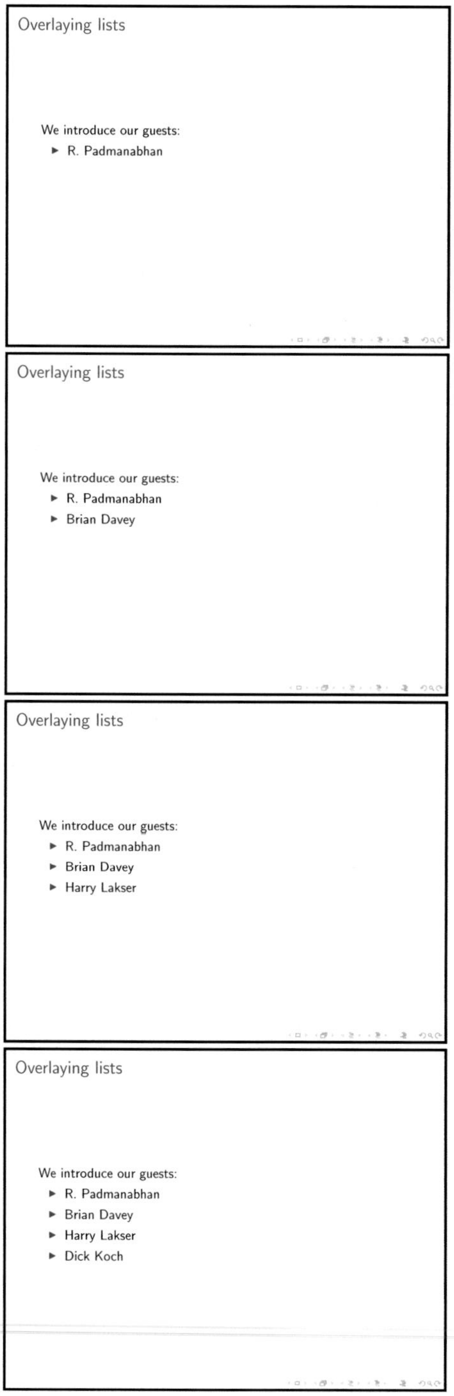

Overlaying lists

We introduce our guests:
- R. Padmanabhan

Overlaying lists

We introduce our guests:
- R. Padmanabhan
- Brian Davey

Overlaying lists

We introduce our guests:
- R. Padmanabhan
- Brian Davey
- Harry Lakser

Overlaying lists

We introduce our guests:
- R. Padmanabhan
- Brian Davey
- Harry Lakser
- Dick Koch

Figure 12.7: `babybeamer4` presentation.

```
\begin{frame}
\frametitle{Overlaying lists}

We introduce our guests:
\begin{itemize}[<+->]
\item R. Padmanabhan
\item Brian Davey
\item Harry Lakser
\item Dick Koch
\end{itemize}
\end{frame}
\end{document}
```

This shorthand allows adding and reordering items without having to change overlay specifications. Of course, if you do not want the items to appear in sequence, you have to use overlay specifications.

12.2.5 Out of sequence overlays

We now present an example of "out of sequence overlays". Look at Figure 12.8. I want to make this part of my presentation. First, I want to show the theorem, then illustrate it with the diagram at the bottom. Finally, I present the proof in the middle. So I need three overlays.

Theorem
Every finite distributive lattice can be embedded in a boolean lattice.

Proof.
Use join-irreducible elements.

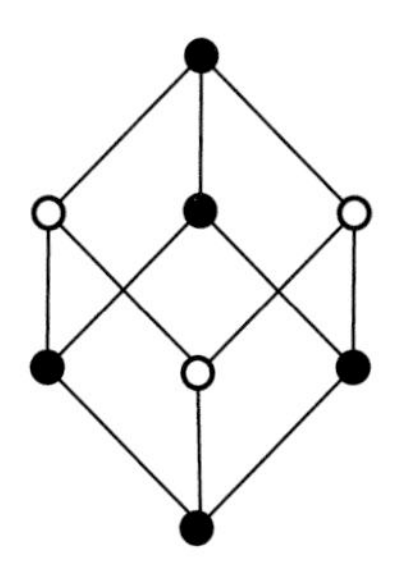

Figure 12.8: The slide to represent.

The theorem is on all three overlays, 1, 2, 3. Its illustration is on overlays 2 and 3, leaving room for the proof that appears only on overlay 3, This is an example of "out of sequence overlays". We code this in `babybeamer6` (in the `samples` folder).

Since declarations, proofs, and the `\includegraphics` command may all have overlay specifications, this seems easy to accomplish.

```
%babybeamer6 presentation, first try
\documentclass{beamer}
\begin{document}

\begin{frame}
\frametitle{Overlaying declarations and graphics}
\begin{theorem}<1->
Every finite distributive lattice can be embedded
in a boolean lattice.
\end{theorem}
\begin{proof}<3->
Use join-irreducible elements.
\end{proof}
\includegraphics<2,->{cube}
\end{frame}
\end{document}
```

This does not work too well. On overlay 1 the theorem appears in the middle and then it jumps up to make room for the illustration. This is the same problem we encountered in the `babybeamer2` presentation in Section 12.2.1 and the solution is also the same, the use of the `\onslide` command. Replace the line

```
\includegraphics<2,->{cube}
```

with

```
\onslide<2->{\includegraphics{cube}}
```

12.2.6 Blocks and overlays

You can think of a theorem in `beamer` as the contents of the `theorem` environment with a heading and, optionally, with an overlay specification, and with most themes—see Section 12.5—colorful visual highlighting, see Figures 12.3 and 12.4.

`beamer` provides the `block` environment that works the same way except that you name the block. The (partial) syntax of the `block` environment is

```
\begin{block}<overlay spec>{title}
source
\end{block}
```

Blocks are shaped as theorems. If there is no title, you still need the braces. The overlay specification is optional.

As an example, save babybeamer6.tex as babybeamer6block.tex in the work folder (also in the samples folder along with the PDF file) and replace the theorem environment with

```
\begin{block}<1->{Theorem}
  Every finite distributive lattice can be embedded
in a boolean lattice.
\end{block}
```

If you want a block of LaTeX code with an overlay specification but with no title and no visual highlighting, use one of the commands, \onslide and \only.

12.2.7 *Links*

A presentation is a PDF file, so it is not surprising that you can set links of various types in a beamer presentation. Just as the hyperref package helps us with hyperlinks in a PDF file (see Section 11.2), the beamer package allows us to conveniently set links in a presentation.

Some links are automatically provided. If you look closer at Figures 12.1 and 12.2, you see that the section titles are shown in the *sidebar*. In fact, the sidebar is a *navigation bar*. First, it shows which section you are in. Second, clicking on a section title takes you to that section.

Creating a link is a two-step process.

1. Name the place you want to link to.

2. Create a button with the property that clicking on it jumps you to the designated place.

To illustrate this process, we modify the presentation babybeamer4. Open the file babybeamer4.tex and save it as babybeamer7.tex in the work folder (the edited version is in the samples folder along with the PDF file).

1. Name the frame you want to link to by adding a label to the \begin{frame} line. In babybeamer7, add a label to the frame fourguests:

   ```
   \begin{frame}[label=fourguests]
   ```

   Labels of frames are also useful for selective typesetting of your presentation, see Section 12.6.

2. Add the following line to babybeamer7:

   ```
   \hyperlink{fourguests<3>}%
             {\beamergotobutton{Jump to third guest}}
   ```

This creates a link to the third overlay of the frame named `fourguests`, and creates a button, with the text `Jump to third guest`. Clicking on this button will jump to the third overlay of the frame `fourguests`.

3. To add variety to linking, include a new first frame:

```
\begin{frame}
\frametitle{First frame with a button}
Button example

Jumping to an overlay of a different frame
\bigskip

\hyperlink{fourguests<3>}%
{\beamergotobutton{Jump to third guest}}
\end{frame}
```

which has a button for jumping to the third overlay of the `fourguests` frame.

4. We also add a new third frame.

```
\begin{frame}
\frametitle{Third frame with a button}
Button example

Jumping to another frame
\bigskip

\hyperlink{fourguests}%
{\beamergotobutton{Jump to guest list}}
\end{frame}
```

with a button, with the text `Jump to guest list`. Clicking on this button will jump to the second frame, overlay not specified (defaults to 1).

5. Add a fourth frame,

```
\begin{frame}
\frametitle{Hidden link}
\hyperlink{fourguests}{Jumping to the guest list}
\end{frame}
```

introducing another version of the `\hyperlink` command:

```
\hyperlink{fourguests}{Jumping to the guest list}
```

which typesets the second argument as regular text, making it an *invisible link*. However, you may notice that the cursor changes when it hovers over the link. For instance, you may want to link the use of a concept to its earlier definition, where you also need a button for the return jump.

Here is babybeamer7:

```
%babybeamer7 presentation
\documentclass{beamer}
\begin{document}

\begin{frame}
\frametitle{First frame with a button}

Button example

Jumping to an overlay of a different frame
\bigskip

\hyperlink{fourguests<3>}%
{\beamergotobutton{Jump to third guest}}
\end{frame}

\begin{frame}[label=fourguests]
\frametitle{Overlaying lists}

We introduce our guests:
\begin{itemize}
\item<1-> R. Padmanabhan
\item<2-> Brian Davey
\item<3-> Harry Lakser
\item<4-> Dick Koch
\end{itemize}

\hyperlink{fourguests<3>}%
{\beamergotobutton{Jump to third guest}}
\end{frame}

\begin{frame}
\frametitle{Third frame with a button}
Button example

Jumping to another frame
\bigskip
```

First frame with a button

Button example
Jumping to an overlay of a different frame

Jump to third guest

Overlaying lists

We introduce our guests:
- R. Padmanabhan

Jump to third guest

Overlaying lists

We introduce our guests:
- R. Padmanabhan
- Brian Davey

Jump to third guest

Third frame with a button

Button example
Jumping to another frame

Jump to guest list

Figure 12.9: `babybeamer7` presentation.

```
\hyperlink{fourguests}%
{\beamergotobutton{Jump to guest list}}
\end{frame}

\begin{frame}
\frametitle{Hidden link}

\hyperlink{fourguests}{Jumping to the guest list}
\end{frame}
\end{document}
```

Figure 12.9 shows all these buttons. We do not show overlays 3 and 4 of frame 2 and frame 4, where the button is invisible.

12.2.8 Columns

Often, it is useful to put the display into columns. A simple illustration is given in `babybeamer8`:

```
%babybeamer8 presentation
\documentclass{beamer}
\begin{document}

\begin{frame}
\frametitle{Columns, top alignment}

\begin{columns}[t]
\begin{column}{2in}
Is it true that there is no new result
on the Congruence Lattice Characterization Problem?
\end{column}
\begin{column}{2in}
F. Wehrung found a distributive algebraic lattice that
cannot be represented as the congruence lattice
of a lattice.
\end{column}
\end{columns}
\end{frame}
\end{document}
```

The environment is `columns`. It has an optional argument for alignment, `t` for top, `c` for center, and `b` for bottom. The columns, usually two, are both in the `column` environment; the width of the column is in the argument; it can be given as a measurement—`2in` in the example—or relative to the width of the whole frame as `0.4\textwidth`.

Figure 12.10 shows the babybeamer8 presentation.

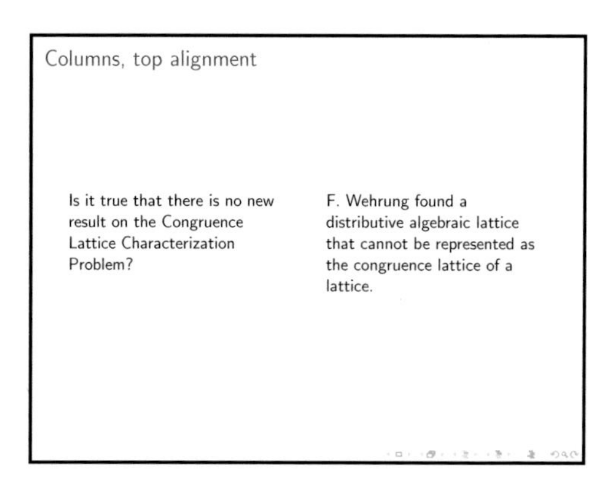

Figure 12.10: babybeamer8 presentation.

12.2.9 Coloring

LaTeX's job is to produce articles and books that contain text, math formulas, and graphics. Such publications—with the exception of textbooks—cannot afford color printing. Presentations are different. If you prepare a color presentation, it will project in color.

Nevertheless, the color commands are of limited use even for presentations. You probably use the color scheme of the chosen theme (see Section 12.5), and have limited opportunity to color things yourself. If you do, be very careful, too much color distracts from the presentation but judicious use of color—say, for highlighting a word or phrase—may be very effective.

beamer uses the sophisticated xcolor package of Uwe Kern. It colors by specifying the color model: rgb (red, green, blue), or cmyk (cyan, magenta, yellow, black), or gray (black and white)—there are many more models to choose from—and how much of each color you want to mix.

So \color[rgb]{0,1,0} paints everything—within its scope—green. You can color some text green with the command

```
\textcolor[rgb]{0,1,0}{This text is green.}
```

There are seventeen predefined colors: red, green, blue, cyan, magenta, yellow, orange, violet, purple, brown, pink, olive, black, darkgray, gray, lightgray, and white. With the proper options, there are hundreds more. So the previous command could also be given as

```
\textcolor{green}{This text is green.}
```

or as

```
{\color{green}This text is green.}
```

To pretty things up, you can use `\colorbox{green}{Green box}`, which puts the argument in a green box and `\fcolorbox{red}{green}{Green box}`, which also adds a red frame.

`xcolor` is automatically loaded by `beamer`. To make sure that `xcolor` is loaded with the options desired, you have to include these options in the preamble in the

```
\documentclass{beamer}
```

line. For instance, to have the `dvipsnam` option for `xcolor`, invoke `beamer` with

```
\documentclass[xcolor=dvipsnam]{beamer}
```

You can also mix predefined colors:

```
{\color{green!40!yellow} This text is of what color?}
```

which sets the text 40% green and 60% yellow.

There are commands for defining colors and color sets, as well as for coloring the background, frames, and hyperlinks (see Uwe Kern, *Extending LaTeX's color facilities: the xcolor package* [46]).

Here is a simple illustration:

```
%babybeamer9 presentation
\documentclass{beamer}
\begin{document}

\setbeamercolor{normal text}{bg=yellow!15}
\begin{frame}
\frametitle{Colors}

\begin{columns}[t]
  \begin{column}{2in}
{\color{red}Is it true that there is no new result
on the Congruence Lattice Characterization Problem?}
  \end{column}
    \begin{column}{2in}
{\color{green}F. Wehrung found a distributive
algebraic lattice that cannot be represented
as the congruence lattice of a lattice.}
  \end{column}
\end{columns}
\end{frame}
\setbeamercolor{normal text}{bg=green!15}
```

```
\begin{frame}
\frametitle{Colors fading out}

We introduce our guests:
\begin{itemize}
\item {\color{red}R. Padmanabhan}
\item {\color{red!60!white}Brian Davey}
\item {\color{red!40!white}Harry Lakser}
\item {\color{red!20!white}Dick Koch}
\end{itemize}
\end{frame}
\end{document}
```

The command

```
\setbeamercolor{normal text}{bg=yellow!15}
```

sets the background color to light (15%) yellow. In the first column, the text is red, in the second, green. Set the foreground with `fg=`.

The background of the second frame is light green. The four participants are in lighter and lighter shades of red.

Figure 12.11 shows the frames of the `babybeamer9` presentation—unfortunately, not in color (unless you have the PDF of the book).

12.3 The structure of a presentation

The structure of your presentation is, by and large, determined by the sectioning commands: `\section` and `\subsection`. For a very long lecture there may also be `\part` commands. The argument of any of these commands may have a short version for the navigational side bar (see Section 12.1.4).

The sectioning commands used in a `beamer` presentation look the same as they do for articles and books, but they play a different role. They do not display a section title, but they add an entry to the table of contents. They also act as place markers in the sense that if you click on the title of a section in a navigation bar, then you will jump to the *frame following* the section command.

Rule ■ Sectioning commands

1. Sectioning commands can only be placed between frames.

2. There must be a frame following the last sectioning command.

3. For a long (sub)section title, use `\breakhere` to break a line.

4. The optional short versions are for the navigation bar.

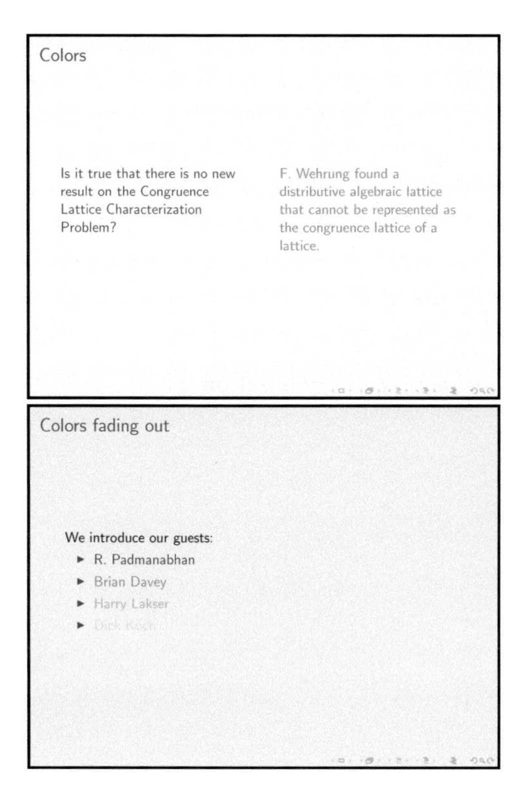

Figure 12.11: `babybeamer9` presentation.

These are illustrated with `beamerstructure1`, see Figure 12.12. The line

```
\tableofcontents[pausesections, pausesubsections]
```

causes the table of contents to appear a line at a time. This command may also be used without an option or only with one, `pausesections`.

The second page shown in Figure 12.12 is the table of contents. The page is about half filled with only five listed items, so no more than 10 sections and subsections would fit. There should be fewer.

```
%beamerstructure1 presentation
\documentclass{beamer}
\usetheme{Berkeley}
\begin{document}

\begin{frame}
\frametitle{Outline}
```

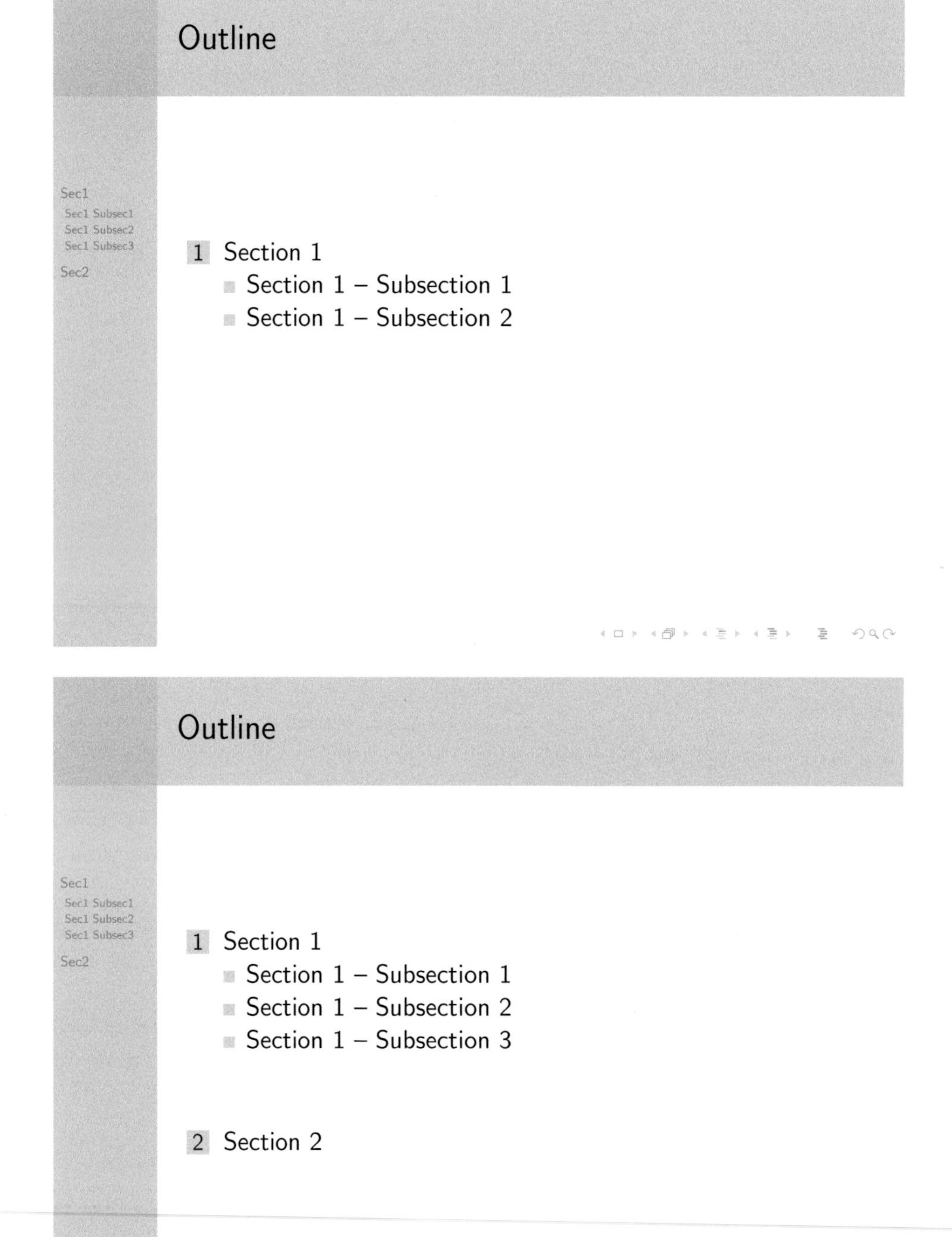

Figure 12.12: `beamerstructure1` presentation, pages 3 and 5.

```
\tableofcontents[pausesections, pausesubsections]
\end{frame}

\section[Sec1]{Section 1}

\begin{frame}
\frametitle{Section 1}

Text of Section 1
\end{frame}

\subsection[Sec1 Subsec1]{Section 1 -- Subsection 1}

\begin{frame}
\frametitle{Section 1\\Subsection 1}

Text of Section 1, Subsection 1
\end{frame}

\subsection[Sec1 Subsec2]{Section 1 -- Subsection 2}
\begin{frame}
\frametitle{Section 1\\Subsection 2}

Text of Section 1, Subsection 2
\end{frame}

\subsection[Sec1 Subsec3]{Section 1 -- Subsection 3}
\begin{frame}
\frametitle{Section 1\\Subsection 3}

Text of Section 1, Subsection 3
\end{frame}

\section[Sec2]{Section 2}

\begin{frame}
\frametitle{Section 2}

Text of Section 2
\end{frame}
```

```
\end{document}
```

12.3.1 Longer presentations

Longer presentations may need parts and a more complicated table of contents. I will not discuss these topics, but the presentation `beamerstructure2` (in the `samples` folder) illustrates the use of parts and some other features. I added some comments to point these out. See Figure 12.13 for two sample pages of this presentation.

12.3.2 Navigation symbols

The more structure you have in a presentation, the more you may appreciate the navigation icons shown by default on each page in the last line on the right. The icons are:

- the slide
- the frame
- the section
- the presentation icons

each surrounded by a left and a right arrow

- the appendix
- the back and forward icons (circular arrows)
- the search icon (a magnifying glass)

If you decide not to have them, as in the presentation `beamerstructure2`, then give the following command in the preamble:

```
\setbeamertemplate{navigation symbols}{}
```

12.4 Notes

You can place notes in your presentation to remind yourself of what you want to say in addition to what is being projected. A note is placed in the presentation as the argument of the `\note` command, as in

```
\note{This is really difficult to compute.}
```

By default, notes are not shown in the presentation. If you invoke `beamer` with

```
\documentclass[notes=show]{beamer}
```

then the notes pages are included. The command

```
\documentclass[notes=show, trans]{beamer}
```

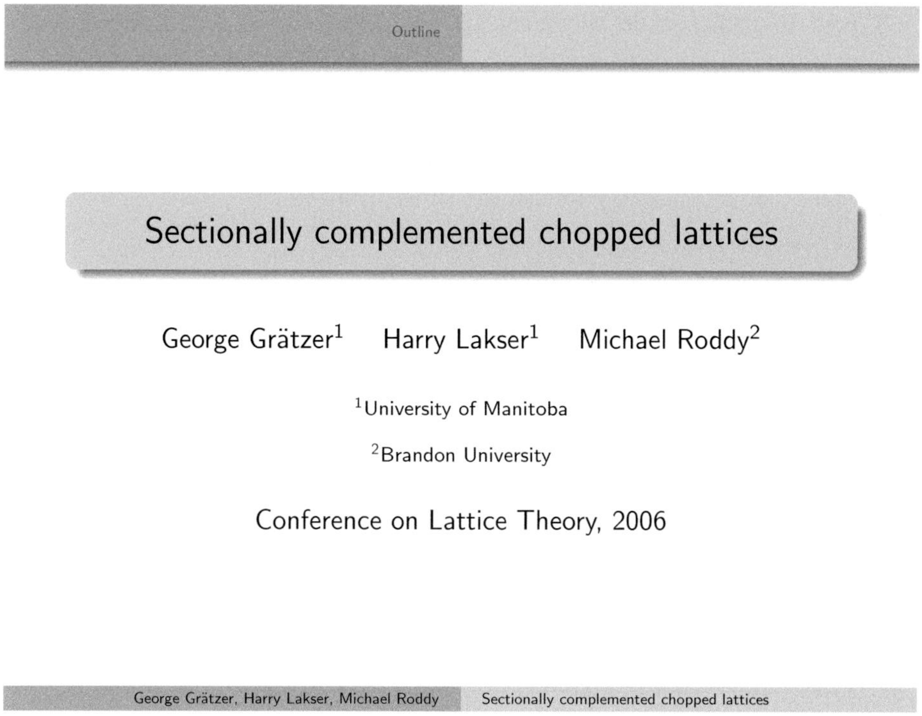

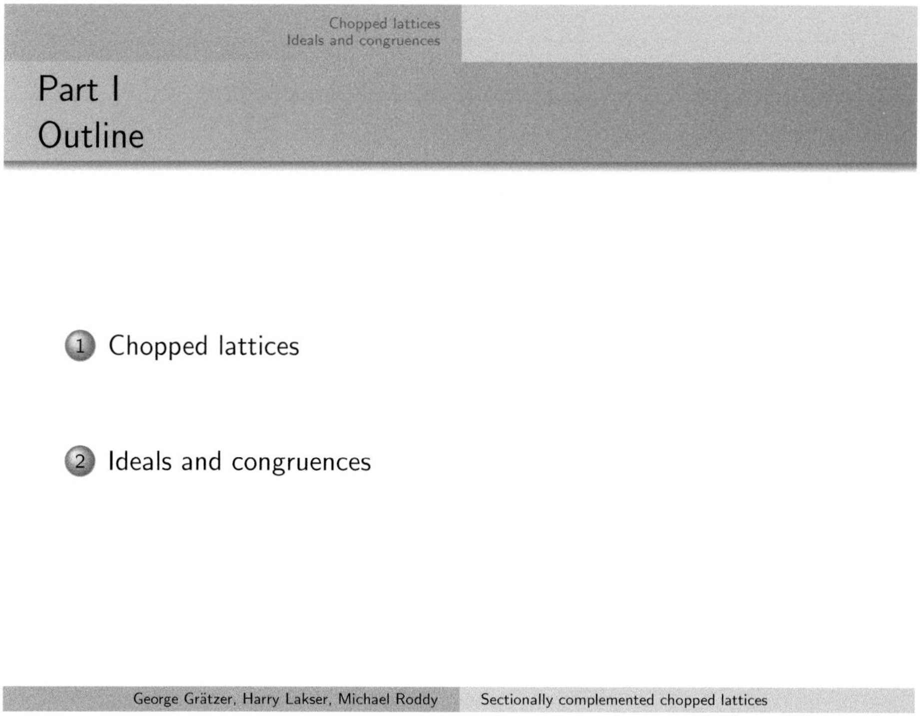

Outline

Sectionally complemented chopped lattices

George Grätzer[1] Harry Lakser[1] Michael Roddy[2]

[1]University of Manitoba

[2]Brandon University

Conference on Lattice Theory, 2006

George Grätzer, Harry Lakser, Michael Roddy Sectionally complemented chopped lattices

Chopped lattices
Ideals and congruences

Part I
Outline

1. Chopped lattices
2. Ideals and congruences

George Grätzer, Harry Lakser, Michael Roddy Sectionally complemented chopped lattices

Figure 12.13: `beamerstructure2` presentation, pages 1 and 10.

produces transparencies with notes, and

```
\documentclass[notes=only]{beamer}
```

produces only the note pages, one note page for every overlay of a frame with a note. To avoid this, print the output of

```
\documentclass[trans, notes=only]{beamer}
```

In addition to these examples, all the notes placed in a single frame are collected together on one note page. And a note between frames becomes a page on its own.

beamer does an excellent job of producing notes pages, for example, see Figure 12.14. In the upper-left corner, it displays precisely where we are in the structure of the presentation. The upper-right corner shows a small picture of the page to which the notes are attached.

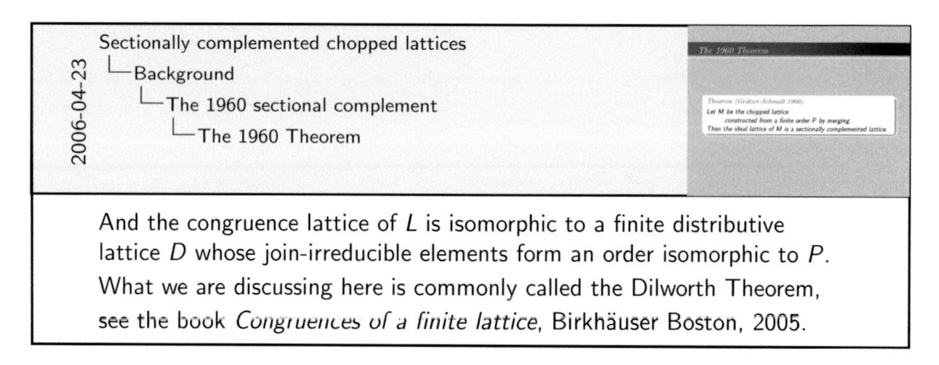

Figure 12.14: A note page.

12.5 Themes

If you look carefully at Figures 12.1, 12.2, and 12.13 (even better, if you look at the PDF files of these presentations), you see how every little detail of the presentation is provided by beamer. Figures 12.1 and 12.2 show a presentation style dominated by a dark blue headline and left sidebar, a complete navigation bar in the left sidebar, the name of the author(s) and the title repeated in every overlay, section numbers in colored squares, and so on.

The presentation in Figure 12.13 has no sidebars, section numbers are in colored circles, the navigation elements are in the headline. The display of lists and theorems (and other similar elements of a presentation) also vary a great deal.

You can achieve such detailed control over your presentation by defining all these elements yourself. beamer places dozens of commands at your disposal to make this possible. Or you can use a *presentation theme* that will do the job for you.

The command to name a presentation theme is \usetheme{}. The presentation beamerstructure2 uses the theme Warsaw (see Figure 12.13), so following the document class line type the command

```
\usetheme{Warsaw}
```

quickbeamer1 uses the theme Berkeley (see Figures 12.1 and 12.2) and so does beamerstructure1 (see Figure 12.12).

The presentation themes are in the theme subfolder of the themes folder of beamer. As of this writing, there are 26 of them, named after cities:

Presentation Themes

Without Navigation Bars default, Bergen, Boadilla, Madrid, AnnArbor, CambridgeUS, Pittsburgh, Rochester

With a Navigation Bar Antibes, JuanLesPins, Montpellier

With a table of contents Sidebar Berkeley, PaloAlto, Goettingen, Marburg, Hannover

With Mini Frame Navigation Berlin, Ilmenau, Dresden, Darmstadt, Frankfurt, Singapore, Szeged

With Section and Subsection Table Copenhagen, Luebeck, Malmoe, Warsaw

How do you choose a presentation theme? After the presentation is finished, try out the various themes. Ask yourself:

- Do sidebars take too much room away from my illustrations?
- Do stronger colors add to the presentation or do they distract?
- Do I want to use a navigation bar?

Answering these questions will narrow your choice.

The presentation theme defines all the colors, but you can alter them with the command \usecolortheme{}. You have a choice of albatross, beetle, crane, fly, and seagull.

For instance,

```
\usetheme{Warsaw}
\usecolortheme{seagull}
```

is a gray version of the Warsaw theme, appropriate for printing in black-and-white. In addition, you can further modify the "inner elements", such as blocks, with

```
\usecolortheme{lily}
```

or `orchid`, or `rose`. You can modify the "outer elements", such as headlines and sidebars, with

```
\usecolortheme{whale}
```

or `seahorse`, or `dolphin`. So you can have, for instance,

```
\usetheme{Warsaw}
\usecolortheme{lily}
\usecolortheme{whale}
```

This gives you 45 "out of the box" color schemes.

Similarly, font themes can also be specified, modifying the presentation theme, with the command `\usefonttheme{}`. You have the default and the following options:
`professionalfonts`
`structurebold`
`structureitalicserif`
`structuresmallcapserif`.

12.6 *Planning your presentation*

Step 1 As a rule, your presentation is based on one or more of your articles. Collect them in one folder. Resolve naming conventions as necessary. There should be only one `Fig1`!

Step 2 Rewrite the article(s) to sketch out your presentation. The pages correspond to frames. A page should not have too many words, say, no more than 40. Replace your numbered theorems with named theorems. Never reference another page. Have few sections and subsections. Add a table of contents, which is a readable overview of the new article.

Step 3 Base the new presentation on a presentation in the `samples` folder, a sample presentation in `beamer`'s `solution` folder, or on one of your own or of a colleague's older presentations. Turn the pages into frames.

Step 4 Design your frames and add frame titles. Completely disregard what we wrote in Section 2.3 (*the idea behind* LaTeX *is that you should concentrate on what you have to say and let* LaTeX *take care of the visual design*). The new principle is: *You are completely responsible for the visual appearance of every frame and overlay.*

This is, of course, in addition to brevity and readability. Do not let LaTeX break your lines. Do it with the `\\` command and keep words that belong together on the same line.

Step 5 Write notes to remind yourself what you want to say in your lecture that is not on the slides. Print the notes for your lecture.

Step 6 Build in flexibility. For instance, if you have four examples to illustrate a definition, put each one on a different frame or overlay, and add a link to each that skips the rest of the examples. Depending on your audience's understanding, show an example or two, and skip the rest. The same way, you may skip proof ideas and even topics.

Step 7 Prepare for the worst—the computer system may fail, but projectors seldom do—so print a set of transparencies for your lecture as a backup by invoking the option `trans` of the documentclass

```
\documentclass[trans]{beamer}
```

To print a *handout*, use the `handout` option

```
\documentclass[handout]{beamer}
```

Open the presentation in Acrobat Reader. In `Printer/Page Setup...` set landscape and 140% magnification. In the `Print` dialogue box in `Layout` choose two pages per sheet and print—assuming, of course, that you have a printer offering these options.

12.7 What did I leave out?

Since the `beamer` reference manual is 245 pages long, it is clear that this chapter covers maybe 10% of it.

For most presentations, you won't even need most of what I have included. If you read Sections 12.1, 12.2.1 and maybe Section 12.3, you should have enough for most math presentations.

If you are in other fields, or if you are more ambitious, you may need more. For example, a computer scientist will want program listings in a `verbatim` environment. This is easy. Start your frame with

```
\begin{frame}{fragile}
```

and then you can use the `verbatim` environment.

If you want to include sounds or movies in your presentation, consult Till Tantau's *User's Guide to the Beamer Class* [69].

You can do very simple animation with what we have covered here. This is illustrated with the `babybeamer10` presentation (in the `samples` folder).

```
babybeamer10 presentation
\documentclass{beamer}
\begin{document}

\begin{frame}

\includegraphics<1>{basem3-1}

\includegraphics<2>{basem3-2}
```

```
\includegraphics<3>{basem3-3}

\includegraphics<4>{basem3-4}
\end{frame}
\end{document}
```

The congruence generated by the dashed red line, see Figure 12.15, spreads in three steps, illustrating an interesting result. The animation is quite effective and instructive.

If you want to place such changing pictures lower in a frame, put them in the `overprint` environment.

I would recommend that you read Section 5 of Till Tantau, *User's Guide*, which has many good pointers about creating presentations.

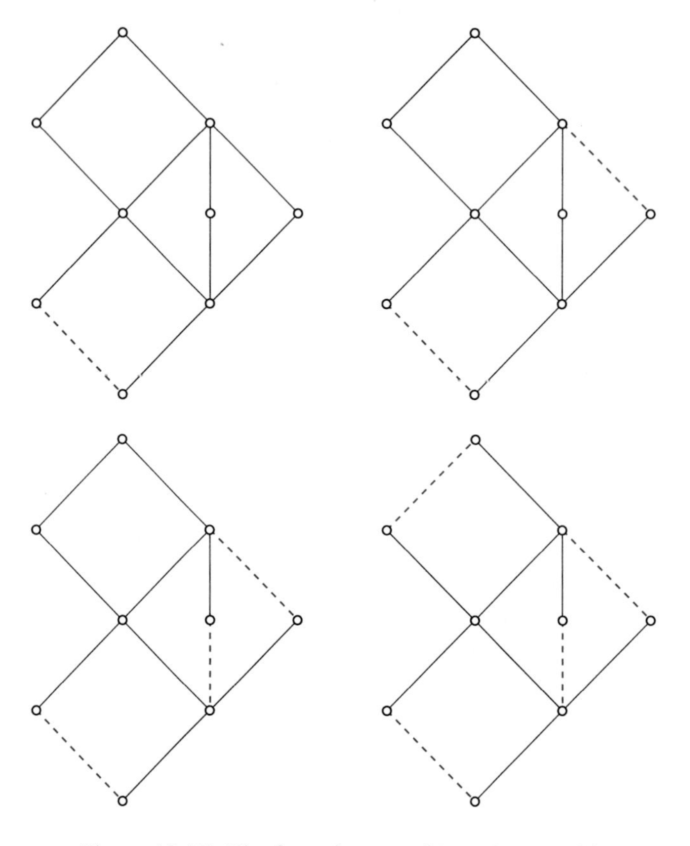

Figure 12.15: The four pictures of `babybeamer10`.

CHAPTER

13

Illustrations

Most illustrations in math are "vector graphics" such as the following example:

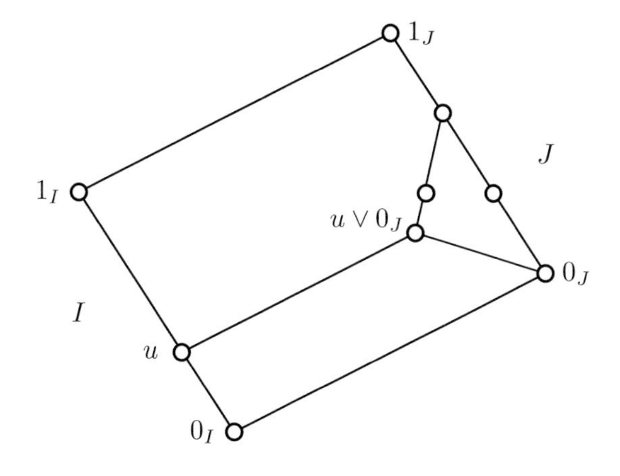

There are lots of circles, connecting lines, arrows, and so on. And the annotations are in LaTeX! Many mathematician used Adobe Illustrator or special purpose software to

G. Grätzer, *More Math Into* LaTeX, DOI 10.1007/978-3-319-23796-1_13

create such diagrams; see Sections 7.8 and 13.4 for commutative diagrams. Since there is no more academic pricing to buy Adobe Illustrator (you have to subscribe to Creative Cloud), some turn to Till Tantau's Ti*k*Z package.

We introduce Ti*k*Z in this chapter with a few commands. We hope they will serve your needs.

This chapter is based on Jacques Crémer's *A very minimal introduction to TikZ*, with his permission, and Michael Doob's detailed suggestions.

13.1 Your first picture

To use the Ti*k*Z package, include
`\usepackage{tikz}`
in the preamble of your document. A picture is in a `tikzpicture` environment, which is, in turn, typically within a `figure` environment (see Section 1.10):

```
\begin{figure}[htb]
{\centering
\begin{tikzpicture}
...
\end{tikzpicture}}
\end{figure}
```

Let us draw the illustration of Figure 13.1.

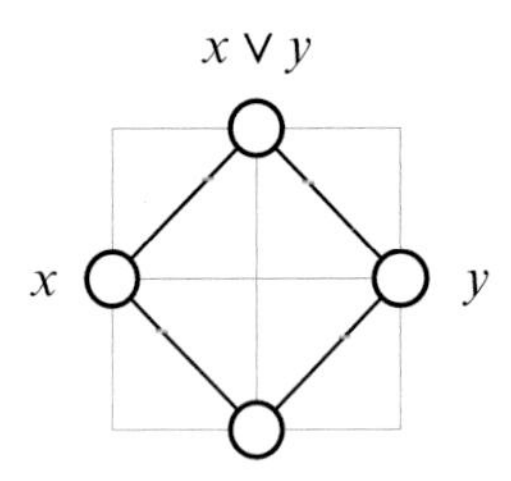

Figure 13.1: Our first Ti*k*Z illustration

Step 1: Draw the grid

The command is `\draw`, the optional argument: `help lines`.

```
\begin{tikzpicture}
\draw[help lines] (0,0) grid (2,2);
\end{tikzpicture}
```

Note the semicolon terminating the line. If you forget, you get the helpful reminder:

```
Package tikz Error: Giving up on this path.
Did you forget a semicolon?
```

Step 2: Draw the four edges

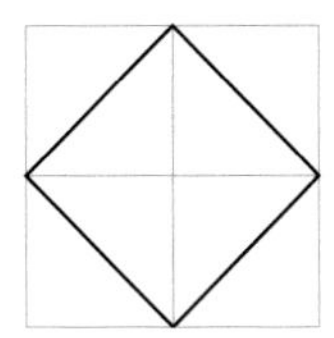

The command is `\draw`; the argument is a series of grid points connected by two dashes, `--`.

```
{\centering\begin{tikzpicture}
\draw[help lines] (0,0) grid (2,2);
\draw (1,0)--(2,1)--(1,2)--(0,1)--(1,0);
\end{tikzpicture}
```

Again, note the semicolon terminating the line. It has to terminate all Ti*k*Z lines!!! No more warnings.

Let me specify the conventions used in my field for such a diagram (note that 1 inch is 2.54 cm and 1 cm is 28.35 points): We use a grid with lines 1 cm apart; the circles have radius 1.8 mm and line width 1 pt; the lines have line width 0.7 pt.

So the `\draw` command we would use for the illustration is

```
\draw[line width=0.7pt] (1,0)--(2,1)--(1,2)--(0,1)--(1,0);
```

to make the lines a little thicker. More about line width soon.

Step 3: Draw the circles

We add a circle with the `\draw` command

```
\draw (1,0) circle[radius=1.8mm];
```

By the conventions (above), we want the line width to be 1pt:

```
\draw[line width=1pt] (1,0) circle[radius=1.8mm];
```

We add four circles at `(1,0)`, `(2,1)`, `(1,2)`, `(0,1)`:

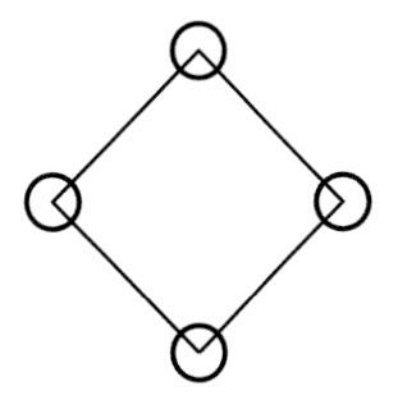

Step 4: Fill the circles

This looks ugly. We should not be seeing the line segments in the circles. Easy to help. Fill the circles with white by adding the `fill=white` option to `\draw`. So we get the illustration of Figure 13.1 with the grid. We comment out the line producing the grid:

```
{\centering\begin{tikzpicture}
\draw[help lines] (0,0) grid (2,2);
\draw[line width=0.7pt](1,0)--(2,1)--(1,2)--(0,1)--(1,0);
\draw[fill=white, line width=1pt] (1,0) circle[radius=1.8mm];
\draw[fill=white, line width=1pt] (2,1) circle[radius=1.8mm];
\draw[fill=white, line width=1pt] (1,2) circle[radius=1.8mm];
\draw[fill=white, line width=1pt] (0,1) circle[radius=1.8mm];
\end{tikzpicture}}
```

and this produces the illustration of Figure 13.1 except for the labels.

Step 5: Add the labels

We add the labels with the `\node at` command. To add the label y to the circle with center at (2,1):

```
\node at (2.5,1) {$y$};
```

You get 2.5 by 2.5 = 2 + 0.18 + a little nudge. Experiment until you like the result. Then proceed to the other circles:

```
\begin{figure}[h!]
{\centering\begin{tikzpicture}
\draw[help lines] (0,0) grid (2,2);
\draw[line width=0.7pt] (1,0)--(2,1)--(1,2)--(0,1)--(1,0);
\draw[fill=white, line width=1pt] (1,0) circle[radius=1.8mm];
\draw[fill=white, line width=1pt] (2,1) circle[radius=1.8mm];
\node at (2.5,1) {$y$};
\draw[fill=white, line width=1pt] (1,2) circle[radius=1.8mm];
\node at (1,2.5) {$x \vee y$};
\draw[fill=white, line width=1pt] (0,1) circle[radius=1.8mm];
\node at (-0.5,1) {$x$};
\end{tikzpicture}}
\caption{Our first \tikzname illustration}\label{Fi:firsttikz}
\end{figure}
```

producing the illustration of Figure 13.1!

Step 6: Remember, this is LaTeX

Section 14.1 introduces custom commands. Let's get a little ahead of ourselves, and define

```
\newcommand{\mycircle}[1]
{\draw[fill=white,line width=1pt] (#1) circle[radius=1.8mm]}
```

Even better. Ti*k*Z allows us to set default values. The command

```
\tikzset{every picture/.style={line width=0.7pt}}
```

sets the default value of line width to 0.7pt. Now the code for Figure 13.1 becomes easier to read (and write):

```
{\centering\begin{tikzpicture}
\draw[help lines] (0,0) grid (2,2);
\draw (1,0)--(2,1)--(1,2)--(0,1)--(1,0);
\mycircle{1,0};
\mycircle{2,1};
\node at (2.5,1) {$y$};
\mycircle{1,2};
\node at (1,2.5) {$x \vee y$};
\mycircle{0,1};
\node at (-0.5,1) {$x$};
\end{tikzpicture}}
```

13.2 The building blocks of an illustration

An illustration is built from components. We discuss some of them: line segments, circles, dots (or vertices), ellipses, rectangles, arcs, smooth curves (Bézier curves), and labels.

Line segments A path drawn with the command `\draw (1,0)--(2,2)--(4,1);`

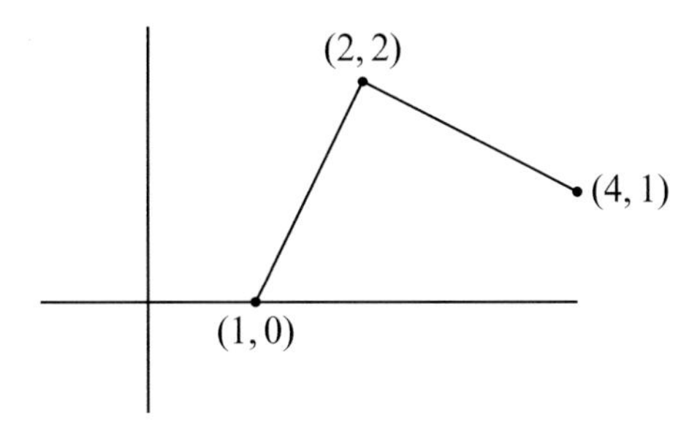

Circles A circle drawn using `\draw (2,2) circle[radius=1];`

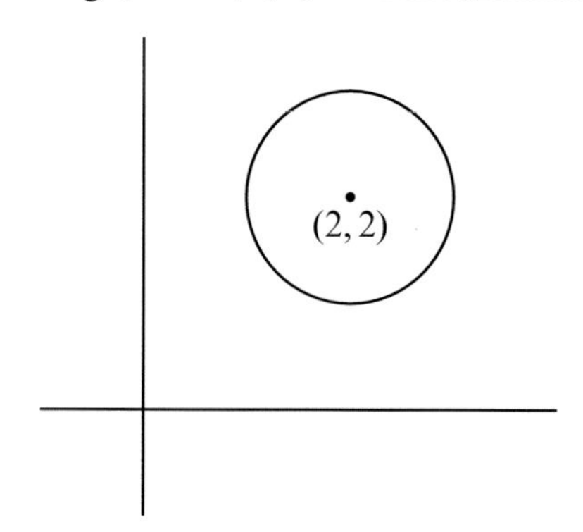

Dots (or vertices) The command `\draw[fill] (2,2) circle[radius=1pt];` draws a dot (or vertex):

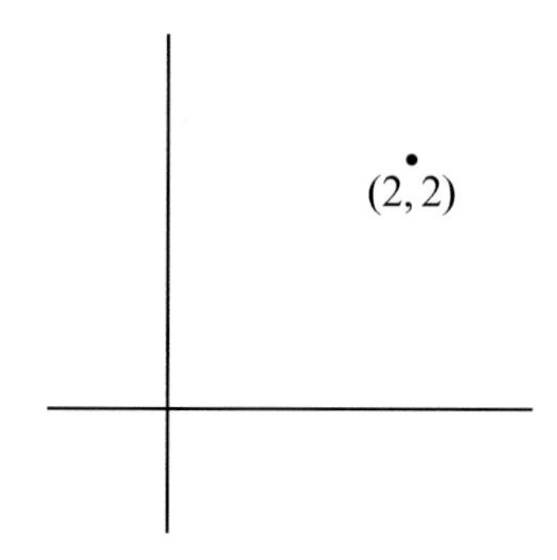

Ellipses Drawn by `\draw (2,2) ellipse[x radius=2, y radius=1];`

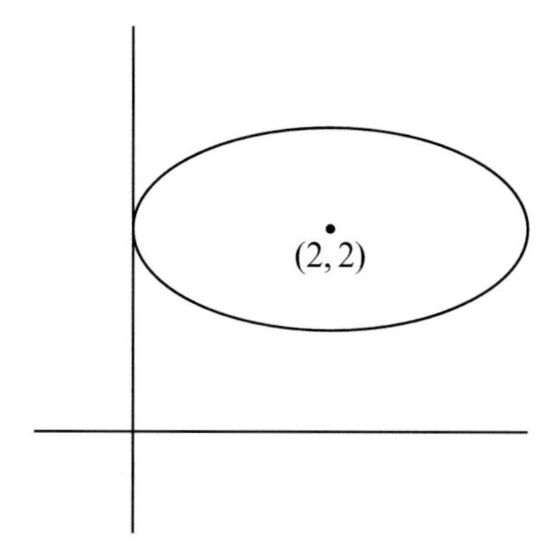

Rectangles A rectangle drawn with `\draw (1,1) rectangle (2,3);`

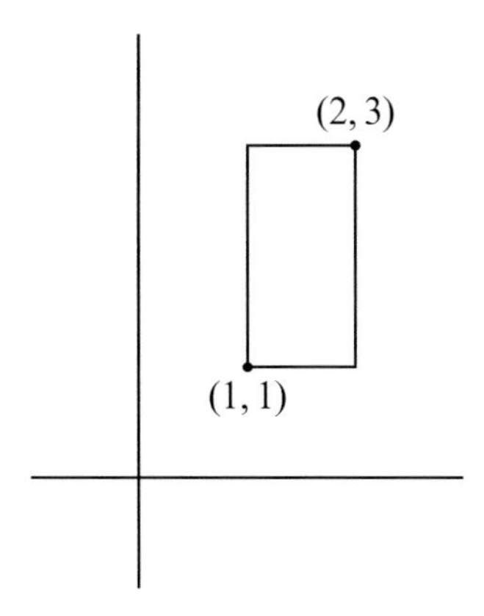

Arcs `\draw (2,1) arc[start angle=0, end angle=90, radius=1];` draws an arc of a circle:

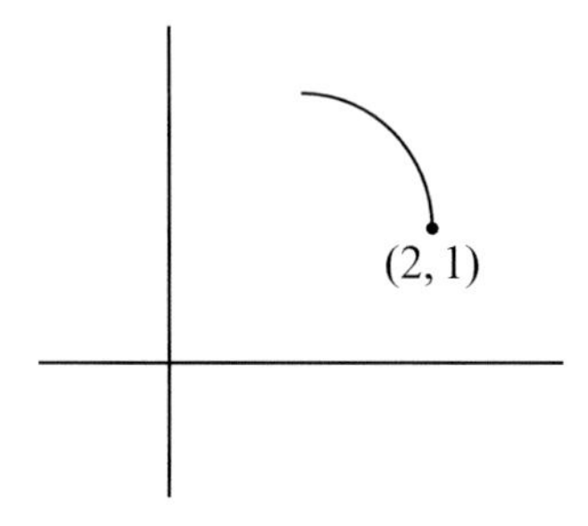

Surprise! The point is not the center of the circle.

Finally, an arc of an ellipse is drawn with the command

```
\draw (2,0) arc[x radius=1cm, y radius=5mm,
            start angle=0, end angle=120];
```

which typesets as

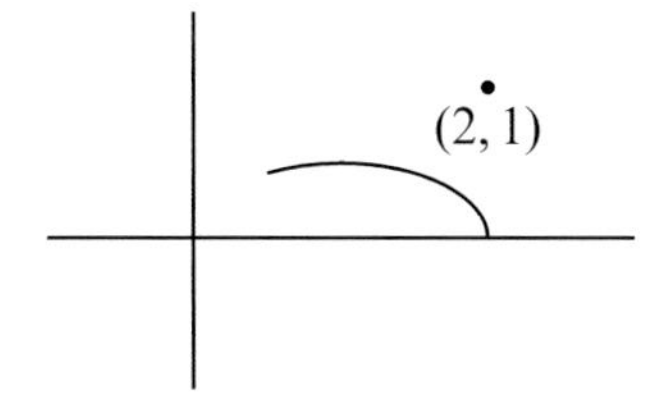

Smooth curves (Bézier curves) Nice curves can be drawn with a single control point (quadratic), as in

```
\draw (2,0)..controls (2,3)..(0,2);
\draw (2,0)..controls (4,2)..(0,2);
```

or with two control points (cubic):

```
\draw[dotted] (2,0)..controls (4,2) and (2,3)..(0,2);
```

These three curves typeset as

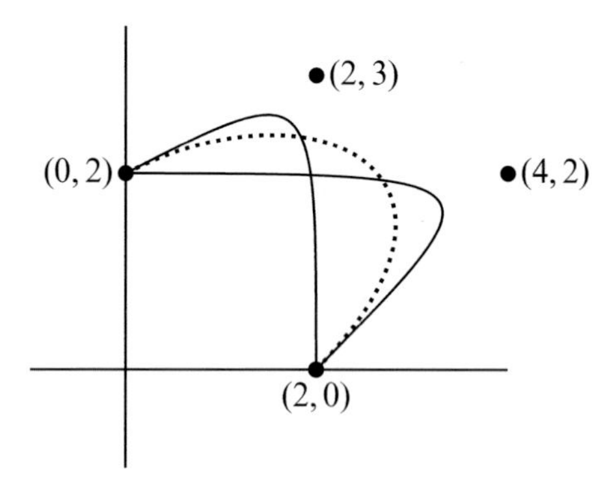

Alternatively, draw the curve defined by two points, A and B, the start and the end, and by the direction it leaves A and the direction it arrives at B with the command: `\draw[very thick] (0,0) to[out=90,in=195] (2,1.5);` This draws a curve from `(0,0)` to `(2,1)` which "leaves" at an angle of 90° and "arrives" at an angle of 195°:

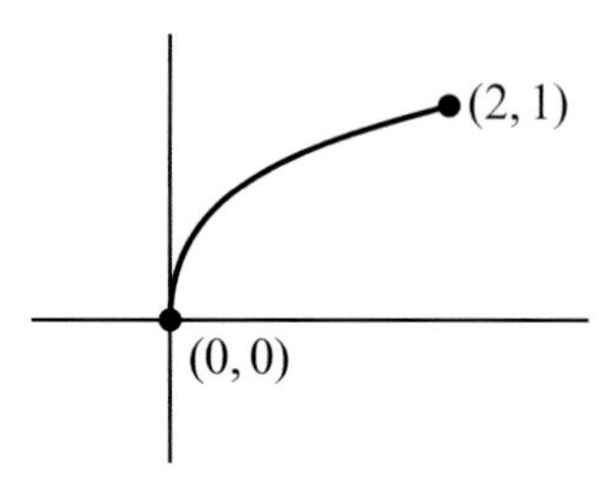

Labels We place text or formula in a picture with `\node at (1,1) {yes};`

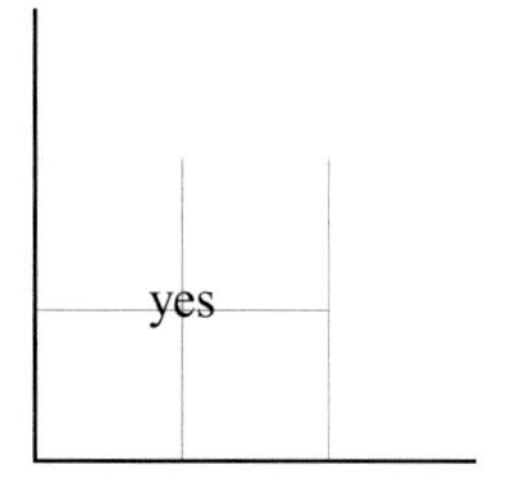

Notice how the "yes" is positioned relative to (1,1).

To place a label *below* a point, use the option `below`:

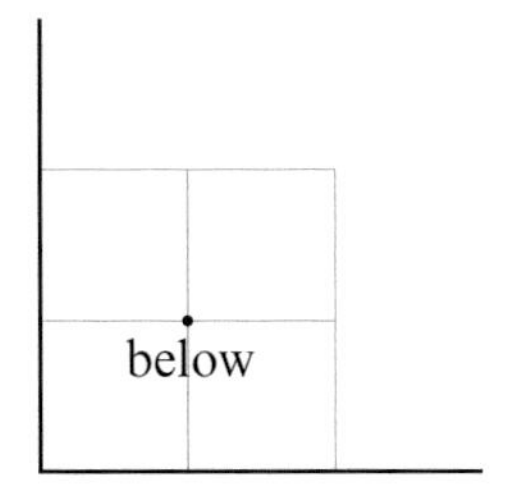

with the command `\node[below] at (1,1) {below};`. You can also use `above`, `left`, and `right`, and their combinations, for instance, `above left`.

If the text is several lines long, break it with \\ and tell Ti*k*Z how to align it:

```
\begin{mypicture}[xscale=1.3]
\draw[thick] (0,0)--(3,0);
\draw (0,-.2)--(0,.2);
\draw (3,-.2)--(3,.2);
\node[align=left, below] at (1.5,-.5)%
    {This text\\ is left justified};
\end{mypicture}
```

which typesets as

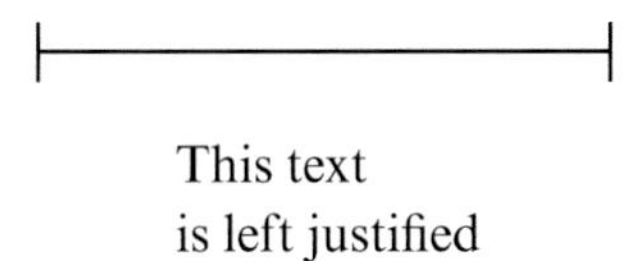

This text
is left justified

Text could be

- left justified, option: `align=left`;
- right justified, option: `align=right`;
- centered, option: `align=center`;

13.3 Transformations

We can rotate, translate, and scale the illustrations.

Rotations The command

`\draw[rotate=30] (1,1) rectangle (2,3);`

rotates the rectangle by 30° (around the origin) and

`\draw (2,2) ellipse[x radius=2, y radius=1, rotate=60];`

rotates an ellipse by 60° (around its center):

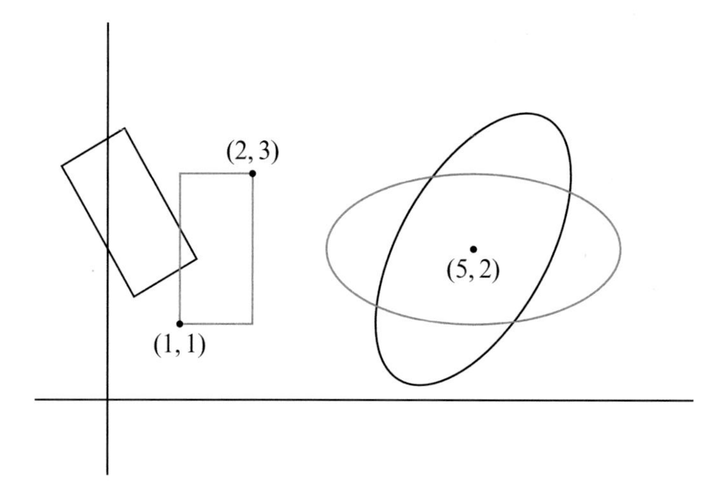

Translations `\draw[shift={(3,1)}] (1,1) rectangle (2,3);` shifts the rectangle:

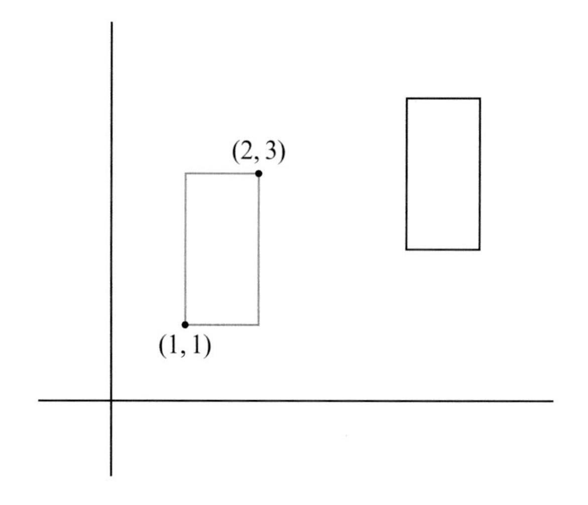

Scaling The command `\draw[scale=0.3] (1,1) rectangle (2,3);` scales the rectangle by 0.3 from the origin and
`\draw[scale around={0.5:(4.5,2)}] (4,1) rectangle (5,3);`
scales the rectangle by 2.5 using around the center of the rectangle, that is, around (4.5, 2):

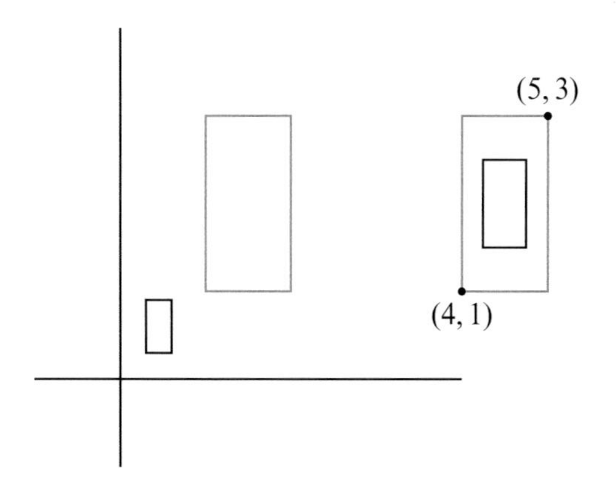

You can scale the two dimensions in different proportions:

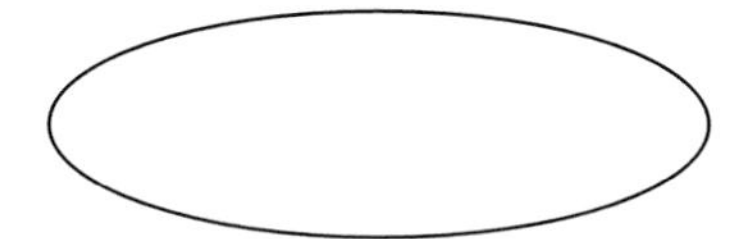

coded with

```
\begin{tikzpicture}[xscale=1.5, yscale=0.5]\\
\draw (2,2) circle[radius=1.5];\\
```

or with

```
\draw[xscale=1.5, yscale=0.5] (2,2) circle[radius=1.5];
```

13.4 Path attributes

A `\draw` command draws a path, with a start point and an end point. The start and end points of `\draw (1,0)--(2,1)--(1,2)--(0,1);` are `(1,0)` and `(0,1)`; of `\draw (2,2) circle[radius=1.5];` are `(2,2)` and `(2,2)`.

We consider now some of the common attributes.

Line width We have already seen the `line width=1pt` option of `\draw`. Ti*k*Z comes with seven additional built in widths: `ultra thin`, `very thin`, `thin`, `semithick`, `thick`, `very thick`, and `ultra thick`.

Dashes and dots You can also make dotted and dashed lines. The commands

```
\draw[dashed] (0,0.5)--(2,0.5);
\draw[dotted, thick] (0,0)--(2,0);
```

make two thick lines, one dashed and one dotted:

Colors Articles, as a rule, are printed black and white. But illustrations in PDF files and in presentations are shown in full color.

Ti*k*Z comes with the following colors ready to use: `red`, `green`, `blue`, `cyan`, `magenta`, `yellow`, `black`, `gray`, `darkgray`, `lightgray`, `brown`, `lime`, `olive`, `orange`, `pink`, `purple`, and `teal`.

The following example uses `white`, `lightgray`, and `gray`.

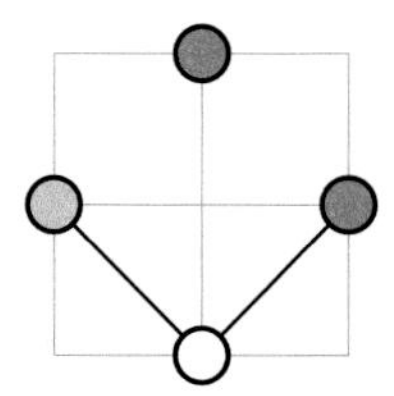

Arrows We can put arrows or bars on one or both ends of a path:

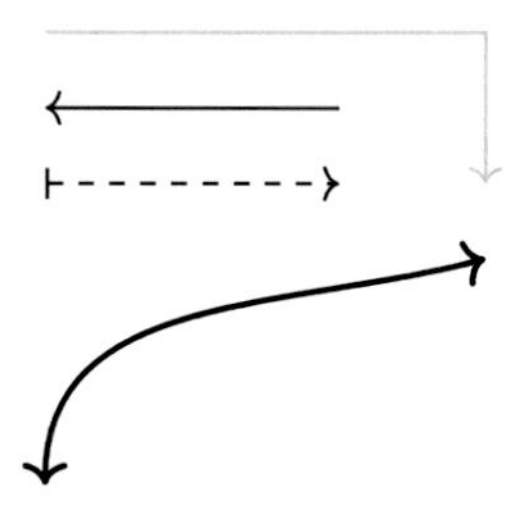

coded as

```
\begin{tikzpicture}
\draw[->, lightgray] (0,0)--(3,0)--(3,-1);
\draw[<-] (0,-0.5)--(2,-0.5);
\draw[|->, dashed] (0,-1)--(2,-1);
\draw[very thick, <->] (0,-3) to[out=90,in=195] (3,-1.5);
\end{tikzpicture}
```

Ti*k*Z provides you with dozens of arrows. You get them by invoking the arrows library with `\usepgflibrary{arrows}` in the preamble. Even better, use the `tikzcd` package by Florêncio Neves invoked by `\usepackage{tikz-cd}`. It is designed to code commutative diagrams and it provides arrows very close to the LaTeX style, see Figure 13.2. It comes with an excellent (and short) manual `tikz-cd-doc.pdf`.

In the `tikzcd` environment, the command `\arrow` produces an arrow. It takes one argument, a character `r`, `l`, `u`, or `d`, for right, left, up and down. A label is placed on an arrow as the second argument.

Here are two examples of commutative diagrams from the `tikz-cd` manual.

A basic example:

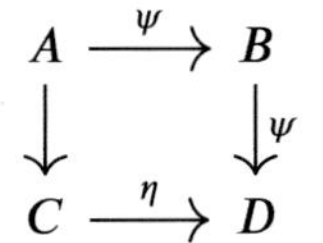

coded as

```
\begin{tikzcd}
  A \arrow{r}{\psi} \arrow{d}
    & B \arrow{d}{\psi} \\
  C \arrow{r}{\eta}
    & D
\end{tikzcd}
```

`rightarrow`	yields $\longrightarrow$
`leftarrow`	yields $\longleftarrow$
`leftrightarrow`	yields $\longleftrightarrow$
`dash`	yields $\relbar\joinrel\relbar$
`Rightarrow`	yields $\Longrightarrow$
`Leftarrow`	yields $\Longleftarrow$
`Leftrightarrow`	yields $\Longleftrightarrow$
`equal`	yields $=\joinrel=$
`mapsto` (or `maps to`)	yields $\longmapsto$
`mapsfrom`	yields $\longmapsfrom$
`hookrightarrow` (or `hook`)	yields $\hookrightarrow$
`hookleftarrow`	yields $\hookleftarrow$
`rightharpoonup`	yields $\rightharpoonup$
`rightharpoondown`	yields $\rightharpoondown$
`leftharpoonup`	yields $\leftharpoonup$
`leftharpoondown`	yields $\leftharpoondown$
`dashrightarrow` (or `dashed`)	yields $\dashrightarrow$
`dashleftarrow`	yields $\dashleftarrow$
`rightarrowtail` (or `tail`)	yields $\rightarrowtail$
`leftarrowtail`	yields $\leftarrowtail$
`twoheadrightarrow` (or `two heads`)	yields $\twoheadrightarrow$
`twoheadleftarrow`	yields $\twoheadleftarrow$
`rightsquigarrow` (or `squiggly`)	yields $\rightsquigarrow$
`leftsquigarrow`	yields $\leftsquigarrow$
`leftrightsquigarrow`	yields $\leftrightsquigarrow$

Figure 13.2: The arrows provided by the `tikz-cd` package

An example with curved and dashed arrows:

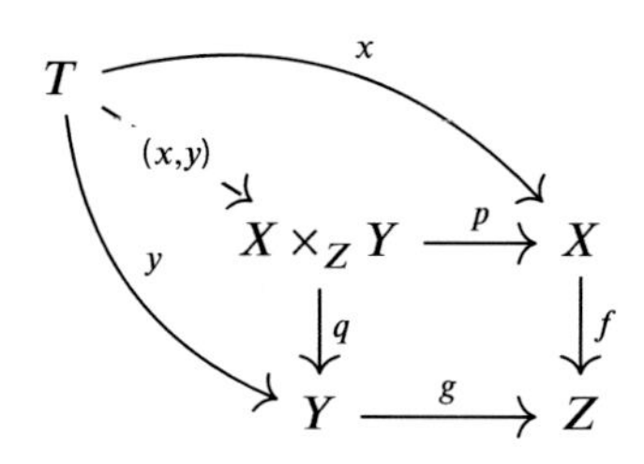


```
\begin{tikzcd}
  T
  \arrow[bend left]{drr}{x}
  \arrow[bend right]{ddr}{y}
  \arrow[dashed]{dr}[description]{(x,y)} & & \\
    & X \times_Z Y \arrow{r}{p} \arrow{d}{q}
      & X \arrow{d}{f} \\
    & Y \arrow{r}{g}
      & Z
\end{tikzcd}
```

13.5 Coding the example

To code the example on page 343, at the start of your article, following the

```
\usepackage{tikz}
```

line, define:

```
\tikzset{every picture/.style={line width=0.7pt}}
\newcommand{\mycircle}[1]
{\draw[fill=white,line width=1pt] (#1) circle[radius=1.0mm]}
```

Then the following code will produce the diagram:

```
\begin{figure}[hbt]
\centerline{\begin{tikzpicture}
%\draw[help lines] (0,0) grid (6,6);
\draw (2,0)--(0,3)--(4,5)--(6,2)--(2,0);
\draw (1.325,1)--(4.325,2.5);
\draw (6,2)--(4.325,2.5)--(4.66,4);
\mycircle{2,0};
\node at (2-.4,0) {$0_I$};
\mycircle{6,2};
\node at (6+.4,2) {$0_J$};
\mycircle{0,3};
```

```
\node at (0-.4,3) {$1_I$};
\mycircle{4,5};
\node at (4+.4,5) {$1_J$};
\mycircle{1.325,1};
\node at (1.325-.4,1) {$u$};
\mycircle{5.325,3};
\mycircle{4.325,2.5};
\mycircle{4.46,3};
\mycircle{4.67,4};
\node at (6,3.5) {$J$};
\node at (0,1.5) {$I$};
\node at (3.7,2.66) {$u \vee 0_J$};
\end{tikzpicture}}
\end{figure}
```

13.6 What did I leave out?

The Ti*k*Z manual is 726 pages, so this chapter covers maybe 2% of it. For most math illustrations, this chapter will do. (I still use Adobe Illustrator for my lattice diagrams. I use maybe 2% of Illustrator's power for my work.)

Ti*k*Z can plot. It can graph many built in functions, has a small programming language, has libraries, for instance, for circuits.

There is an excellent book on Ti*k*Z: Gérard Tisseau and Jacques Duma's Ti*k*Z *pur l'impatient* [70]. Although the book is in French, the Ti*k*Z code is in English; this book should be easy enough to read.

PART V

Customization

CHAPTER

14

Commands and environments

In Section C.1.2, we discuss how Donald E. Knuth designed TEX as a platform on which *convenient work environments* could be built. One such work environment, LATEX, predominates today, and it is indeed convenient.

Nevertheless, LATEX is designed for all of us, so it is not surprising that we could improve on it for our personal use. There are many reasons to customize LATEX:

Goal 1 to enhance the readability of the source file

Goal 2 to make notational and terminological changes easier

Goal 3 to redefine names used by LATEX

Goal 4 to introduce consistent layouts

G. Grätzer, *More Math Into* LATEX, DOI 10.1007/978-3-319-23796-1_14

There are many techniques to accomplish these.

Technique 1 Define commands and environments in order to enhance LaTeX to meet your particular needs (see Sections 14.1, 14.1.7, and 14.2).

Technique 2 Utilize delimited commands to write LaTeX documents in a more readable fashion (see Section 14.1.9).

Technique 3 Collect your frequently used commands into a command file (see Section 14.3).

Technique 4 Manipulate *counters*, integers—for instance, equation numbers and section numbers—and *length commands*, distance measurements—the `\voffset` command is an example (see Section 14.5).

Technique 5 Create customized list environments with the `list` environment (see Section 14.6).

Of course, there are many more reasons to customize and many more techniques to employ. We cover them in detail in this chapter.

We dedicate the last section to the pitfalls of customization (see Section 14.7). While the benefits of customization are great, there are many practices to avoid.

14.1 Custom commands

LaTeX provides hundreds of commands. Chances are good, however, that you still have specific needs that are not directly addressed by these commands. By judiciously adding *custom commands* (or *macros*) you can make your work more productive.

Custom commands follow the same rules as regular LaTeX commands (see Section 3.3.1).

14.1.1 Examples and rules

Commands to enhance readability

Let us start with a few examples of custom commands as shorthand for longer command(s) or text in order to enhance readability of the source file (Goal 1).

1. If you use the `\leftarrow` command a lot, you could define

   ```
   \newcommand{\larr}{\leftarrow}
   ```

   Then you would only have to type `\larr` to obtain a left arrow.

2. Instead of

   ```
   \widetilde{a}
   ```

you could simply type `\wtilda` after defining

```
\newcommand{\wtilda}{\widetilde{a}}
```

I show you how to define a generalized version of such a command in Section 14.1.2.

3. If you want to suppress the ligature in iff (see Section 3.4.6), you would normally have to type

   ```
   if\textcompwordmark f
   ```

   By defining a command `\Iff`,

   ```
   \newcommand{\Iff}{if\textcompwordmark f}
   ```

   you can type `\Iff` to get iff. We name this command `\Iff` because `\iff` is the symbol $\iff$ (see Section A.4).

4. If you use the construct $D^{[2]} \times D^{[3]}$ often, you could introduce the `\DxD` (D times D) command,

   ```
   \newcommand{\DxD}{D^{[2]}\times D^{[3]}}
   ```

   and then type `\DxD` instead of the longer, and hard to read, version throughout your document—serves also Goal 2.

5. If you want to get a backslash in typewriter style, you would normally have to type (see Section 3.4.4)

   ```
   \texttt{\symbol{92}}
   ```

   Instead, you can introduce the `\bsl` command,

   ```
   \newcommand{\bsl}{\texttt{\symbol{92}}}
   ```

   and `\bsl` typesets as `\`.

6. You can also use commands as a shorthand for text. For instance, if you use the phrase `subdirectly irreducible` many times in your document, you could define

   ```
   \newcommand{\subdirr}{subdirectly irreducible}
   ```

   `\subdirr` is now shorthand for `subdirectly irreducible`, which typesets as subdirectly irreducible.

Tip With modern editors, the need to have custom commands as shorthand is reduced. Most editors have "command completion" or "phrase completion". For instance, in `TeXShop`, type the first few letters of a word and hit the escape key. The remaining letters are entered to match the first entry in the completion dictionary. Hitting escape again cycles through all possible completions. To make this feature useful, you have to customize the completion dictionary. Note that command completion minimizes the number of characters you have to type, but it does not improve the readability of the source.

Rule ■ Custom commands

1. Issue the `\newcommand` command.
2. In braces, type the name of your new command, for example, `\subdirr`, including the backslash (\).
3. In a second pair of braces, define the command, in this example, `subdirectly irreducible`.
4. Use the command as `\subdirr\␣` or `\subdirr{}` before a space, before an alphabetical character as `\subdirr{}`, and `\subdirr` otherwise.

Examples for Rule 4. For subdirectly irreducible lattice type

```
\subdirr{} lattice
```

or

```
\subdirr\ lattice
```

and not `\subdirr lattice`. Indeed, typesetting `\subdirr lattice` results in subdirectly irreduciblelattice. By the first spacing rule, `\subdirr␣lattice` is not any better (see Section 3.2.1). If you want subdirectly irreducibles, you must use the `\subdirr{}` form. Indeed, `\subdirr{}s` typesets as subdirectly irreducibles.

Using new commands

It is good practice to place custom commands in the preamble of your document or in a command (style) file you load with a `\usepackage` command (see Section 14.3)—provided that you do not submit to a journal that does not allow this. Then you always know where to look for the command definitions. An exception is a custom command that you want to restrict to a part of the document. Delimit the segment with braces and define the custom command within those braces (see Section 3.3.2). Instead of a pair of braces, you can use `\begingroup` and `\endgroup`, which is easier to see. Section 14.2.5 recommends yet another approach.

Tip

- If errors occur, isolate the problem. Comment out the custom commands and reintroduce them one at a time.
- LaTeX only checks whether the braces match in the command definition. Other mistakes are found only when the command is used.

For instance, if you define a command with a spelling error

```
\newcommand{\bfA}{\textf{A}}
```

then at the first use of `\bfA` you get the message

```
! Undefined control sequence.
\bfA ->\textf
              {A}
```

Note that LaTeX is not complaining about `\bfA` but about the misspelled `\textbf` command in the definition of `\bfA`.

Be careful not to define a custom command with a name that is already in use. If you do, you get a message such as

```
! LaTeX Error: Command \larr already defined.
```

To correct the error, replace the command name with a new one. On the other hand, if you need to replace an existing command, you have to *redefine* it. See Section 14.1.5 for how to do so.

Tip Use spaces to make your source files more readable, but avoid them in definitions.

For example, you may type

```
$D^{ \langle 2 \rangle } + 2 = x^{ \mathbf{a} }$
```

This may help you see how the braces match, easily identify relations and operations, and so on. *Do not add these spaces in command definitions* because it may result in unwanted spaces in your typeset document. You may start a new line to increase the readability of a command definition, provided that you terminate the previous line with %. For instance, borrowing an example from page 370:

```
\newcommand{\Xquotphi}[2]{%
   \dfrac{\varphi \cdot X_{n, #1}}%
   {\varphi_{#2}\times \varepsilon_{#1}}}
```

Tip In the definition of a new command, command declarations need an extra pair of braces (see Section 3.3.3).

Say you want to define a command that typesets the warning: *Do not redefine this variable!* It is very easy to make the following mistake:

```
\newcommand{\Warn}{\em Do not redefine this variable!}
```

`\Warn` typesets the warning emphasized, but everything that follows the warning is also emphasized (more precisely, until the end of the `\Warn` command's scope). Indeed, `\Warn` is replaced by `\em Do not redefine this variable!` so the effect of `\em` goes beyond the sentence to the next closing brace.

The correct definition is

```
\newcommand{\Warn}{{\em Do not redefine this variable!}}
```

Even simpler, you could use a command with an argument

```
\newcommand{\Warn}{\emph{Do not redefine this variable!}}
```

Tip There are limits as to what can be done with custom commands. For instance, you cannot introduce `\bal` for `\begin{align}` and `\eal` for `\end{align}`. So if you want to introduce a new type of custom command, do just one and try it out.

The `xspace` *package*

Rule 4 (on page 364) is the source of many annoying problems in LaTeX. David Carlisle's `xspace` package (see Section 10.3.1) helps eliminate such problems. In the preamble, load the package with

```
\usepackage{xspace}
```

Whenever you define a command that may have such problems, add the `\xspace` command to the definition. For instance, define `\subdirr` as

```
\newcommand{\subdirr}{subdirectly irreducible\xspace}
```

Then all the following typesets `subdirectly irreducible lattice` correctly:

```
\subdirr\␣lattice
\subdirr{}␣lattice
\subdirr␣lattice
```

Note that \xspace does not add space if followed by a punctuation mark, so to get

the lattice is subdirectly irreducible.

type

```
the lattice is \subdirr.
```

Tip Be careful not to use \xspace twice in a definition.

For instance, if you define

```
\newcommand{\tex}{\TeX\xspace}
\newcommand{\bibtex}{\textsc{Bib}\kern-.1em\tex\xspace}% Bad!!!
```

then

```
\bibtex, followed by a comma
```

typesets as

BibTeX , followed by a comma

The correct definitions are

```
\newcommand{\tex}{\TeX\xspace}
\newcommand{\bibtex}{\textsc{Bib}\kern-.1em\TeX\xspace}% Correct!
```

Of course, if you want to get TeXbook, you cannot use the \xspace variant definition: \tex.

Ensuring math

The \ensuremath command is useful for defining commands for both text and math mode. Suppose you want to define a command for $D^{\langle 2\rangle}$. If you define it as

```
\newcommand{\Dsq}{D^{\langle2\rangle}}
```

then you can use the command in math mode, but not in text mode. If you define it as

```
\newcommand{\Dsq}{$D^{\langle2\rangle}$}
```

then it works in text mode, but not in math mode. Instead, define this command as

```
\newcommand{\Dsq}{\ensuremath{D^{\langle2\rangle}}}
```

Then \Dsq works correctly in both contexts.

This example also shows the editorial advantages of custom commands. Suppose the referee suggests that you change the notation to $D^{[2]}$. To carry out the change you only have to change one line:

```
\newcommand{\Dsq}{\ensuremath{D^{[2]}}}
```

It is hard to overemphasize the importance of this example. You may want to change notation because:

- you found a better notation;
- your coauthor insists;
- your article appears in a conference proceedings, and the editor wants to unify the notation;
- you are reusing the code from this article in another one or in a book, where the notation is different.

See also the discussion of the \TextOrMath command on page 293.

14.1.2 Arguments

Arguments of custom commands work the same way as for LaTeX commands (see page 52). Define

```
\newcommand{\fsqAB}{(f^2)^{[[\frac{A^2}{B-1}]]}}
```

Then \fsqAB typesets as $(f^2)^{[[\frac{A^2}{B-1}]]}$ in a math formula. If you want to use \fsqAB in math and also by itself in text, define it with \ensuremath, as

```
\newcommand{\fsqAB}{\ensuremath{(f^2)^
                {[[\frac{A^2}{B-1}]]}}}
```

However, if you use this construct for many functions f, then you may need a generalized command, such as

```
\newcommand{\sqAB}[1]{\ensuremath{(#1^2)^
                   {[[\frac{A^2}{B-1}]]}}}
```

Now \sqAB{g} typesets $(g^2)^{[[\frac{A^2}{B-1}]]}$. The form of this \newcommand is the same as before, except that after the name of the command in braces, {\sqAB}, we specify the number of arguments in brackets (in this example, [1]). Then we can use #1 in the definition of the command. When the command is invoked, the argument you provide replaces #1 in the definition. Typing $\sqAB{q}$ results in the formula $(q^2)^{[[\frac{A^2}{B-1}]]}$, while $\sqAB{r}$ gives $(r^2)^{[[\frac{A^2}{B-1}]]}$.

Notice how these examples disrupt the normal spacing between lines—a practice to avoid!

A custom command may have up to nine arguments, numbered 1–9.

The simplest examples just allow you to invoke an existing command under a new name. For instance, the `\eqref` command introduced in Section 5.3 to reference equations (the equation number upright, enclosed in parentheses), would also be useful to reference items for the custom list environment `enumeratei` introduced in Section 14.2.1—see Example 3. Indeed, for the `enumeratei` environment, we want references to items to be typeset upright, enclosed in parentheses. So if the first item has label `First`, we could reference it with `\eqref{First}`, which typesets as (i). But `\eqref{First}` seems awkward and inappropriate; we are referencing an item not an equation.

So define

```
\newcommand{\itemref}[1]{\eqref{#1}}
```

and now we can reference the first item with `\itemref{First}`, which typesets as (i).

Following are some simple examples of custom commands with arguments.

1. In the preamble of the source file for this book, I defined

   ```
   \newcommand{\env}[1]{\textnormal{\texttt{#1}}}
   ```

   In this example, the `\env` command is used to typeset environment names. So the environment name `center` is typed as

   ```
   \env{center}
   ```

   Again the editorial advantage is obvious. If the editor wants the environment names set in sans serif, only one line in the book has to be changed to alter every occurrence of a typeset environment name:

   ```
   \newcommand{\env}[1]{\textsf{#1}}
   ```

2. An argument (e.g., `#1`) may occur more than once in a definition. A natural example is provided by the `\index` command (see Section 16.1). Typically, if you wanted to include a phrase, say subdirectly irreducible lattice, in your index, you would have to type

   ```
   this proves that $L$ is a subdirectly irreducible lattice
   ```

   You could instead define an "index it" command such as

   ```
   \newcommand{\indexit}[1]{#1\index{#1}}
   ```

   The argument of this command is a phrase to be both typeset and included in the index. Using this command, you can type

   ```
   it is a \indexit{subdirectly irreducible lattice}
   ```

If you want all such index entries to be typeset in italics, then \indexit should be defined as

```
\newcommand{\indexit}[1]{#1\index{#1@\textit{#1}}}
```

in which #1 occurs three times. (See Chapter 16 for more information about index commands.)

3. Let us define a command with three arguments for congruences:

```
\newcommand{\congr}[3]{#1\equiv#2\pod{#3}}
```

Now type $\congr{a}{b}{\theta}$ to typeset $a \equiv b \pod{\theta}$. In Section 14.1.9, I present another command for typesetting congruences.

4. In the secondarticle.tex article (see Section 9.3), there are a lot of vectors with only one nonzero entry:

$$\langle\dots,0,\dots,\overset{i}{d},\dots,0,\dots\rangle$$

the i above the d indicates that it is the ith component of the vector. A command \vectsup, a vector with a superscript, producing this symbol can be defined as

```
\newcommand{\vectsup}[2]{\langle\dots,0,\dots,
\overset{#1}{#2},\dots,0,\dots\rangle}
```

\vectsup{i}{d} in a math formula now produces $\langle\dots,0,\dots,\overset{i}{d},\dots,0,\dots\rangle$.

Formula 20 of the *Formula Gallery* (Section 5.10),

$$\mathbf{A} = \begin{pmatrix} \dfrac{\varphi \cdot X_{n,1}}{\varphi_1 \times \varepsilon_1} & (x+\varepsilon_2)^2 & \cdots & (x+\varepsilon_{n-1})^{n-1} & (x+\varepsilon_n)^n \\ \dfrac{\varphi \cdot X_{n,1}}{\varphi_2 \times \varepsilon_1} & \dfrac{\varphi \cdot X_{n,2}}{\varphi_2 \times \varepsilon_2} & \cdots & (x+\varepsilon_{n-1})^{n-1} & (x+\varepsilon_n)^n \\ \hdotsfor{5} \\ \dfrac{\varphi \cdot X_{n,1}}{\varphi_n \times \varepsilon_1} & \dfrac{\varphi \cdot X_{n,2}}{\varphi_n \times \varepsilon_2} & \cdots & \dfrac{\varphi \cdot X_{n,n-1}}{\varphi_n \times \varepsilon_{n-1}} & \dfrac{\varphi \cdot X_{n,n}}{\varphi_n \times \varepsilon_n} \end{pmatrix} + \mathbf{I}_n$$

is a good candidate for custom commands. By defining

```
\newcommand{\Xquotphi}[2]{%
   \dfrac{\varphi \cdot X_{n, #1}}%
   {\varphi_{#2}\times \varepsilon_{#1}}}
\newcommand{\exn}[1]{(x+\varepsilon_{#1})^{#1}}
```

the two new commands,

```
\[
   \Xquotphi{2}{3} \qquad \exn{n}
\]
```

are typeset as

$$\frac{\varphi \cdot X_{n,2}}{\varphi_3 \times \varepsilon_2} \qquad (x + \varepsilon_n)^n$$

With these custom commands, you can rewrite Formula 20 as follows:

```
\[
   \mathbf{A} =
   \begin{pmatrix}
      \Xquotphi{1}{1} & \exn{2} & \cdots & \exn{n - 1}+
                      & \exn{n}\\[10pt]
      \Xquotphi{1}{2} & \Xquotphi{2}{2}  & \cdots
       & \exn{n - 1}  &\exn{n}\\
      \hdotsfor{5}\\
      \Xquotphi{1}{n} & \Xquotphi{2}{n}  & \cdots &
  \Xquotphi{n - 1}{n} & \Xquotphi{n}{n}
   \end{pmatrix}
    + \mathbf{I}_{n}
\]
```

Observe how much shorter this form is than the version shown in the *Formula Gallery* and how much easier it is to read. It is also easier to reuse in a subsequent article.

14.1.3 Short arguments

There are three ways of defining new commands:

`\newcommand` `\renewcommand` `\providecommand`

We take up the last two in Section 14.1.5. They define commands that can take any number of paragraphs as arguments. The ∗-ed versions of these commands define *short* commands (see Section 3.3.3) that take a block of text that contains no paragraph break as an argument. For instance,

```
\newcommand{\bigbold}[1]{{\large\bfseries#1}}
```

makes its argument large and bold. So

```
\bigbold{First paragraph.

Second paragraph.

}
```

prints

First paragraph.
Second paragraph.

as expected. On the other hand, if you define

```
\newcommand*{\bigbold}[1]{{\large\bfseries#1}}
```

and then attempt to typeset the previous example, you get the message

```
Paragraph ended before \bigbold was complete.
<to be read again>
                   \par
                   \par
                   \par
```

Short commands are often preferable because of their improved error checking.

14.1.4 *Optional arguments*

You can define a command whose first argument is *optional,* and provide a *default value* for this optional argument. To illustrate, let us define the command

```
\newcommand{\SimpleSum}{a_{1}+a_{2}+\dots+a_{n}}
```

`$\SimpleSum$` now produces $a_1 + a_2 + \dots + a_n$. Now we change this command so that we can sum from 1 to m if necessary, with n as the default:

```
\newcommand{\BetterSum}[1][n]{a_{1}+a_{2}+\dots+a_{#1}}
```

`$\BetterSum$` still produces $a_1 + a_2 + \dots + a_n$, but `$\BetterSum[m]$` typesets as $a_1 + a_2 + \dots + a_m$.

A `\newcommand` may have up to nine arguments, but *only the first* may be optional. The following command has two arguments, one optional:

```
\newcommand{\BestSum}[2][n]{#2_{1}+#2_{2}+\dots+#2_{#1}}
```

Now

`$\BestSum{a}$`	typesets as	$a_1 + a_2 + \dots + a_n$
`$\BestSum{b}$`	typesets as	$b_1 + b_2 + \dots + b_n$
`$\BestSum[m]{c}$`	typesets as	$c_1 + c_2 + \dots + c_m$

14.1.5 Redefining commands

LaTeX makes sure that you do not inadvertently define a new command with the same name as an existing command (see, for example, page 365). Assuming that you have already defined the \larr command as in Section 14.1.1 (to typeset $\leftarrow$), to *redefine* \larr, use \renewcommand:

```
\renewcommand{\larr}{\Longleftarrow}
```

and now \larr typesets as $\Longleftarrow$.

Tip Use the \renewcommand command sparingly and make sure that you understand the consequences of redefining an existing command. Redefining LaTeX commands may cause LaTeX to behave in unexpected ways, or even crash.

Blind redefinition is the route to madness.

See also the discussion in Section 14.7.

You can also use \renewcommand to redefine commands defined by LaTeX or any package. For instance, the end of proof symbol, \qedsymbol, used by the proof environment, can be changed to the solid black square some people prefer (defined in the amssymb package) with the command

```
\renewcommand{\qedsymbol}{$\blacksquare$}
```

Even better, define

```
\renewcommand{\qedsymbol}{\ensuremath{\blacksquare}}
```

so that you can use \qedsymbol in both text and math mode. Section 14.1.7 has more on redefining names.

\renewcommand has a companion, \providecommand. If the command it defines has already been defined, the original command is left unchanged. Otherwise, the \providecommand command acts exactly like \newcommand. For instance, the \bysame command (see Section 8.5.1, page 250) is defined in some document classes as

```
\newcommand{\bysame}{\makebox[3em]{\hrulefill}\thinspace}
```

If you want to use the \bysame command in your bibliography and include this definition in your document, LaTeX generates a message when you typeset your document using a document class that already defines \bysame (all AMS document classes do). However, if you define \bysame in your document using \providecommand:

```
\providecommand{\bysame}%
{\makebox[3em]{\hrulefill}\thinspace}
```

the \bysame command typesets correctly whether or not the document class defines it.

14.1.6 Defining operators

The powerful `\DeclareMathOperator` command defines a new operator:

```
\DeclareMathOperator{\opCommand}{opName}
```

Invoke the new operator with `\opCommand`, which is then typeset with *opName*.

Rule ■ The `\DeclareMathOperator` command must be placed in the preamble.

For example, to define the operator Truncat, invoked by the command `\Trunc`, place this in the preamble:

```
\DeclareMathOperator{\Trunc}{Truncat}
```

An operator is typeset in math roman with a little space after it, so `$\Trunc A$` typesets as $\operatorname{Truncat} A$.

The second argument is typeset in math mode but - and * are typeset as they would be in text. Here are some more examples. Define in the preamble two operators:

```
\DeclareMathOperator{\Trone}{Trunc_{1}}
\DeclareMathOperator{\Ststar}{Star-one*}
```

Then in the body of the article
`$\Trone A$` is typeset as $\operatorname{Truncat}_1 A$
`$\Ststar A$` is typeset as $\operatorname{Star-one*} A$

To define an operator with limits, use the *-ed form

```
\DeclareMathOperator*{\doublesum}{\sum\sum}
```

and then (see Section 5.6.4 for multiline subscripts)

```
\[
    \doublesum_{\begin{subarray}{l}
            i^2+j^2 = 50\\
            i,\ j \leq 10
            \end{subarray}}
            \frac{x^i + y^j}{(i + j)!}
\]
```

typesets as

$$\sum\sum_{\begin{subarray}{l} i^2+j^2 = 50\\ i,\ j \leq 10 \end{subarray}} \frac{x^i + y^j}{(i + j)!}$$

14.1.7 Redefining names

A number of names, such as Table, List of Tables, Abstract, and so on, are typeset in your document by LaTeX. You can easily change these names.

For instance, if you are preparing your manuscript for the proceedings of a meeting, and Abstract has to be changed to Summary, you can do so with

```
\renewcommand{\abstractname}{Summary}
```

Table 14.1 lists the commands that define such names in various document classes, along with their default definitions and the major document classes using the commands. It is easy to check whether your document class defines such a command, simply open the appropriate `cls` file and search for the command.

If your document has photographs rather than figures, you could redefine

```
\renewcommand{\figurename}{Photograph}
\renewcommand{\listfigurename}{List of Photographs}
```

14.1.8 Showing the definitions of commands

If you are defining a new command with `\newcommand` and an error message informs you that the command name is already in use, then it may be useful to find out the existing definition of the command. For instance, the `\vectsup` command is defined in `secondarticleccom.tex` (in the `samples` folder and in Section 14.4). If you called this new command `\vec`, you would get the message

```
! LaTeX Error: Command \vec already defined.
```

You can find out the definition of the `\vec` command by getting into interactive mode (see Section C.3) and typing

```
*\show \vec
```

LaTeX responds with

```
> \vec=macro:
->\mathaccent "017E .
<*> \show \vec
```

informing you that `\vec` is a command, and, specifically, a math accent (see Sections 5.7 and A.8). Now try `\hangafter` (see Section 3.7.2):

```
*\show \hangafter
```

```
> \hangafter=\hangafter.
```

Command	Default Value	Defined by Document Class
\abstractname	Abstract	aa, ab, ap, a, p, r
\appendixname	Appendix	aa, ab, ap, a, b, r
\bibname	Bibliography	ab, b, r
\ccname	Cc	l
\chaptername	Chapter	ab, b, r
\contentsname	Contents	aa, ab, ap, a, b, r
\datename	Date	aa, ab, ap
\enclname	Enclosure	l
\figurename	Figure	aa, ab, ap, a, b, r
\headtoname	To	l
\indexname	Index	aa, ab, ap, a, b, r
\keywordsname	Key words and phrases	aa, ab, ap
\listfigurename	List of Figures	aa, ab, ap, a, b, r
\listtablename	List of Tables	aa, ab, ap, a, b, r
\pagename	Page	l, p
\partname	Part	aa, ab, ap, a, b, r
\proofname	Proof	aa, ab, ap
\refname	References	aa, ap, a
\see	see	aa, ab, ap
\seealso	see also	aa, ab, ap
\subjclassname	1991 Mathematics Subject Classification	aa, ab, ap
\subjclassname[2010]	2010 Mathematics Subject Classification	aa, ab, ap
\tablename	Table	aa, ab, ap, a, b, r

Document class codes: aa amsart, ab amsbook, ap amsproc, a article, b book, l letter, p proc, and r report

Table 14.1: Redefinable name commands in LaTeX.

The response indicates that \hangafter is a *primitive,* defined by TeX itself. Redefining a primitive is not a good idea.

Try one more command, \medskip (see Section 3.8.2), to find out how large it is:

```
*\show \medskip
> \medskip=macro:
->\vspace \medskipamount .
```

The third line indicates that the length is stored in `\medskipamount`. If we use `\show` to ask what `\medskipamount` is defined to be:

```
*\show \medskipamount
> \medskipamount=\skip14.
```

we do not get a very useful answer. `\medskipamount` is unlike most of the commands you have seen so far. It is a *length command* (see Section 14.5.2), containing the value of `\medskip`. You can ask for the value of a length command (or parameter) with the `\showthe` command:

```
*\showthe \medskipamount

> 6.0pt plus 2.0pt minus 2.0pt.
```

So `\medskip` is a vertical space of 6 points that can stretch or shrink by up to 2 points.

LaTeX has many registers that contain numbers:

- counters containing integers, such as `3`
- dimensions such as `10.2pt`, for example, `\textwidth` (see Section 8.6)
- lengths, written in the form `6.0pt plus 2.0pt minus 2.0pt`, also called a *glue* or a *rubber length* (see Sections 14.5.2 and C.2.2)

Use the `\showthe` command to display the value for any of these registers.

You can also type the `\show` and `\showthe` commands directly into your document rather than go into interactive mode. LaTeX's response appears in the `log` window, and is saved into the `log` file.

14.1.9 Delimited commands

You can define new commands in TeX using characters and symbols to delimit arguments. Such *delimited commands* provide a way to write more readable source documents.

First we have to learn how to define a command using TeX's `\def` command. Type `\def`, followed by the new command name (not in braces), then the definition in braces. For example, the first command defined in Section 14.1.1,

```
\newcommand{\larr}{\leftarrow}
```

could be typed

```
\def\larr{\leftarrow}
```

TeX's `\def` command does not check whether a new command name is already in use, so `\def` behaves differently from the LaTeX's `\newcommand`, `\renewcommand`, and `\providecommand` (see Section 14.1.5). If the `\larr` command was defined previously, the original definition is overwritten.

Tip It is your responsibility to ensure that your command name is unique when you define a command using \def. LaTeX provides no protection. Use the techniques introduced in Section 14.1.8 to check a name before you define a command with \def.

Now we can start discussing delimited commands with a simple example, defining a command for vectors:

```
\def\vect<#1>{\langle#1\rangle}
```

Note that \vect is a command with one argument, #1. When invoked, it typesets $\langle$, the argument, and then $\rangle$.

In the definition of \vect, the argument #1 is delimited by < and >. When the command is invoked, the argument must be delimited the same way. So to typeset the vector $\langle a, b\rangle$, we invoke \vect with

```
\vect<a,b>
```

This looks somewhat like a vector, and the name \vect serves as a reminder.

You have to be careful with delimited commands because the math spacing rules (see Section 5.2) do not hold in either the definition or the invocation. So if there is a space before #1, in the definition of \vect,

```
\def\vect< #1>{\langle#1\rangle}
```

then $\vect<a,b>$ results in the message

```
! Use of \vect doesn't match its definition.
l.12 $\vect<a
             ,b>$
```

which is clear enough. If the space is on the other side of the #1, as in

```
\def\vect<#1 >{\langle#1\rangle}
```

the message is slightly more confusing:

```
Runaway argument?
a,b>$
! Paragraph ended before \vect was complete.
<to be read again>
                   \par
```

The moral is that if you use delimited commands, you must be very careful that each invocation exactly matches the definition. /Users/gratzer-new/Dropbox/MiL5/Sample files/newlattice.sty In Example 3 of Section 14.1.2, we introduced a command with three arguments for typing congruences:

```
\newcommand{\congr}[3]{#1\equiv#2\pod{#3}}
```

`$\congr{a}{b}{\theta}$` produces $a \equiv b \ (\theta)$. This command is easy to remember, but it does not make the source file more readable. For that, we use a delimited command. Let us redo the congruence example with a delimited command

```
\def\congr#1=#2(#3){#1\equiv#2\pod{#3}}
```

so that `$\congr a=b(\theta)$` produces $a \equiv b \ (\theta)$. In the source document, the formula `\congr a=b(\theta)` looks a bit like the typeset congruence and it is easier to read. I included this definition in the `newlattice.sty` command file (see Section 14.3).

There is only one catch. Suppose you want to typeset the formula

$$x = a \equiv b \quad (\theta)$$

If you type `$\congr x=a=b(\theta)$`, LaTeX typesets it as $x \equiv a = b \ (\theta)$. Indeed, `x` is delimited on the right by the first =, so LaTeX believes that the first argument is `x`. The second argument is delimited by the first = and the left parenthesis, so it is `a=b`. In such cases, you can help LaTeX find the correct first argument by enclosing it in braces:

```
$\congr{x=a}=b(\theta)$
```

Here is our final example. In Section 3.3.1 we discuss the problem of typing a command such as `\TeX` (the example there was `\today`) in the form `\TeX\␣` so that it is typeset as a separate word. The problem is that if you type `\TeX` without the trailing `\␣`, TeX is merged with the next word, and there is no message to warn you. One solution is to use a delimited command:

```
\def\tex/{\TeX}
```

Now to get TeX, type `\tex/`. If a space is needed after it, type `\tex/␣`. If you forget the closing `/`, you get a message.

A better solution to this problem is the use of the `xspace` package—provided you do not want to typeset something like TeXbook (see Section 14.1.1). However, many documents use the delimited construct (including the AMS documentation), so you should be familiar with it.

14.2 Custom environments

Most custom commands are new commands. *Custom environments,* as a rule, are built on existing environments. We start with such custom environments (Section 14.2.1) and then proceed to investigate

- arguments (Section 14.2.2)
- optional arguments (Section 14.2.3)
- short arguments (Section 14.2.4)

Finally, we discuss how to define brand-new environments (Section 14.2.5).

14.2.1 Modifying existing environments

If you do not like the name of the proof environment and would prefer to use the name demo, define

```
\newenvironment{demo}
   {\begin{proof}}
   {\end{proof}}
```

Note that this does not change how the environment is typeset, only the way it is invoked.

To modify an existing environment, oldenv, type

```
\newenvironment{name}
   {begin_text}
   {end_text}
```

where *begin_text* contains the command \begin{*oldenv*} and *end_text* contains the command \end{*oldenv*}.

Tip Do not give a new environment the name of an existing command or environment.

For instance, if you define

```
\newenvironment{parbox}
   {...}
   {...}
```

you get the message

```
! LaTeX Error: Command \parbox already defined.
```

If there is an error in such a custom environment, the message generated refers to the environment that was modified, not to your environment. For instance, if you misspell proof as prof when you define

```
\newenvironment{demo}
   {\begin{prof}}
   {\end{proof}}
```

then *at the first use* of the demo environment you get the message

```
! LaTeX Error: Environment prof undefined.

l.13 \begin{demo}
```

If you define

```
\newenvironment{demo}
   {\begin{proof}\em}
   {\end{prof}}
```

at the first use of demo you get the message

```
! LaTeX Error: \begin{proof} on input line 5
        ended by \end{prof}.
l.14 \end{demo}
```

Here are four more examples of modified environments.

1. The command

   ```
   \newenvironment{demo}
      {\begin{proof}\em}
      {\end{proof}}+
   ```

   defines a demo environment that typesets an emphasized proof. Note that the scope of \em is the demo environment.

2. The following example defines a very useful environment. It takes an argument to be typeset as the name of a theorem:

   ```
   \newtheorem*{namedtheorem}{\theoremname}
   \newcommand{\theoremname}{testing}
   \newenvironment{named}[1]{
      \renewcommand{\theoremname}{#1}
      \begin{namedtheorem}}
      {\end{namedtheorem}}
   ```

   For example,

   ```
   \begin{named}{Name of the theorem}
   Body of theorem.
   \end{named}
   ```

   produces

   Name of the theorem. *Body of theorem.*

   in the style appropriate for the \newtheorem* declaration. This type of environment is often used to produce an unnumbered **Main Theorem** (see Section 14.4) or when typesetting an article or book in which the theorem numbering is already fixed, for instance, when publishing a book in LaTeX that was originally typeset by another typesetting system.

3. In Sections 4.2.4 and 10.3.1, we came across the enumerate package, which allows you to customize the enumerate environment. If the enumerate package is loaded, you can invoke the enumerate environment with an optional argument specifying how the counter should be typeset, for instance, with the option [\upshape (i)],

```
\begin{enumerate}[\upshape (i)]
   \item First item\label{First}
\end{enumerate}
```

items are numbered (i), (ii), and so on. So now we define

```
\newenvironment{enumeratei}{\begin{enumerate}%
                [\upshape (i)]}%
                            {\end{enumerate}}
```

and we can invoke the new environment with (see Sections 14.3 and 14.4)

```
\begin{enumeratei}
   \item \label{ }
\end{enumeratei}
```

Reference items in the enumeratei environment with the \itemref command introduced in Section 14.1.2.

4. If you want to define an environment for displaying text that is numbered as an equation, you might try

```
\newenvironment{texteqn}
   {\begin{equation} \begin{minipage}{0.9\linewidth}}
   {\end{minipage} \end{equation}}
```

But there is a problem. If you use this environment in the middle of a paragraph, an interword space appears at the beginning of the first line after the environment. To remove this unwanted space, use the \ignorespacesafterend command, as in

```
\newenvironment{texteqn}
  {\begin{equation} \begin{minipage}{0.9\linewidth}}
  {\end{minipage} \end{equation} \ignorespacesafterend}
```

Examples 2 and 3 are included in the newlattice.sty command file (see Section 14.3). See the sample article, secondarticleccom.tex in Section 14.4, for some instances of their use.

See Section 14.6.3 for custom lists as custom environments.

Redefine an existing environment with the \renewenvironment command. It is similar to the \renewcommand command (see Section 14.1.5).

There are some environments you cannot redefine; for instance, verbatim and all the AMS multiline math environments.

14.2.2 Arguments

An environment defined by the `\newenvironment` command can take arguments (see Example 2 in Section 14.2.1), but they can only be used in the *begin_text* argument of the `\newenvironment` command. Here is a simple example. Define a `theorem` proclamation in the preamble (see Section 4.4), and then define a theorem that can be referenced:

```
\newenvironment{theoremRef}[1]
   {\begin{theorem}\label{T:#1}}
   {\end{theorem}}
```

This is invoked with

```
\begin{theoremRef}{label}
```

The `theoremRef` environment is a modified environment. It is a `theorem` that can be referenced (with the `\ref` and `\pageref` commands, of course) and it invokes the `theorem` environment when it defines `T:`*label* to be the label for cross-referencing.

14.2.3 Optional arguments with default values

The first argument of an environment created with the `\newenvironment` command may be an *optional argument with a default value*. For example,

```
\newenvironment{narrow}[1][3in]
   {\noindent\begin{minipage}{#1}}
   {\end{minipage}}
```

creates a `narrow` environment. By default, it sets the body of the environment in a 3-inch wide box, with no indentation. So

```
\begin{narrow}
This text was typeset in a \texttt{narrow}
  environment, in a 3-inch wide box, with no indentation.
\end{narrow}
```

typesets as

This text was typeset in a `narrow` environment, in a 3-inch wide box, with no indentation.

You can also give an optional argument to specify the width. For example,

```
\begin{narrow}[3.5in]
  This text was typeset in a \texttt{narrow} environment,
  in a 3-inch wide box, with no indentation.
\end{narrow}
```

which produces the following false statement:

This text was typeset in a narrow environment, in a 3-inch wide box, with no indentation.

14.2.4 Short contents

We have discussed two commands that define new environments,

\newenvironment and \renewenvironment

These commands allow you to define environments whose contents (*begin_text* and *end_text*; see page 380) can include any number of paragraphs. The *-ed versions of these commands define *short* environments whose contents cannot contain a paragraph break (a blank line or a \par command).

14.2.5 Brand-new environments

Some custom environments are not modifications of existing environments. Here are two examples:

1. A command remains effective only within its scope (see Section 3.3.2). Now suppose that you want to make a change, say redefining a counter, for only a few paragraphs. You could simply place braces around these paragraphs, but they are hard to see. So define

   ```
   \newenvironment{exception}
     {\relax}
     {\relax}
   ```

   and then

   ```
   \begin{exception}
       new commands
       body
   \end{exception}
   ```

   The environment stands out better than a pair of braces, reminding you later about the special circumstances. The \relax command does nothing, but it is customary to include a \relax command in such a definition to make it more readable.

2. In this example, we define a new environment that centers its body vertically on a new page:

   ```
   \newenvironment{vcenterpage}
                  {\newpage\vspace*{\fill}}
                  {\vspace*{\fill}\par\pagebreak}
   ```

For \vspace, see Section 3.8.2 and for \fill, see the last subsection in Section 14.5.

14.3 A custom command file

Custom commands, of course, are a matter of individual need and taste. I have collected some commands for writing papers in lattice theory in the newlattice.sty file, which you can find in the samples folder (see page 5). I hope that this model helps you to develop a command file of your own. Please remember that everything we discuss in this section is a reflection of *my* work habits. Many experts disagree with one or another aspect of the way I define the commands, so take whatever suits your needs. And keep in mind the dangers of customization discussed in Section 14.7.

Tip Some journals do not permit the submission of a separate custom command file. For such journals, just copy the needed custom commands into the preamble of the article.

This file is named newlattice.sty. It can be loaded with \usepackage. This has a number of advantages.

Your command names should be mnemonic. If you cannot easily remember a command's name, rename it. The implication here is that your command file should not be very large unless you have an unusual ability to recall abbreviations.

Here are the first few lines of the newlattice.sty command file:

```
% newlattice.sty
% New command file for lattice papers
\NeedsTeXFormat{LaTeX2e}[2011/01/30]
\ProvidesPackage{newlattice}[2015/03/15 v1.6
         Commands for lattices based on LTF]
\RequirePackage{amsmath}
\RequirePackage{amssymb}
\RequirePackage{latexsym}
\RequirePackage[mathscr]{eucal}
\RequirePackage{verbatim}
\RequirePackage{enumerate}
\RequirePackage{xspace}
```

The line

```
\NeedsTeXFormat{LaTeX2e}[2005/12/01]
```

gives a message if a document loading the newlattice package is typeset with LaTeX 2.09 or with an older version of the standard LaTeX. The next line provides information that is written in your log file.

The next seven lines declare what packages are required. If some of these packages have not yet been loaded, then the missing packages are loaded. A package already loaded is not read in again by \RequirePackage.

Being able to specify the packages we need is one of the great advantages of command files. When I write a document, the packages are there if I need them.

You may want some justification for the inclusion of two of these packages in this list. The `verbatim` package is on the list so that I can use the `comment` environment to comment out large blocks of text (see Section 3.5.1), which is useful for finding errors and typesetting only parts of a longer document—but do not forget to remove your comments before submission. The `enumerate` package is on the list because the `enumeratei` and `enumeratea` environments, defined in `newlattice.sty`, require it.

If you start your article with

```
\documentclass{amsart}
\usepackage{newlattice}
```

then the `\listfiles` command (see Section C.2.4) produces the following list when your document is typeset:

```
 *File List*
  amsart.cls    2009/07/02 v2.20.1
 amsmath.sty    2013/01/14 v2.14 AMS math features
 amstext.sty    2000/06/29 v2.01
  amsgen.sty    1999/11/30 v2.0
  amsbsy.sty    1999/11/29 v1.2d
  amsopn.sty    1999/12/14 v2.01 operator names
    umsa.fd     2013/01/14 v3.01 AMS symbols A
amsfonts.sty    2013/01/14 v3.01 Basic AMSFonts support
newlattice.sty    2011/07/14 Commands for lattices based on LTF
fixltx2e.sty    2014/05/13 v1.1q fixes to LaTeX
 amssymb.sty    2013/01/14 v3.01 AMS font symbols
latexsym.sty    1998/08/17 v2.2e Standard LaTeX package
               (lasy symbols)
   eucal.sty    2009/06/22 v3.00 Euler Script fonts
verbatim.sty    2003/08/22 v1.5q LaTeX2e package
                for verbatim enhancements
enumerate.sty   1999/03/05 v3.00 enumerate extensions (DPC)
  xspace.sty    2009/10/20 v1.13
                Space after command names (DPC,MH)
microtype.sty   2013/05/23 v2.5a
                Micro-typographical refinements (RS)
  keyval.sty    2014/05/08 v1.15 key=value parser (DPC)
microtype-pdftex.def     2013/05/23 v2.5a
                Definitions specific to pdftex (RS)
microtype.cfg   2013/05/23 v2.5a microtype
                main configuration file (RS)
    umsa.fd    2013/01/14 v3.01 AMS symbols A
    umsb.fd    2013/01/14 v3.01 AMS symbols B
```

```
  ulasy.fd    1998/08/17 v2.2e LaTeX symbol font definitions
 mt-cmr.cfg   2013/05/19 v2.2 microtype config. file:
              Computer Modern Roman (RS)
***********
```

This provides a list of all packages already invoked.

Now we continue with `newlattice.sty`. After the introductory section dealing with LaTeX and the packages, we define some commands for writing about lattices and sets:

```
% Lattice operations
\newcommand{\jj}{\TextOrMath{$\vee$\xspace}{\vee}}
\newcommand{\mm}{\TextOrMath{$\wedge$\xspace}{\wedge}}
\newcommand{\JJ}{\bigvee}% big join
\newcommand{\MM}{\bigwedge}% big meet
\newcommand{\JJm}[2]{\JJ(\,#1\mid#2\,)}% big join with a middle
\newcommand{\MMm}[2]{\MM(\,#1\mid#2\,)}% big meet with a middle

% Set operations
\newcommand{\uu}{\cup}% union
\newcommand{\ii}{\cap}% intersection
\newcommand{\UU}{\bigcup}% big union
\newcommand{\II}{\bigcap}% big intersection
\newcommand{\UUm}[2]{\UU(\,#1\mid#2\,)}% big union with a middle
\newcommand{\IIm}[2]{\II(\,#1\mid#2\,)}
   % big intersection with a middle

% Sets
\newcommand{\ci}{\subseteq}% contained in with equality
\newcommand{\nc}{\nsubseteq}% not \ci
\newcommand{\sci}{\subset}% strictly contained in
\newcommand{\nci}{\nc}% not \ci
\newcommand{\ce}{\supseteq}% containing with equality
\newcommand{\nce}{\nsupseteq}% not \ce
\newcommand{\nin}{\notin}% not \in
\newcommand{\es}{\varnothing}% the empty set
\newcommand{\set}[1]{\{#1\}}% set
\newcommand{\setm}[2]{\{\,#1\mid#2\,\}}% set with a middle
\def\vv<#1>{\langle#1\rangle}% vector

% Partial ordering
\newcommand{\nle}{\nleq}% not \leq
```

So `$a \jj b$` produces $a \vee b$ and `$A \contd B$` produces $A \subseteq B$, and so on. The original commands are not redefined, so if a coauthor prefers `$a \vee b$` to `$a \jj b$`, the `\vee` command is available.

The commands with a "middle" are exemplified by `\setm`:

```
$\setm{x \in R}{x^2 \leq 2}$
```

which typesets as $\{x \in R \mid x^2 \leq 2\}$.

Using the `\set` command, we can type the set $\{a, b\}$ as `$\set{a,b}$`, which is easier to read than `$\{a,b\}$`. Similarly, we type `$\vect<a,b>$` for the vector $\langle a, b\rangle$, so it looks like a vector.

Next in `newlattice.sty`, I map the Greek letters and bold Greek letters to easy to remember commands. For some, I prefer to use the variants, but that is a matter of individual taste. It is also a matter of taste whether or not to change the commands for the Greek letters at all, and how far one should go in changing commonly used commands.

```
% Greek letters
\newcommand{\ga}{\TextOrMath{$\alpha$\xspace}{\alpha}}
\newcommand{\gb}{\TextOrMath{$\beta$\xspace}{\beta}}
\newcommand{\gc}{\TextOrMath{$\chi$\xspace}{\chi}}
\newcommand{\gd}{\TextOrMath{$\delta$\xspace}{\delta}}
\renewcommand{\ge}{\TextOrMath{$\varepsilon$\xspace}{\varepsilon}}
\newcommand{\gf}{\TextOrMath{$\varphi$\xspace}{\varphi}}
\renewcommand{\gg}{\TextOrMath{$\gamma $\xspace}{\gamma}}
\newcommand{\gh}{\TextOrMath{$\eta$\xspace}{\eta}}
\newcommand{\gi}{\TextOrMath{$\iota$\xspace}{\iota}}
\newcommand{\gk}{\TextOrMath{$\kappa$\xspace}{\kappa}}
\newcommand{\gl}{\TextOrMath{$\lambda$\xspace}{\lambda}}
\newcommand{\gm}{\TextOrMath{$\mu$\xspace}{\mu}}
\newcommand{\gn}{\TextOrMath{$\nu$\xspace}{\nu}}
\newcommand{\go}{\TextOrMath{$\omega$\xspace}{\omega}}
\newcommand{\gp}{\TextOrMath{$\pi$\xspace}{\pi}}
\newcommand{\gq}{\TextOrMath{$\theta$\xspace}{\theta}}
\newcommand{\gr}{\TextOrMath{$\varrho$\xspace}{\varrho}}
\newcommand{\gs}{\TextOrMath{$\sigma$\xspace}{\sigma}}
\newcommand{\gt}{\TextOrMath{$\tau$\xspace}{\tau}}
\newcommand{\gu}{\TextOrMath{$\upsilon$\xspace}{\upsilon}}
\newcommand{\gv}{\TextOrMath{$\vartheta$\xspace}{\vartheta}}
\newcommand{\gx}{\TextOrMath{$\xi$\xspace}{\xi}}
\newcommand{\gy}{\TextOrMath{$\psi$\xspace}{\psi}}
\newcommand{\gz}{\TextOrMath{$\gz$\xspace}{\gz}}
\newcommand{\gG}{\TextOrMath{$\Gamma$\xspace}{\Gamma}}
\newcommand{\gD}{\TextOrMath{$\Delta$\xspace}{\Delta}}
\newcommand{\gF}{\TextOrMath{$\Phi$\xspace}{\Phi}}
```

```
\newcommand{\gL}{\TextOrMath{$\Lambda$\xspace}{\Lambda}}
\newcommand{\gO}{\TextOrMath{$\Omega$\xspace}{\Omega}}
\newcommand{\gP}{\TextOrMath{$\Pi$\xspace}{\Pi}}
\newcommand{\gQ}{\TextOrMath{$\Theta$\xspace}{\Theta}}
\newcommand{\gS}{\TextOrMath{$\Sigma$\xspace}{\Sigma}}
\newcommand{\gU}{\TextOrMath{$\Upsilon$\xspace}{\Upsilon}}
\newcommand{\gX}{\TextOrMath{$\Xi$\xspace}{\Xi}}
\newcommand{\gY}{\TextOrMath{$\Psi$\xspace}{\Psi}}

% Bold Greek letters
\newcommand{\bga}{\TextOrMath{$\boldsymbol{\alpha}$\xspace}
          {\boldsymbol\alpha}}
\newcommand{\bgb}{\TextOrMath{$\boldsymbol{\beta}$\xspace}
          {\boldsymbol\beta}}
\newcommand{\bgc}{\TextOrMath{$\boldsymbol{\chi}$\xspace}
          {\boldsymbol\chi}}
\newcommand{\bgd}{\TextOrMath{$\boldsymbol{\delta}$\xspace}
          {\boldsymbol\delta}}
\newcommand{\bge}{\TextOrMath{$\boldsymbol{\varepsilon}$\xspace}
          {\boldsymbol\varepsilon}}
\newcommand{\bgf}{\TextOrMath{$\boldsymbol{\varphi}$\xspace}
          {\boldsymbol\varphi}}
\newcommand{\bgg}{\TextOrMath{$\boldsymbol{\gamma}$\xspace}
          {\boldsymbol\gamma}}
\newcommand{\bgh}{\TextOrMath{$\boldsymbol{\eta}$\xspace}
          {\boldsymbol\eta}}
\newcommand{\bgi}{\TextOrMath{$\boldsymbol{\iota}$\xspace}
          {\boldsymbol\iota}}
\newcommand{\bgk}{\TextOrMath{$\boldsymbol{\kappa}$\xspace}
          {\boldsymbol\kappa}}
\newcommand{\bgl}{\TextOrMath{$\boldsymbol{\lambda}$\xspace}
          {\boldsymbol\lambda}}
\newcommand{\bgm}{\TextOrMath{$\boldsymbol{\mu}$\xspace}
          {\boldsymbol\mu}}
\newcommand{\bgn}{\TextOrMath{$\boldsymbol{\nu}$\xspace}
          {\boldsymbol\nu}}
\newcommand{\bgo}{\TextOrMath{$\boldsymbol{\omega}$\xspace}
          {\boldsymbol\omega}}
\newcommand{\bgp}{\TextOrMath{$\boldsymbol{\pi}$\xspace}
          {\boldsymbol\pi}}
\newcommand{\bgq}{\TextOrMath{$\boldsymbol{\theta}$\xspace}
          {\boldsymbol\theta}}
```

```
\newcommand{\bgr}{\TextOrMath{$\boldsymbol{\varrho}$\xspace}
          {\boldsymbol\varrho}}
\newcommand{\bgs}{\TextOrMath{$\boldsymbol{\sigma}$\xspace}
          {\boldsymbol\sigma}}
\newcommand{\bgt}{\TextOrMath{$\boldsymbol{\tau}$\xspace}
          {\boldsymbol\tau}}
\newcommand{\bgu}{\TextOrMath{$\boldsymbol{\upsilon}$\xspace}
          {\boldsymbol\upsilon}}
\newcommand{\bgv}{\TextOrMath{$\boldsymbol{\vartheta}$\xspace}
          {\boldsymbol\vartheta}}
\newcommand{\bgx}{\TextOrMath{$\boldsymbol{\xi}$\xspace}
          {\boldsymbol\xi}}
\newcommand{\bgy}{\TextOrMath{$\boldsymbol{\psi}$\xspace}
          {\boldsymbol\psi}}
\newcommand{\bgz}{\TextOrMath{$\boldsymbol{\gz}$\xspace}
          {\boldsymbol\gz}}
\newcommand{\bgL}{\TextOrMath{$\boldsymbol{\gL}$\xspace}
          {\boldsymbol\gL}}
\newcommand{\bgF}{\TextOrMath{$\boldsymbol{\gF}$\xspace}
          {\boldsymbol\gF}}
```

I also introduce some new names for text font commands by abbreviating `text` to `t` (so that `\textbf` becomes `\tbf`) and for math font commands by abbreviating `math` to `m` (so that `\mathbf` becomes `\mbf`).

```
% Font commands
\newcommand{\tbf}{\textbf}% text bold
\newcommand{\tit}{\textit}% text italic
\newcommand{\tsl}{\textsl}% text slanted
\newcommand{\tsc}{\textsc}% text small cap
\newcommand{\ttt}{\texttt}% text typewriter
\newcommand{\trm}{\textrm}% text roman
\newcommand{\tsf}{\textsf}% text sans serif
\newcommand{\tup}{\textup}% text upright

\newcommand{\mbf}{\mathbf}% math bold
\providecommand{\mit}{\mathit}% math italic
\newcommand{\msf}{\mathsf}% math sans serif
\newcommand{\mrm}{\mathrm}% math roman
\newcommand{\mtt}{\mathtt}% math typewriter
```

The math alphabets are invoked as commands with arguments: `\Bold` for bold, `\Cal` for calligraphic, `\DD` for blackboard bold (double), and `\Frak` for fraktur (German

Gothic—see Section 6.4.2. Notice that \Cal and \Euler are different because of the option mathscr of the eucal package (see Section 6.4.1).

```
\newcommand{\B}{\boldsymbol}
   % Bold math symbol, use as \B{a}
\newcommand{\C}[1]{\mathcal{#1}}
   % Euler Script - only caps, use as \C{A}
\newcommand{\D}[1]{\mathbb{#1}}
   % Doubled - blackboard bold - only caps, use as \D{A}
\newcommand{\E}[1]{\mathcal{#1}}% same as \C
   % Euler Script - only caps, use as \E{A}
\newcommand{\F}[1]{\mathfrak{#1}}% Fraktur, use as \F{a}

%Sansserif, special lattices
%Chains and Boolean lattces

\newcommand{\SC}[1]{\msf{C}_{#1}}
\newcommand{\SB}[1]{\msf{B}_{#1}}
\newcommand{\SD}[1]{\msf{D}_{#1}}
\newcommand{\SL}[1]{\msf{L}_{#1}}
\newcommand{\SM}[1]{\msf{M}_{#1}}
\newcommand{\SMb}[1]{\msf{M}_{3}[#1]_\text{bal}}
\newcommand{\SN}[1]{\msf{N}_{#1}}
\newcommand{\SH}[1]{\msf{H}_{#1}}
\newcommand{\SV}[1]{\msf{V}_{#1}}
\newcommand{\SfC}[1]{\msf{C}_{#1}}
\newcommand{\SfB}[1]{\msf{B}_{#1}}
\newcommand{\SfM}[1]{\msf{M}_{#1}}
\newcommand{\SfN}[1]{\msf{N}_{#1}}
\newcommand{\SfS}[1]{\msf{S}_{#1}}
```

Here are some commands of importance in lattice theory:

```
% Constructs
\DeclareMathOperator{\Id}{Id}
\DeclareMathOperator{\Fi}{Fi{}l}
\DeclareMathOperator{\Con}{Con}
\DeclareMathOperator{\Aut}{Aut}
\DeclareMathOperator{\Sub}{Sub}
\DeclareMathOperator{\Pow}{Pow}
\DeclareMathOperator{\Part}{Part}
\DeclareMathOperator{\Ker}{Ker}
\DeclareMathOperator{\Joinir}{Join}
```

```
\DeclareMathOperator{\Meetir}{Meet}
\DeclareMathOperator{\Down}{Down}
\DeclareMathOperator{\Ji}{Ji}
\DeclareMathOperator{\Mi}{Mi}

% Generated by
\newcommand{\con}[1]{\tup{con}(#1)}
\newcommand{\consub}[2]{\tup{con}_{#1}(#2)}
\newcommand{\sub}[1]{\tup{sub}(#1)}
\newcommand{\id}[1]{\tup{id}(#1)}
\newcommand{\fil}[1]{\tup{f{}il}(#1)}

% Miscellaneous
\newcommand{\nl}{\newline}
\newcommand{\ol}[1]{\overline{#1}}
\newcommand{\ul}[1]{\underline{#1}}
\providecommand{\bysame}{\makebox[3em]{\hrulefill}\thinspace}
\newcommand{\q}{\quad}% spacing
\newcommand{\qq}{\qquad}% more spacing
\newcommand{\iso}{\cong}% isomorphic
\newenvironment{enumeratei}{\begin{enumerate}[\upshape (i)]}%
                           {\end{enumerate}}
        %produces (i), (ii), etc. Cross-reference with \eqref.
\newenvironment{enumeratea}{\begin{enumerate}[\upshape (a)]}%
                           {\end{enumerate}}
        %produces (a), (b), etc. Cross-reference with \eqref.
\theoremstyle{plain}
\newtheorem*{namedtheorem}{\theoremname}
\newcommand{\theoremname}{testing}
\newenvironment{named}[1]{\renewcommand{\theoremname}{#1}
   \begin{namedtheorem}}
   {\end{namedtheorem}}
   %use it as \begin{named}{Name of theorem}
   %Body of theorem \end{named}

\newcommand{\Dg}{\downarrow\!}% down-set generated by congruences
\newcommand{\per}{\sim}% perspective ~
\newcommand{\pu}{\stackrel{\textrm{u}}{\sim}}
          % perspective up ~ with u on top
\newcommand{\pd}{\stackrel{\textrm{d}}{\sim}}
          % perspective down ~ with d on top
\newcommand{\proj}{\approx}% projective
```

```
\newcommand{\cpu}{\nearrow}
          % congruence perspective up -- up arrow
\newcommand{\cpd}{\searrow}
          % congruence perspective down -- down arrow
\newcommand{\cper}{\hookrightarrow}
          % congruence perspective onto-- hooked right arrow
\newcommand{\cproj}{\Rightarrow}
          % congruence projective into -- double headed arrow
\newcommand{\cprojboth}{\Leftrightarrow}
          % congruence projective both ways
          %-- two headed double arrow

%perspective
\newcommand{\perspsymb}{\thicksim}% perspective symbol
\newcommand{\persp}{\perspsymb}% perspective
\newcommand{\perspup}{\stackrel{\textrm{up}}{\perspsymb}}
    % perspective up
\newcommand{\perspdn}{\stackrel{\textrm{dn}}{\perspsymb}}
    % perspective down

% c-perspective
\newcommand{\cperspsymb}{\hookrightarrow}
          % c-perspective symbol
\newcommand{\cpersp}{\cperspsymb}% c-perspective
\newcommand{\cperspup}{\stackrel{\textrm{up}}{\cperspsymb}}
          % c-perspective up
\newcommand{\cperspdn}{\stackrel{\textrm{dn}}{\cperspsymb}}
          % c-perspective dn


\newcommand{\lp}{\tup{(}}
\newcommand{\rp}{\tup{)}\xspace}
\newcommand{\up}[1]{\tup{(}#1\tup{)}}
\newcommand{\one}{\mathbf{1}}
\newcommand{\zero}{\mathbf{0}}
\newcommand{\restr}{\rceil}

\def\cng#1=#2(#3){#1\equiv#2\pmod{#3}}
   %congruence, use it as \cng a=b(\theta)%
\def\cngd#1=#2(#3){#1\equiv#2\!\pmod{#3}}
   %congruence for display, use it as \cngd a=b(\theta)%
\def\ncng#1=#2(#3){#1\not\equiv#2\pmod{#3}}
   %negate cng
```

```
\def\ncngd#1=#2(#3){#1\not\equiv#2\!\pmod{#3}}
    %negate dcng

\endinput
```

See Section 14.1.9 for the `\congr` command. The `enumeratei` and `named` environments are discussed in Section 14.2.1. The `enumeratea` environment is similar.

This command file, like all command files, is terminated with the `\endinput` command. In Section 17.3.2, we discuss the same rule for files that are `\include`-d.

My `newlattice.sty` evolves with time. I keep a copy in the folder of every article I write. This way, even years later, with the command file changed, I can typeset the article with no problem.

The `\TextOrMath` command (see Section 10.3) is very useful for command files. For instance, we can use it to define our Greek letters, such as

```
\newcommand{\ga}{\TextOrMath{$\alpha$}{\alpha}}
```

Then we can use \ga both in text and math to produce α.

14.4 The sample article with custom commands

In this section, we look at the `secondarticleccom.tex` sample article, which is a rewrite of the `secondarticle.tex` sample article utilizing the custom commands collected in the command file `newlattice.sty` (for these files, see Section 14.3 and the `samples` folder).

```
% Sample file: secondarticleccom.tex
% The sample article with custom commands and environments

\documentclass{amsart}
\usepackage{newlattice}

\theoremstyle{plain}
\newtheorem{theorem}{Theorem}
\newtheorem{corollary}{Corollary}
\newtheorem{lemma}{Lemma}
\newtheorem{proposition}{Proposition}

\theoremstyle{definition}
\newtheorem{definition}{Definition}

\theoremstyle{remark}
\newtheorem*{notation}{Notation}
```

```
\numberwithin{equation}{section}

\newcommand{\Prodm}[2]{\GrP(\,#1\mid#2\,)}
   % product with a middle
\newcommand{\Prodsm}[2]{\GrP^{*}(\,#1\mid#2\,)}
   % product * with a middle
\newcommand{\vectsup}[2]{\vect<\dots,0,\dots,\overset{#1}{#2},%
\dots,0,\dots>}% special vector
\newcommand{\Dsq}{D^{\langle2\rangle}}

\begin{document}
\title[Complete-simple distributive lattices]
      {A construction of complete-simple\\
       distributive lattices}
\author{George~A. Menuhin}
\address{Computer Science Department\\
         University of Winnebago\\
         Winnebago, Minnesota 23714}
\email{menuhin@ccw.uwinnebago.edu}
\urladdr{http://math.uwinnebago.edu/homepages/menuhin/}
\thanks{Research supported by the NSF under grant number~23466.}
\keywords{Complete lattice, distributive lattice, complete
   congruence, congruence lattice}
\subjclass[2000]{Primary: 06B10; Secondary: 06D05}
\date{March 15, 2006}

\begin{abstract}
   In this note we prove that there exist \emph{complete-simple
   distributive lattices,} that is, complete distributive
   lattices in which there are only two complete congruences.
\end{abstract}
\maketitle

\section{Introduction}\label{S:intro}
In this note we prove the following result:

\begin{named}{Main Theorem}
   There exists an infinite complete distributive lattice
   $K$ with only the two trivial complete congruence relations.
\end{named}
\section{The $\Dsq$ construction}\label{S:Ds}
For the basic notation in lattice theory and universal algebra,
```

```
see Ferenc~R. Richardson~\cite{fR82} and George~A.
Menuhin~\cite{gM68}. We start with some definitions:

\begin{definition}\label{D:prime}
   Let $V$ be a complete lattice, and let $\Frak{p} = [u, v]$ be
   an interval of $V$. Then $\Frak{p}$ is called
   \emph{complete-prime} if the following three conditions
   are satisfied:
   \begin{enumeratei}
      \item $u$ is meet-irreducible but $u$ is \emph{not}
         completely meet-irreducible;\label{m-i}
      \item $v$ is join-irreducible but $v$ is \emph{not}
         completely join-irreducible;\label{j-i}
      \item $[u, v]$ is a complete-simple lattice.\label{c-s}
   \end{enumeratei}
\end{definition}

Now we prove the following result:

\begin{lemma}\label{L:Dsq}
   Let $D$ be a complete distributive lattice satisfying
   conditions \itemref{m-i} and~\itemref{j-i}.
   Then $\Dsq$ is a sublattice of $D^{2}$; hence $\Dsq$ is
   a lattice, and $\Dsq$ is a complete distributive lattice
   satisfying conditions \itemref{m-i} and~\itemref{j-i}.
\end{lemma}

\begin{proof}
   By conditions~\itemref{m-i} and \itemref{j-i}, $\Dsq$ is a
   sublattice of $D^{2}$. Hence, $\Dsq$ is a lattice.

   Since $\Dsq$ is a sublattice of a distributive lattice,
   $\Dsq$ is a distributive lattice. Using the characterization
   of standard ideals in Ernest~T. Moynahan~\cite{eM57},
   $\Dsq$ has a zero and a unit element, namely,
   $\vect<0, 0>$ and $\vect<1, 1>$. To show that $\Dsq$ is
   complete, let $\empset \ne A \contd \Dsq$, and let $a = \JJ A$
   in $D^{2}$. If $a \in \Dsq$, then
   $a = \JJ A$ in $\Dsq$; otherwise, $a$ is of the form
   $\vect<b, 1>$ for some $b \in D$ with $b < 1$. Now
   $\JJ A = \vect<1, 1>$ in $D^{2}$, and
   the dual argument shows that $\MM A$ also exists in
```

```
   $D^{2}$. Hence $D$ is complete. Conditions \itemref{m-i}
   and~\itemref{j-i} are obvious for $\Dsq$.
\end{proof}
\begin{corollary}\label{C:prime}
   If $D$ is complete-prime, then so is $\Dsq$.
\end{corollary}

The motivation for the following result comes from Soo-Key
Foo~\cite{sF90}.

\begin{lemma}\label{L:ccr}
   Let $\gQ$ be a complete congruence relation of $\Dsq$ such
   that
   \begin{equation}\label{E:rigid}
      \congr \vect<1, d>=\vect<1, 1>(\gQ),
   \end{equation}
   for some $d \in D$ with $d < 1$. Then $\gQ = \gi$.
\end{lemma}

\begin{proof}
   Let $\gQ$ be a complete congruence relation of $\Dsq$
   satisfying \itemref{E:rigid}. Then $\gQ = \gi$.
\end{proof}

\section{The $\gp^{*}$ construction}\label{S:P*}
The following construction is crucial to our proof of the
Main~Theorem:

\begin{definition}\label{D:P*}
   Let $D_{i}$, for $i \in I$, be complete distributive
   lattices satisfying condition~\itemref{j-i}. Their $\gp^{*}$
   product is defined as follows:
   \[
     \Prodsm{ D_{i} }{i \in I} = \Prodm{ D_{i}^{-} }{i \in I}+1;
   \]
   that is, $\Prodsm{ D_{i} }{i \in I}$ is
   $\Prodm{ D_{i}^{-} }{i \in I}$ with a new unit element.
\end{definition}


\begin{notation}
   If $i \in I$ and $d \in D_{i}^{-}$, then
```

```
   \[
      \vectsup{i}{d}
   \]
   is the element of $\Prodsm{ D_{i} }{i \in I}$ whose
   $i$-th component is $d$ and all the other
   components are $0$.
\end{notation}

See also Ernest~T. Moynahan~\cite{eM57a}. Next we verify:

\begin{theorem}\label{T:P*}
   Let $D_{i}$, for $i \in I$, be complete distributive
   lattices satisfying condition~\itemref{j-i}. Let $\gQ$
   be a complete congruence relation on
   $\Prodsm{ D_{i} }{i \in I}$. If there exist
   $i \in I$ and $d \in D_{i}$ with $d < 1_{i}$ such
   that for all $d \leq c < 1_{i}$,
   \begin{equation}\label{E:cong1}
      \congr\vectsup{i}{d}=\vectsup{i}{c}(\gQ),
   \end{equation}
   then $\gQ = \gi$.
\end{theorem}

\begin{proof}
   Since
   \begin{equation}\label{E:cong2}
      \congr\vectsup{i}{d}=\vectsup{i}{c}(\gQ),
   \end{equation}
   and $\gQ$ is a complete congruence relation, it follows
   from condition~\itemref{c-s} that
   \begin{equation}\label{E:cong}
   \begin{split}
       &\langle \dots, \overset{i}{d}, \dots, 0,
        \dots \rangle\\
       &\equiv \bigvee ( \langle \dots, 0, \dots,
        \overset{i}{c},\dots, 0,\dots \rangle \mid d \leq c < 1)
         \equiv 1 \pmod{\Theta}.
   \end{split}
   \end{equation}

   Let $j \in I$, for $j \neq i$, and let
   $a \in D_{j}^{-}$. Meeting both sides of the congruence
```

```
    \itemref{E:cong} with $\vectsup{j}{a}$, we obtain
    \begin{equation}\label{E:comp}
       \begin{split}
           0 &= \vectsup{i}{d} \mm \vectsup{j}{a}\\
             &\equiv \vectsup{j}{a}\pod{\gQ}.
      \end{split}
    \end{equation}
   Using the completeness of $\gQ$ and \itemref{E:comp}, we get:
    \begin{equation}\label{E:cong3}
      \congr{0=\JJm{ \vectsup{j}{a} }{ a \in D_{j}^{-} }}={1}(\gQ),
    \end{equation}
    hence $\gQ = \gi$.
\end{proof}

\begin{theorem}\label{T:P*a}
    Let $D_{i}$, for $i \in I$, be complete distributive
    lattices satisfying
    conditions \itemref{j-i} and~\itemref{c-s}. Then
    $\Prodsm{ D_{i} }{i \in I}$ also satisfies
    conditions~\itemref{j-i} and \itemref{c-s}.
\end{theorem}

\begin{proof}
    Let $\gQ$ be a complete congruence on
    $\Prodsm{ D_{i} }{i \in I}$. Let $i \in I$. Define
    \begin{equation}\label{E:dihat}
      \widehat{D}_{i} = \setm{ \vectsup{i}{d} }{ d \in D_{i}^{-} }
        \uu \set{1}.
    \end{equation}
    Then $\widehat{D}_{i}$ is a complete sublattice of
    $\Prodsm{ D_{i} }{i \in I}$, and $\widehat{D}_{i}$
    is isomorphic to $D_{i}$. Let $\gQ_{i}$ be the
    restriction of $\gQ$ to $\widehat{D}_{i}$. Since
    $D_{i}$ is complete-simple, so is $\widehat{D}_{i}$,
    hence $\gQ_{i}$ is $\go$ or $\gi$. If $\gQ_{i} = \go$
    for all $i \in I$, then $\gQ = \go$.
    If there is an $i \in I$, such that $\gQ_{i} = \gi$,
    then $\congr0=1(\gQ)$, and hence $\gQ = \gi$.
\end{proof}

The Main Theorem follows easily from Theorems~\ref{T:P*} and
\ref{T:P*a}.
```

```
\begin{thebibliography}{9}

   \bibitem{sF90}
      Soo-Key Foo, \emph{Lattice Constructions}, Ph.D. thesis,
      University of Winnebago, Winnebago, MN, December, 1990.

   \bibitem{gM68}
      George~A. Menuhin, \emph{Universal algebra}. D.~van
      Nostrand, Princeton, 1968.
   \bibitem{eM57}
      Ernest~T. Moynahan, \emph{On a problem of M. Stone},
      Acta Math. Acad. Sci. Hungar. \tbf{8} (1957), 455--460.

   \bibitem{eM57a}
     \bysame, \emph{Ideals and congruence relations in
     lattices}.~II, Magyar Tud. Akad. Mat. Fiz. Oszt. K\"{o}zl.
     \tbf{9} (1957), 417--434  (Hungarian).

   \bibitem{fR82}
      Ferenc~R. Richardson, \emph{General lattice theory}. Mir,
      Moscow, expanded and revised ed., 1982 (Russian).

\end{thebibliography}
\end{document}
```

14.5 *Numbering and measuring*

LaTeX stores integers in *counters*. For example, the `section` counter contains the current section number. Distance measurements are saved in *length commands*. For instance, the `\textwidth` command contains the width of the text. For this book, the length command `\textwidth` is set to 345.0 points.

In this section, we take a closer look at counters and length commands.

14.5.1 *Counters*

Counters may be defined by LaTeX, by document classes, by packages, or by the user.

Standard LaTeX *counters*

LaTeX automatically generates numbers for equations, sections, theorems, and so on. Each such number is stored in a *counter*. Table 14.2 shows the standard LaTeX counters. Their names are more or less self-explanatory. In addition, for every proclamation *name*, there is a matching counter called `name` (see Section 4.4).

Setting counters

The command for setting a counter's value is \setcounter. When LaTeX generates a number, it first increments the appropriate counter, so if you want the next chapter to be numbered 3, you should set the chapter counter to 2 by typing

```
\setcounter{chapter}{2}
```

before the \chapter command. The only exception to this rule is the page number, which is first used to number the current page, and then incremented. If you wanted to set the current page number to 63, you would include the command

```
\setcounter{page}{63}
```

somewhere in the page.

LaTeX initializes and increments its standard counters automatically. Sometimes you may want to manipulate them yourself. To typeset only chapter3.tex, the third chapter of your book, start with

```
\setcounter{chapter}{2}
\include{chapter3}
```

and when chapter3.tex is typeset, the chapter is properly numbered. You can also type

```
\setcounter{page}{63}
```

if the first page of this chapter is supposed to be 63. Of course, the preferred way to typeset parts of a larger document is with the \includeonly command (see Section 17.3.2).

Tip If you need to manipulate counters, always look for solutions in which LaTeX does the work for you.

Defining new counters

You can define your own counters. For example,

```
\newcounter{mycounter}
```

makes mycounter a new counter. In the definition, you can use an optional argument, the name of another counter:

```
\newcounter{mycounter}[basecounter]
```

equation	part	enumi
figure	chapter	enumii
footnote	section	enumiii
mpfootnote	subsection	enumiv
page	subsubsection	
table	paragraph	
	subparagraph	

Table 14.2: Standard LaTeX counters.

Style	Command	Sample
Arabic	\arabic{*counter*}	1, 2, ...
Lowercase Roman	\roman{*counter*}	i, ii, ...
Uppercase Roman	\Roman{*counter*}	I, II, ...
Lowercase Letters	\alph{*counter*}	a, b, ..., z
Uppercase Letters	\Alph{*counter*}	A, B, ..., Z

Table 14.3: Counter styles.

which automatically resets `mycounter` to 0 if `basecounter` changes value. This command has the same form as the command LaTeX uses internally for tasks such as numbering theorems and subsections within sections.

Rule ■ New counters

New counters should be defined in the preamble of the document. They should not be defined in a file read in with an `\include` command (see Section 17.3.2).

Let us suppose that you define a new counter, `mycounter`, in `chapter5.tex`, which is made part of your whole document with an `\include` command. When you typeset your document with `\includeonly` commands not including `chapter5.tex`, you get a message, such as

```
! LaTeX Error: No counter 'mycounter' defined.
```

Counter styles

The value of `counter` can be displayed in the typeset document with the command

```
\thecounter
```

If you want to change the counter's appearance when typeset, issue the command

```
\renewcommand{\thecounter}{new_style}
```

where *new_style* specifies the counter modified as shown in Table 14.3. The default style is arabic. For instance, if you give the command

```
\renewcommand{\thetheorem}{\Alph{theorem}}
```

then the theorems appear as **Theorem A**, **Theorem B**, ...

Here is a more complicated example for a book:

```
\renewcommand{\thechapter}{\arabic{chapter}}
\renewcommand{\thesection}{\thechapter-\arabic{section}}
\renewcommand{\thesubsection}
    {\thechapter-\arabic{section}.\arabic{subsection}}
```

With these definitions, Section 1 of Chapter 3 is numbered in the form 3-1 and Subsection 2 of Section 1 of Chapter 3 is numbered in the form 3-1.2.

The \pagenumbering command is a shorthand method for setting the page numbering in a given style. For instance, \pagenumbering{roman} numbers pages as i, ii, and so on.

The subequations environment (see Section 6.6) uses parentequation as the counter for the whole equation group and it uses equation as the counter for the subequations. To change the default format of the equation numbers from (2a), (2b), and so forth, to (2i), (2ii), and so on, type the following line inside the subequations environment

```
\renewcommand{\theequation}
            {\theparentequation\roman{equation}}
```

If you want equation numbers like (2.i), (2.ii), and so on, type

```
\renewcommand{\theequation}%
              {\theparentequation.\roman{equation}}
```

Counter arithmetic

The \stepcounter{counter} command increments counter and sets all the counters that were defined with the optional argument counter to 0. The variant

```
\refstepcounter{counter}
```

does the same, and also sets the value for the next \label command.

You can do some arithmetic with the command

```
\addtocounter{counter}{n}
```

where n is an integer. For example,

```
\setcounter{counter}{5}
\addtocounter{counter}{2}
```

sets `counter` to 7.

The value stored in a counter can be accessed using the `\value` command, which is mostly used with the `\setcounter` or `\addtocounter` commands. For instance, you can set `counter` to equal the value of another counter, `oldcounter`, by typing

```
\setcounter{counter}{\value{oldcounter}}
```

Here is a typical example of counter manipulation. You have a theorem (invoked in a `theorem` environment) and you want it followed by several corollaries (each in a `corollary` environment) starting with Corollary 1. In other words, Theorem 1 should be followed by Corollary 1, Corollary 2, and so forth and so should Theorem 3. By default, LaTeX numbers the next corollary as Corollary 3, even if it follows another theorem. To tell LaTeX to start numbering the corollaries from 1 again, issue the command

```
\setcounter{corollary}{0}
```

after each theorem. But such a process is error-prone, and goes against the spirit of LaTeX.

Instead, follow my advice on page 401, and let LaTeX do the work for you. In the preamble, type the proclamations

```
\newtheorem{theorem}{Theorem}
\newtheorem{corollary}{Corollary}[theorem]
```

We are almost there. Theorem 1 now is followed by Corollary 1.1, Corollary 1.2 and Theorem 3 by Corollary 3.1. If we redefine `\thecorollary`,

```
\renewcommand{\thecorollary}{\arabic{corollary}}
```

then Theorem 1 is followed by Corollary 1 and Corollary 2, and Theorem 3 is also followed by Corollary 1.

If you need to perform more complicated arithmetic with counters, use Kresten K. Thorup and Frank Jensen's `calc` package (see Section 10.3.1). This package is discussed in Section A.3.1 of *The LaTeX Companion*, 2nd edition [56].

Two special counters

The `secnumdepth` and `tocdepth` counters control which sectional units are numbered and which are listed in the table of contents, respectively. For example,

```
\setcounter{secnumdepth}{2}
```

sets `secnumdepth` to 2. As a result, chapters—if they are present in the document class—sections, and subsections are numbered, but subsubsections are not. This command must be placed in the preamble of the document. `tocdepth` is similar.

14.5.2 Length commands

While a counter contains integers, a length command contains a *real number* and a *dimensional unit.*

LaTeX recognizes many different dimensional units. We list five *absolute* units:

- `cm` centimeter
- `in` inch
- `pc` pica (`1 pc = 12 pt`)
- `tip` point (`1 in = 72.27 pt`)
- `mm` millimeter

and two *relative* units:

- `em`, approximately the width of the letter M in the current font
- `ex`, approximately the height of the letter x in the current font

LaTeX defines many length commands. For instance, Section 4.1 of *The LaTeX Companion,* 2nd edition [56] lists 17 length commands for page layout alone. You can find some of them in Figure 8.4. A list environment sets about a dozen additional length commands (see Figure 14.2). Length commands are defined for almost every aspect of LaTeX's work, including displayed math environments—a complete list would probably contain a few hundred. Many are listed in Leslie Lamport's LaTeX*: A Document Preparation System,* 2nd edition [53] and in *The LaTeX Companion,* 2nd edition [56]. Many more are hidden in packages such as `amsmath`.

The most common length commands are:

- `\parindent`, the amount of indentation at the beginning of a paragraph
- `\parskip`, the extra vertical space inserted between paragraphs
- `\textwidth`, the width of the text on a page

A more esoteric example is `\marginparpush`, the minimum vertical space between two marginal notes. Luckily, you do not have to be familiar with many length commands because LaTeX and the document class set them for you.

Defining new length commands

You can define your own length commands. For example,

```
\newlength{\mylength}
```

makes `\mylength` a new length command with a value of 0 points. Note that while you have to type

```
\newcounter{counter}
```

to get a new counter, typing

```
\newlength{mylength}
```

results in a message such as

```
! Missing control sequence inserted.
<inserted text>
                \inaccessible
l.3 \newlength{mylength}
```

Setting length

The `\setlength` command sets or resets the value of a length command. So

```
\setlength{\textwidth}{3in}
```

creates a very narrow page. The first argument of `\setlength` must be a length command, not simply the command name, that is

```
\setlength{textwidth}{3in} % Bad
```

is incorrect. The second argument of `\setlength` must be a real number with a dimensional unit, for instance, `3in`, and *not simply a real number.* In other words,

```
\setlength{\textwidth}{3} % Bad
```

is also incorrect. You can also use `\setlength` as an environment, as in Section 7.3.

Tip A common mistake is to type a command such as

```
\setlength{\marginpar}{0}
```

Instead, type

```
\setlength{\marginpar}{0pt}
```

Always be sure to include a dimensional unit.

The `\addtolength` command adds a quantity to the value of a length command. For instance,

```
\addtolength{\textwidth}{-10pt}
```

narrows the page width by 10 points.

If you define

```
\newlength{\shorterlength}
\setlength{\shorterlength}{\mylength}
\addtolength{\shorterlength}{-.5in}
```

then `\parbox{\shorterlength}{...}` always typesets its second argument in a box 1/2 inch narrower than the parboxes set to be of width `\mylength`.

When LaTeX typesets some text or math, it creates a box. Three measurements are used to describe the size of the box:

- the width
- the height, from the baseline to the top
- the depth, from the baseline to the bottom

as illustrated in Figure 14.1. For instance, the box typesetting "aa" has a width of 10.00003 pt, a height of 4.30554 pt, and a depth of 0 pt. The box typesetting "ag" has the same width and height, but a depth of 1.94444 pt. The box "Ag" (see Figure 14.1) has a width of 12.50003 pt, a height of 6.83331 pt, and a depth of 1.94444 pt.

The commands

```
\settowidth
\settoheight
\settodepth
```

each take two arguments. The first argument is a length command, the second is text (or math) to be measured by LaTeX. The corresponding measurement of the box in which the second argument is typeset is assigned to the length command in the first argument. For example, if `\mylength` is a length command, then

```
\settowidth{\mylength}{Ag}
```

assigns 12.50003 pt to `\mylength`. It should be clear from this example how the `\phantom` and `\hphantom` commands (see Section 3.8.1) are related to this command.

To perform more complicated arithmetic with length commands, use the `calc` package.

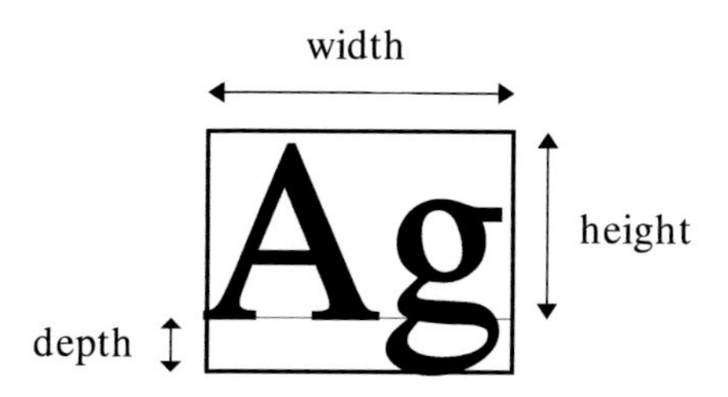

Figure 14.1: The measurements of a box.

Rubber lengths

In addition to rigid lengths, such as `3in`, LaTeX can also set *rubber lengths,* that is, lengths that are allowed to stretch and shrink. Here is an example:

```
\setlength{\stretchspace}{3in plus 10pt minus 8pt}
```

Assuming that `\stretchspace` is a length command, this command assigns it a value of 3 inches that can stretch by 10 points or shrink by 8 points, if necessary. So a box of width `\stretchspace` is 3 inches wide, plus up to 10 points, or minus up to 8 points.

Stretchable vertical spaces are often used before and after displayed text environments. LaTeX adjusts these spaces to make the page look balanced. An example can be found in Section 14.1.8. `\medskipamount` is defined as

```
6.0pt plus 2.0pt minus 2.0pt
```

See Section 14.6.3 for more examples.

The `\fill` command is a special rubber length that can stretch any amount. The stretching is done evenly if there is more than one `\fill` present. See the second example of brand-new environments in Section 14.2.5.

14.6 Custom lists

Although there are three ready-made list environments provided by LaTeX (see Section 4.2), it is often necessary to create one of your own using LaTeX's `list` environment. In fact, LaTeX itself uses the `list` environment to define many of its standard environments, including:

- The three list environments (Section 4.2)
- The `quote`, `quotation`, and `verse` environments (Section 4.8)
- Proclamations (Section 4.4)
- The style environments `center`, `flushleft`, and `flushright` (Section 4.3)
- The `thebibliography` environment (Section 8.5.1)
- The `theindex` environment (Section 8.5.2)

14.6.1 Length commands for the `list` environment

The general layout of a list is shown in Figure 14.2. It uses six horizontal measurements and three vertical measurements. I now list these length commands.

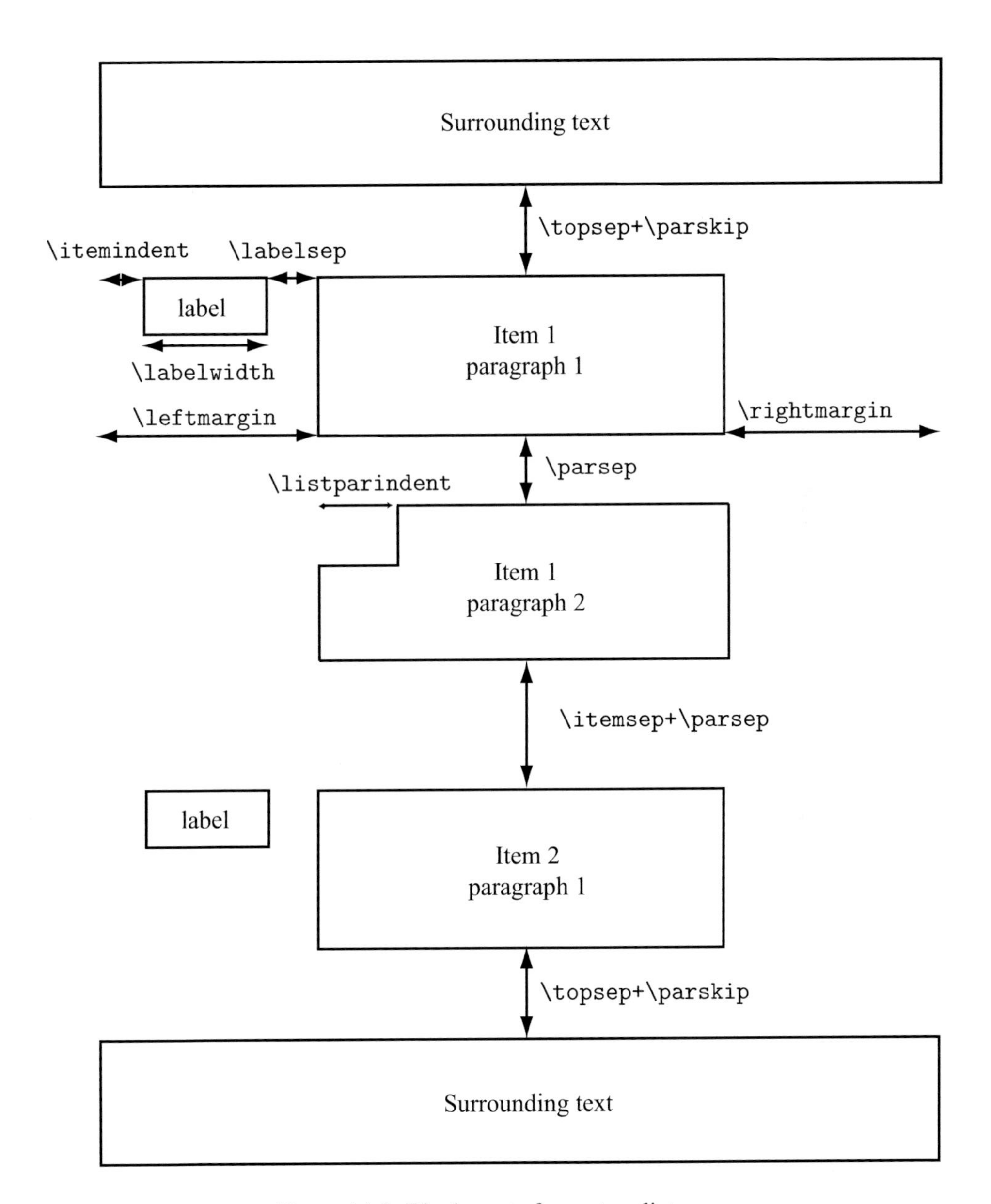

Figure 14.2: The layout of a custom list.

Vertical length commands

\topsep is most of the vertical space between the first item and the preceding text, and also between the last item and the following text. This space also includes \parskip, the extra vertical space inserted between paragraphs, and optionally, \partopsep, provided that the list environment starts a new paragraph.

\parsep is the space between paragraphs of the same item.

\itemsep is the space between items. Like \topsep, the actual gap is the sum of \itemsep and \parsep.

All of these vertical length commands are rubber lengths (see Section 14.5.2).

Horizontal length commands

By default, the margins of a list environment are the same as the margins of the surrounding text. If the list is nested within a list, the margins are wider and so the text is narrower.

The \leftmargin and \rightmargin length commands specify the distance between the edge of the item box and the left and right margins of the page.

The label is the text provided by the optional argument of an \item command or provided as a default in the definition of the list environment. It is typeset in a box of width \labelwidth, which is indented \itemindent units from the left margin, and separated by a space of \labelsep units from the text box. If the label is too wide to fit in the box, it is typeset at its full natural width, and the first line in the text box is indented.

The second and subsequent paragraphs of an item are typeset with their first lines indented by \listparindent units.

14.6.2 *The* list *environment*

Custom lists are created with the list environment, which is invoked as follows:

```
\begin{list}{default_label}{declarations}
   \item item1
   \item item2
   ...
\end{list}
```

The arguments are

- *default_label*, the label for any items that do not specify their own, similar to the optional argument of the \item command
- *declarations*, the vertical and horizontal length commands and any other required parameters for the list

Here is a very simple example:

Here are the most important LaTeX rules about spaces in text, sentences, and paragraphs:

- ◇ **Rule 1:** Two or more spaces in text are the same as one.
- ◇ **Rule 2:** A blank line (that is, two end-of-line characters separated only by blanks and tabs) indicates the end of a paragraph.

Rules 1 and 2 make typing and copying very convenient.

I have used the ◇ math symbol (`$\diamondsuit$`) as a default label, and I set the item box 0.5 inch from either margin. So this example is typed as follows:

```
\noindent Here are the most important \la\ rules about
spaces in text, sentences, and paragraphs:
\begin{list}{$\diamondsuit$}{\setlength{\leftmargin}%
            {.5in}\setlength{\rightmargin}{.5in}}
\item \textbf{Rule 1:} Two or more spaces in text
are the same as one.
\item \textbf{Rule 2:} A blank line (that is, two
end-of-line characters separated only by blanks and tabs)
indicates the end of a paragraph.
\end{list}
Rules 1 and~2 make typing and copying very convenient.
```

Here is a second variant:

Here are the most important LaTeX rules about spaces in text, sentences, and paragraphs:

Rule 1: Two or more consecutive spaces in text are the same as one.

Rule 2: A blank line (that is, two end-of-line characters separated only by blanks and tabs) indicates the end of a paragraph.

Rules 1 and 2 make typing and copying very convenient.

In this example, I dropped the optional *default_label* and typed **Rule 1:** and **Rule 2:** as (optional) arguments of the `\item` commands:

```
\noindent Here are the most important \la\ rules about
spaces in text, sentences, and paragraphs:
\begin{list}{}{\setlength{\leftmargin}{.5in}%
               \setlength{\rightmargin}{.5in}}
```

```
\item[\textbf{Rule 1:}] Two or more consecutive spaces in
text are the same as one.
\item[\textbf{Rule 2:}] A blank line (that is,
two end-of-line characters separated only by blanks and
tabs) indicates the end of a paragraph.
\end{list}
Rules 1 and~2 make typing and copying very convenient.
```

For further simple examples, you can look at various document class files to see how standard environments such as `verse`, `quote`, and so on, are defined.

Using counters

It is not very LaTeX-like to provide the numbers for the rules in the examples above. It would be more logical for LaTeX to do the numbering. The following is a more LaTeX-like coding of the second example:

```
\noindent Here are the most important \la\ rules about
spaces in text, sentences, and paragraphs:
\newcounter{spacerule}
\begin{list}{\textbf{Rule \arabic{spacerule}:}}
            {\setlength{\leftmargin}{.5in}
             \setlength{\rightmargin}{.5in}
             \usecounter{spacerule}}
   \item Two or more consecutive spaces in text are the
   same as one.\label{Li:Twoor}
   \item A blank line (that is, two end-of-line
    characters separated only by blanks and tabs)
    indicates the end of a paragraph. \label{Li:blankline}
\end{list}
Rules \ref{Li:Twoor} and~\ref{Li:blankline} make typing
and copying very convenient.
```

Note that

1. I declared the counter before the `list` environment with the line

   ```
   \newcounter{spacerule}
   ```

2. I defined the *default_label* as

   ```
   \textbf{Rule \arabic{spacerule}:}
   ```

3. In the *declarations*, I specified that the list should use the `spacerule` counter with the command

   ```
   \usecounter{spacerule}
   ```

14.6.3 Two complete examples

In the previous examples, I set the values of `\leftmargin` and `\rightmargin`. The other length commands were not redefined, so their values remained the values set by the document class. In the following examples, I set the values of many more length commands.

Example 1 To get the following list,

Here are the most important LaTeX rules about spaces in text, sentences, and paragraphs:

Rule 1: *Two or more consecutive spaces in text are the same as one.*

Rule 2: *A blank line—that is, two end-of-line characters separated only by blanks and tabs—indicates the end of a paragraph.*

Rules 1 and 2 make typing and copying very convenient.

we type

```
\noindent Here are the most important \la\ rules about
spaces in text, sentences, and paragraphs:
\newcounter{spacerule}
\begin{list}{\upshape\bfseries Rule \arabic{spacerule}:}
            {\setlength{\leftmargin}{1.5in}
             \setlength{\rightmargin}{0.6in}
             \setlength{\labelwidth}{1.0in}
             \setlength{\labelsep}{0.2in}
             \setlength{\parsep}{0.5ex plus 0.2ex
                                   minus 0.1ex}
             \setlength{\itemsep}{0ex plus 0.2ex
                                    minus 0ex}
             \usecounter{spacerule}
             \itshape}
   \item Two or more consecutive spaces in text are the
   same as one.\label{Li:Twoor}
   \item A blank line---that is, two end-of-line
    characters separated only by blanks and
                           tabs---indicates
    the end of a paragraph.\label{Li:blankline}
\end{list}
Rules \ref{Li:Twoor} and~\ref{Li:blankline} make typing
and copying very convenient.
```

Note that

1. I declared the counter as in the previous example.
2. The last item in *declarations* is \itshape, which typesets the entire list in italics.
3. The *default_label* is defined as

   ```
   \upshape\bfseries Rule \arabic{spacerule}
   ```

   My first attempt was to define it as

   ```
   \bfseries Rule \arabic{spacerule}
   ```

   which typesets Rule in bold italics (because in Step 2 we set the whole list in italics). To force the label to be typeset upright, I start the *default_label* with the \upshape command.
4. The left margin is set to 1.5 inches and the right margin to 0.6 inches:

   ```
   \setlength{\leftmargin}{1.5in}
   \setlength{\rightmargin}{0.6in}
   ```

5. Next I set the width of the label to 1 inch, and the space between the label and the item to 0.2 inches:

   ```
   \setlength{\labelwidth}{1.0in}
   \setlength{\labelsep}{0.2in}
   ```

6. Finally, I set the paragraph separation to 0.5 ex, allowing stretching by 0.2 ex and shrinking by 0.1 ex, and the item separation to 0 ex, allowing stretching by 0.2 ex and no shrinking, by

   ```
   \setlength{\parsep}{0.5ex plus 0.2ex minus 0.1ex}
   \setlength{\itemsep}{0ex plus 0.2ex minus 0ex}
   ```

   The actual amount of item separation is calculated by adding the values specified for \parsep and \itemsep.

A complicated list such as this should be defined as a new environment. For example, you could define a myrules environment:

```
\newenvironment{myrules}
  {\begin{list}
    {\upshape \bfseries Rule \arabic{spacerule}:}
    {\setlength{\leftmargin}{1.5in}
      \setlength{\rightmargin}{0.6in}
      \setlength{\labelwidth}{1.0in}
      \setlength{\labelsep}{0.2in}
      \setlength{\parsep}{0.5ex plus 0.2ex minus 0.1ex}
```

```
        \setlength{\itemsep}{0ex plus 0.2ex minus 0ex}
        \usecounter{spacerule}
        \itshape} }
   {\end{list}}
```

and then use it anywhere, as in

```
\begin{myrules}
   \item Two or more consecutive spaces in text are the
    same as one.\label{Li:Twoor}
   \item A blank line---that is, two end-of-line
    characters separated only by blanks and
    tabs---indicates the end of a paragraph.
                        \label{Li:blankline}
Rules \ref{Li:Twoor} and~\ref{Li:blankline} make typing
\end{myrules}
and copying very convenient.
```

which typesets as the first example shown on page 413.

Example 2 In Section 3.7.2, we discussed the formatting of the following type of glossary:

sentence a group of words terminated by a period, exclamation point, or question mark.

paragraph a group of sentences terminated by a blank line or by the new paragraph command.

Now we can create the glossary as a custom list:

```
\begin{list}{}
   {\setlength{\leftmargin}{30pt}
     \setlength{\rightmargin}{0pt}
     \setlength{\itemindent}{14pt}
     \setlength{\labelwidth}{40pt}
     \setlength{\labelsep}{5pt}
     \setlength{\parsep}{0.5ex plus 0.2ex minus 0.1ex}
     \setlength{\itemsep}{0ex plus 0.2ex minus 0ex}}
   \item[\textbf{sentence}\hfill] is a group of words
   terminated by a period, exclamation point,
   or question mark.
   \item[\textbf{paragraph}\hfill] is a group of sentences
    terminated by a blank line or by the \com{par} command.
\end{list}
```

There is nothing new in this example except the `\hfill` commands in the optional arguments to left adjust the labels. With the long words in the example this adjustment is not necessary, but it would be needed for shorter words.

See Section 3.3 of *The LaTeX Companion,* 2nd edition [56] on how to customize the three standard list environments and also for more complicated custom lists.

14.6.4 *The* `trivlist` *environment*

LaTeX also provides a `trivlist` environment, meant more for programmers than users. The environment is invoked in the form

```
\begin{trivlist}
   body
\end{trivlist}
```

It is similar to the `list` environment except that there are no arguments, and all the length commands are trivially set, most to 0 points, except for `\listparindent` and `\parsep`, which are set to equal `\parindent` and `\parskip`, respectively. For instance, LaTeX defines the `center` environment as follows:

```
\begin{trivlist}
   \centering \item[]
\end{trivlist}
```

14.7 The dangers of customization

We can customize LaTeX in so many ways. We can add packages to expand its power and define new commands that better suit our work habits. These enhance LaTeX and make it easier to work with. But they also introduce difficulties. Let us start with the obvious.

Whoever introduced the command `\textcompwordmark` knew that—even if we use command completion—we are not going to type

```
if\textcompwordmark f
```

to avoid having a ligature (see Section 3.4.6). It is a lot of typing, and the source file becomes hard to read. This cries out for a custom command, say, `\Iff`, which is short and *readable* (see Section 14.1.1).

When introducing custom commands, watch out for the following traps.

PART VI

Long Documents

CHAPTER

15

BibTeX

The BibTeX application, written by Oren Patashnik, assists LaTeX users in compiling bibliographies, especially long ones. Short bibliographies can easily be placed in the document directly (see Section 8.5.1).

It takes a little effort to learn BibTeX. But in the long run, the advantages of building bibliographic databases that can be reused and shared outweigh the disadvantage of a somewhat steep learning curve. The *bibliographic database files*, the bib files, contain the *bibliographic entries*. We discuss the format of these entries in Section 15.1, and then describe how to use BibTeX to create bibliographies in Section 15.2.

BibTeX uses a style, called a *bibliographic style,* or bst file, to format entries. On the next two pages we show the bibliography of the secondarticleb.tex sample article typeset with six different style files.

To simplify our discussion, in the rest of this chapter I discuss only one style, the AMS plain style, amsplain.bst, version 2.0. All of the examples shown are in this style, and several of the comments I make are true only for the AMS plain style. If you choose to use a different style, you should check its documentation for special rules.

G. Grätzer, *More Math Into* LaTeX, DOI 10.1007/978-3-319-23796-1_15

[1] Soo-Key Foo. *Lattice Constructions.* PhD thesis, University of Winnebago, Winnebago, MN, December 1990.

[2] George A. Menuhin. *Universal Algebra.* D. van Nostrand, Princeton, 1968.

[3] Ernest T. Moynahan. Ideals and congruence relations in lattices. II. *Magyar Tud. Akad. Mat. Fiz. Oszt. Közl.*, 7:417–434, 1957.

[4] Ernest T. Moynahan. On a problem of M. Stone. *Acta Math. Acad. Sci. Hungar.*, 8:455–460, 1957.

[5] Ferenc R. Richardson. *General Lattice Theory.* Mir, Moscow, expanded and revised edition, 1982.

`plain.bst`

[Foo90] Soo-Key Foo. *Lattice Constructions.* PhD thesis, University of Winnebago, Winnebago, MN, December 1990.

[Men68] George A. Menuhin. *Universal Algebra.* D. van Nostrand, Princeton, 1968.

[Moy57a] Ernest T. Moynahan. Ideals and congruence relations in lattices. II. *Magyar Tud. Akad. Mat. Fiz. Oszt. Közl.*, 7:417–434, 1957.

[Moy57b] Ernest T. Moynahan. On a problem of M. Stone. *Acta Math. Acad. Sci. Hungar.*, 8:455–460, 1957.

[Ric82] Ferenc R. Richardson. *General Lattice Theory.* Mir, Moscow, expanded and revised edition, 1982.

`alpha.bst`

1. Soo-Key Foo, *Lattice constructions*, Ph.D. thesis, University of Winnebago, Winnebago, MN, December 1990.
2. George A. Menuhin, *Universal algebra*, D. van Nostrand, Princeton, 1968.
3. Ernest T. Moynahan, *Ideals and congruence relations in lattices.* II, Magyar Tud. Akad. Mat. Fiz. Oszt. Közl. **7** (1957), 417–434 (Hungarian).
4. ———, *On a problem of M. Stone*, Acta Math. Acad. Sci. Hungar. **8** (1957), 455–460.
5. Ferenc R. Richardson, *General lattice theory*, expanded and revised ed., Mir, Moscow, 1982 (Russian).

`amsplain.bst`

[Foo90] Soo-Key Foo, *Lattice constructions*, Ph.D. thesis, University of Winnebago, Winnebago, MN, December 1990.

[Men68] George A. Menuhin, *Universal algebra*, D. van Nostrand, Princeton, 1968.

[Moy57a] Ernest T. Moynahan, *Ideals and congruence relations in lattices.* II, Magyar Tud. Akad. Mat. Fiz. Oszt. Közl. **7** (1957), 417–434 (Hungarian).

[Moy57b] Ernest T. Moynahan, *On a problem of M. Stone*, Acta Math. Acad. Sci. Hungar. **8** (1957), 455–460.

[Ric82] Ferenc R. Richardson, *General lattice theory*, expanded and revised ed., Mir, Moscow, 1982 (Russian).

`amsalpha.bst`

[1] S.-K. Foo, *Lattice Constructions*, PhD thesis, University of Winnebago, Winnebago, MN, Dec. 1990.
[2] G. A. Menuhin, *Universal Algebra*, D. van Nostrand, Princeton, 1968.
[3] E. T. Moynahan, *Ideals and congruence relations in lattices*. II, Magyar Tud. Akad. Mat. Fiz. Oszt. Közl., 7 (1957), pp. 417–434.
[4] ———, *On a problem of M. Stone*, Acta Math. Acad. Sci. Hungar., 8 (1957), pp. 455–460.
[5] F. R. Richardson, *General Lattice Theory*, Mir, Moscow, expanded and revised ed., 1982.

`siam.bst`

[1] F. R. Richardson, *General Lattice Theory*. Moscow: Mir, expanded and revised ed., 1982.
[2] G. A. Menuhin, *Universal Algebra*. Princeton: D. van Nostrand, 1968.
[3] E. T. Moynahan, "On a problem of M. Stone," *Acta Math. Acad. Sci. Hungar.*, vol. 8, pp. 455–460, 1957.
[4] S.-K. Foo, *Lattice Constructions*. PhD thesis, University of Winnebago, Winnebago, MN, Dec. 1990.
[5] E. T. Moynahan, "Ideals and congruence relations in lattices. II," *Magyar Tud. Akad. Mat. Fiz. Oszt. Közl.*, vol. 7, pp. 417–434, 1957.

`ieeetr.bst`

15.1 The database

A BibTeX database is a text file containing bibliographic entries. To use BibTeX, you first have to learn how to assemble a database. This section explains how to do that.

There may be special tools available for your computer system that assist you in building and maintaining your bibliographic data. Such tools make compiling the data easier and may minimize formatting errors.

You can find all the examples in this section in the `template.bib` file in the `samples` folder (see page 5).

15.1.1 Entry types

A bibliographic entry is given in pieces called *fields*. The style (see Section 15.2.2) specifies how these fields are typeset. Here are two typical entries:

```
@BOOK{gM68,
   author = "George A. Menuhin",
   title = "Universal Algebra",
   publisher = "D.~Van Nostrand",
   address = "Princeton",
   year = 1968,
   }

@ARTICLE{eM57,
   author = "Ernest T. Moynahan",
   title = "On a Problem of {M. Stone}",
```

```
journal = "Acta Math. Acad. Sci. Hungar.",
pages = "455-460",
volume = 8,
year = 1957,
}
```

The start of an entry is indicated with an at sign (@) followed by the *entry type*. In the first example, the entry type is BOOK, while in the second, it is ARTICLE. The entry type is followed by a left brace ({). The matching right brace (}) indicates the end of the entry. BIBTEX also allows you to use parentheses as delimiters for an entry. In this book, however, we use braces to enclose an entry.

The string @BOOK{ is followed by a *label*, gM68, which designates the name of the entry. Refer to this entry in your document with \cite{gM68}. The label is followed by a comma and a series of fields. In this example, there are five fields, author, title, publisher, address, and year. Each field starts with the field name, followed by = and the value of the field enclosed in double quotes ("). Be sure to use " and *not* LATEX double quotes (“ or ”). Alternatively, BIBTEX also allows you to use braces to enclose the field value. In this book, we use double quotes to enclose a field.

Numeric field values, that is, fields consisting entirely of digits, do not need to be enclosed in double quotes or braces, for instance, year in the examples above, volume in the second example, and number in some of the examples that follow. Page ranges, such as 455-460, are not numeric field values since they contain -, so they must be enclosed in double quotes or braces.

There *must* be a comma before each field. The comma before the first field is placed after the label.

There are many standard entry types, including

ARTICLE an article in a journal or magazine

BOOK a book with an author (or editor) and a publisher

BOOKLET a printed work without a publisher

INBOOK a part of a book, such as a chapter or a page range that, in general, is not titled or authored separately

INCOLLECTION a part of a book with its own title and perhaps author

INPROCEEDINGS an article in a conference proceedings with its own title and author

MANUAL technical documentation

MASTERSTHESIS a master's thesis

MISC an entry that does not fit in any other category

PHDTHESIS a Ph.D. thesis

PROCEEDINGS the proceedings of a conference

`TECHREPORT` a report published by a school or institution

`UNPUBLISHED` an unpublished paper

Each entry includes a number of *fields* from the following list:

```
address        institution    pages
author         journal        publisher
booktitle      key            school
chapter        language       series
crossref       month          title
edition        note           type
editor         number         volume
howpublished   organization   year
```

The style you choose determines which of the fields within an entry are actually used. All the others are ignored. You may also add fields for your own use. For example, you may want to add a `mycomments` field for personal comments. Such fields are ignored unless you have a bibliography style that uses them.

Commonly used examples of new field names include `URL`, `abstract`, `ISBN`, `keywords`, `mrnumber`, and so on. The `language` field is used by the AMS styles but not by any of the other styles mentioned in this chapter.

Tip

1. BIBTEX does not care whether you use uppercase or lowercase letters (or mixed) for the names of entry types and fields. In this book, the entry types are shown in uppercase and field names in lowercase.
2. Placing a comma after the last field is optional. I recommend that you put it there so that when you append a new field to the entry, the required comma separating the fields is present.

For each entry type there are both required and optional fields. Later in this section, I give two examples of each entry type. The first example of an entry type uses a small set of fields, while the second example is a maximal one, showing a large number of optional fields.

15.1.2 *Typing fields*

Make sure you type the field names correctly. If you misspell one, BIBTEX ignores the field. BIBTEX also warns you if a required field is missing. The `author` and `editor` fields require a name.

Rule ■ Names

1. Most names can be typed as usual, `"Ernest T. Moynahan"` or `"Moynahan, Ernest T."`, with one comma separating the family name from the given names.

2. Type two or more names separated by `and`. For instance,

```
author= "George Blue and Ernest Brown and Soo-Key Foo",
```

3. The family name of Miguel Lopez Fernandez is Lopez Fernandez, so type it as `"Lopez Fernandez, Miguel"`. This informs BIBTEX that Lopez is not a middle name.

4. Type Orrin Frink, Jr. as `"Frink, Jr., Orrin"`.

Rules 3 and 4 are seldom needed. In a bibliography of about 1,500 items, I found fewer than 10 names that could not be typed as usual. Note that you can type John von Neumann as `"John von Neumann"` or `"von Neumann, John"`. Because BIBTEX knows about `von`, it handles the name properly.

There are a few rules concerning the `title` field.

Rule ■ Title

1. You should not put a period at the end of a title. The style supplies the appropriate punctuation.

2. Many styles, including the AMS styles, convert titles, except for the first letter of the title, to lowercase for all entry types. If you want a letter to appear in uppercase, put it—or the entire word—in braces. The same rule applies to the `edition` field. Some other styles only do this conversion for the titles of non-book-like entries.

3. To maximize the portability of your database, you should type titles with each important word capitalized:

```
title = "On a Problem of {M. Stone}",
```

The style used in this book, `amsplain.bst`, converts `Problem` to `problem`, so it makes no difference, but some styles do not. To be on the safe side, you should capitalize all words that may have to be capitalized.

For the record, here are the complete rules for titles:

Rule ■ Capitalize:

1. the first word;
2. the first word in a subtitle (BIBTEX assumes that a subtitle follows a colon, so it capitalizes the first word after a colon—a colon not introducing a subtitle should be typed in braces);
3. all other words except articles, unstressed conjunctions, and unstressed prepositions.

Words that should never be converted to lowercase, for example proper names such as `Hilbert`, should be enclosed in braces to prevent them from being converted to lowercase. In the example above, two letters in the title should not be converted to lowercase, so we enclosed `M. Stone` in braces. We could also have typed

```
{M. S}tone+ or \verb+{M.} {S}tone
```

BIBTEX and the style automatically handle a number of things for you that you would have to handle yourself when typing text.

1. You do not have to mark periods in abbreviations, as `.\␣` in the names of journals (see Section 3.2.2). So

   ```
   journal = "Acta Math. Acad. Sci. Hungar.",
   ```

   typesets correctly.

2. You can type a single hyphen for a page range instead of the usual `--` in the `pages` field (see Section 3.4.2). So

   ```
   pages = "455-460",
   ```

   typesets correctly with an en dash.

3. You do not have to type nonbreakable spaces with `~` in the `author` or `editor` fields (see Section 3.4.3):

   ```
   author = "George A. Menuhin",
   ```

   is correct. Normally you would type `George~A. Menuhin`.

Finally, we state a rule about accented characters.

Rule ■ Accents

Put accented characters in braces: `{\"{a}}`.

This rule means that

```
 author = "Paul Erd\H{o}s",
```

is not recommended. Instead, type

```
author = "Paul Erd{\H{o}}s",
```

This rule is, again, about portability. Some styles, e.g., alpha and amsalpha, create a citation for an article from the first three letters of the name and the last two digits of the year.

```
author = "Kurt G{\"{o}}del",
year = 1931,
```

creates the citation: [Göd31]. The accent is used only if the accents rule has been followed.

The downside of this rule is that the braces suppress kerning.

15.1.3 Articles

Entry type	ARTICLE
Required fields	author, title, journal, year, pages
Optional fields	volume, number, language, note

Examples:

1. Ernest T. Moynahan, *On a problem of M. Stone*, Acta Math. Acad. Sci. Hungar. **8** (1957), 455–460.
2. Ernest T. Moynahan, *On a problem of M. Stone*, Acta Math. Acad. Sci. Hungar. **8** (1957), no. 5, 455–460 (English), Russian translation available.

typed as

```
@ARTICLE{eM57,
   author = "Ernest T. Moynahan",
   title = "On a Problem of {M. Stone}",
   journal = "Acta Math. Acad. Sci. Hungar.",
   pages = "455-460",
   volume = 8,
   year = 1957,
   }

@ARTICLE{eM57a,
   author = "Ernest T. Moynahan",
   title = "On a Problem of {M. Stone}",
   journal = "Acta Math. Acad. Sci. Hungar.",
   pages = "455-460",
```

```
volume = 8,
number = 5,
year = 1957,
note = "Russian translation available",
language = "English",
}
```

15.1.4 Books

Entry type	BOOK
Required fields	author (or editor), title, publisher, year
Optional fields	edition, series, volume, number, address, month, language, note

Examples:

1. George A. Menuhin, *Universal algebra*, D. Van Nostrand, Princeton, 1968.
2. George A. Menuhin, *Universal algebra*, second ed., University Series in Higher Mathematics, vol. 58, D. Van Nostrand, Princeton, March 1968 (English), no Russian translation.

typed as

```
@BOOK{gM68,
   author = "George A. Menuhin",
   title = "Universal Algebra",
   publisher = "D.~Van Nostrand",
   address = "Princeton",
   year = 1968,
   }

@BOOK{gM68a,
   author = "George A. Menuhin",
   title = "Universal Algebra",
   publisher = "D.~Van Nostrand",
   address = "Princeton",
   year = 1968,
   month = mar,
   series = "University Series in Higher Mathematics",
   volume = 58,
   edition = "Second",
   note = "no Russian translation",
   language = "English",
   }
```

Abbreviations, such as mar, are discussed in Section 15.1.9.

A second variant of book has an `editor` instead of an `author`:

15. Robert S. Prescott (ed.), *Universal algebra*, D. Van Nostrand, Princeton, 1968.

typed as

```
@BOOK{rP68,
   editor = "Robert S. Prescott",
   title = "Universal Algebra",
   publisher = "D.~Van Nostrand",
   address = "Princeton",
   year = 1968,
   }
```

15.1.5 *Conference proceedings and collections*

Entry type	`INPROCEEDINGS`
Required fields	`author`, `title`, `booktitle`, `year`
Optional fields	`address`, `editor`, `series`, `volume`, `number`, `organization`, `publisher`, `month`, `note`, `pages`, `language`

Examples:

7. Peter A. Konig, *Composition of functions.* Proceedings of the Conference on Universal Algebra, 1970.
8. Peter A. Konig, *Composition of functions.* Proceedings of the Conference on Universal Algebra (Kingston, ON) (G. H. Birnbaum, ed.), vol. 7, Canadian Mathematical Society, Queen's Univ., December 1970, available from the Montreal office, pp. 1–106 (English).

typed as

```
@INPROCEEDINGS{pK69,
   author = "Peter A. Konig",
   title = "Composition of Functions".
   booktitle = "Proceedings of the Conference on
      Universal Algebra",
   year = 1970,
   }

@INPROCEEDINGS{pK69a,
   author = "Peter A. Konig",
   title = "Composition of Functions".
   booktitle = "Proceedings of the Conference on
```

```
    Universal Algebra",
  address = "Kingston, ON",
  publisher = "Queen's Univ.",
  organization = "Canadian Mathematical Society",
  editor = "G. H. Birnbaum",
  pages = "1-106",
  volume = 7,
  year = 1970,
  month = dec,
  language = "English",
  }
```

The `address` field provides the location of the meeting. The address of the publisher should be in the `publisher` field and the address of the organization in the `organization` field.

Entry type	`INCOLLECTION`
Required fields	`author`, `title`, `booktitle`, `publisher`, `year`
Optional fields	`editor`, `series`, `volume`, `number`, `address`, `edition`, `month`, `note`, `pages`, `language`

Examples:

1. Henry H. Albert, *Free torsoids*, Current Trends in Lattices, D. Van Nostrand, 1970.
2. Henry H. Albert, *Free torsoids*, Current Trends in Lattices (George Burns, ed.), vol. 2, D. Van Nostrand, Princeton, January 1970, new edition is due next year, pp. 173–215 (German).

is typed as

```
\noindent\verb+@INCOLLECTION{hA70,
  author = "Henry H. Albert",
  title = "Free Torsoids",
  booktitle = "Current Trends in Lattices".
  publisher = "D.~Van Nostrand",
  year = 1970,
  }


@INCOLLECTION{hA70a,
  author = "Henry H. Albert",
  editor = "George Burns",
  title = "Free Torsoids",
  booktitle = "Current Trends in Lattices".
  publisher = "D.~Van Nostrand",
  address = "Princeton",
```

```
    pages = "173-215",
    volume = 2,
    year = 1970,
    month = jan,
    note = "new edition is due next year",
    language = "German",
    }
```

The `address` field contains the address of the publisher.

Cross-referencing

If your database has several articles from the same conference proceedings and collections, you may prefer to make an entry for the entire volume, and cross-reference individual articles to that entry. For instance,

```
@PROCEEDINGS{UA69,
    title = "Proceedings of the Conference on,
       Universal Algebra",
    booktitle = "Proceedings of the Conference on
       Universal Algebra",
    address = "Kingston, ON",
    publisher = "Canadian Mathematical Society",
    editor = "G. H. Birnbaum",
    volume = 7,
    year = 1970,
    }
```

may be the entry for the proceedings volume as a whole, and

```
@INPROCEEDINGS{pK69a,
    author = "Peter A. Konig",
    title = "Composition of Functions",
    booktitle = "Proceedings of the Conference on
       Universal Algebra",
    pages = "1-106",
    crossref = "UA69",
    }
```

is the cross-referencing entry for a specific article. These two entries produce the following:

1. G. H. Birnbaum (ed.), *Proceedings of the conference on universal algebra*, vol. 7, Kingston, ON, Canadian Mathematical Society, 1970.
2. Peter A. Konig, *Composition of functions*, in Birnbaum [1], pp. 1–106.

Rule ■ Cross-references

1. All the required fields of the cross-referencing entry must appear in either that entry or in the cross-referenced entry.

2. The cross-referenced entry should have both a `title` and a `booktitle` field.

3. The cross-referenced entry must appear in the `bib` file later than any entry that cross-references it.

15.1.6 *Theses*

Entry type `MASTERSTHESIS` or `PHDTHESIS`
Required fields `author`, `title`, `school`, `year`
Optional fields `type`, `address`, `month`, `note`, `pages`
Examples:

1. Soo-Key Foo, *Lattice constructions*, Ph.D. thesis, University of Winnebago, 1990.
2. Soo-Key Foo, *Lattice constructions*, Ph.D. dissertation, University of Winnebago, Winnebago, MN, December 1990, final revision not yet available, pp. 1–126.

is typed as

```
@PHDTHESIS{sF90,
   author = "Soo-Key Foo",
   title = "Lattice Constructions",
   school = "University of Winnebago",
   year = 1990,
   }

@PHDTHESIS{sF90a,
   author = "Soo-Key Foo",
   title = "Lattice Constructions",
   school = "University of Winnebago",
   address = "Winnebago, MN",
   year = 1990,
   month = dec,
   note = "final revision not yet available",
   type = "Ph.D. dissertation",
   pages = "1-126",
   }
```

If the `type` field is present, its content takes the place of the phrase Ph.D. thesis (or Master's thesis).

15.1.7 Technical reports

Entry type `TECHREPORT`
Required fields `author`, `title`, `institution`, `year`
Optional fields `type`, `number`, `address`, `month`, `note`
Examples:

1. Grant H. Foster, *Computational complexity in lattice theory*, tech. report, Carnegie Mellon University, 1986.
2. Grant H. Foster, *Computational complexity in lattice theory*, Research Note 128A, Carnegie Mellon University, Pittsburgh, PA, December 1986, in preparation.

is typed as

```
@TECHREPORT{gF86,
   author = "Grant H. Foster",
   title = "Computational Complexity in Lattice Theory",
   institution = "Carnegie Mellon University",
   year = 1986,
   }

@TECHREPORT{gF86a,
   author = "Grant H. Foster",
   title = "Computational Complexity in Lattice Theory",
   institution = "Carnegie Mellon University",
   year = 1986,
   month = dec,
   type = "Research Note",
   address = "Pittsburgh, PA",
   number = "128A",
   note = "in preparation",
   }
```

15.1.8 Manuscripts and other entry types

Entry type `UNPUBLISHED`
Required fields `author`, `title`, `note`
Optional fields `month`, `year`

Examples:

1. William A. Landau, *Representations of complete lattices*, manuscript, 55 pages.
2. William A. Landau, *Representations of complete lattices*, manuscript, 55 pages, December 1975.

is typed as

```
@UNPUBLISHED{wL75,
   author = "William A. Landau",
   title = "Representations of Complete Lattices",
   note = "manuscript, 55~pages",
   }

@UNPUBLISHED{wL75a,
   author = "William A. Landau",
   title = "Representations of Complete Lattices",
   year = 1975,
   month = dec,
   note = "manuscript, 55~pages",
   }
```

Other standard entry types include

Entry type BOOKLET
Required field title
Optional fields author, howpublished, address, month, year, note

Entry type INBOOK
Required fields author or editor, title, chapter or pages, publisher, year
Optional fields series, volume, number, type, address, edition, month, pages, language, note

Entry type MANUAL
Required field title
Optional fields author, organization, address, edition, month, year, note

Entry type MISC
Required field at least one of the optional fields must be present
Optional fields author, title, howpublished, month, year, note, pages

Entry type PROCEEDINGS
Required fields title, year
Optional fields editor, series, volume, number, address, organization, publisher, month, note

15.1.9 Abbreviations

You may have noticed the field month = dec in some of the examples. This field uses an abbreviation. Most BIBTEX styles, including the AMS styles, include abbreviations for the months of the year: jan, feb, …, dec. When an abbreviation is used, it is not enclosed in quotes (") or braces ({ }). The style defines what is actually to be typeset. Most styles typeset dec as either Dec. or December.

The name of the abbreviation, such as `dec`, is a string of characters that starts with a letter, does not contain a space, an equal sign (=), a comma, or any of the special characters listed in Section 3.4.4.

You may define your own abbreviations using the command `@STRING`. For example,

```
@STRING{au = "Algebra Universalis"}
```

A string definition can be placed anywhere in a `bib` file, as long as it precedes the first use of the abbreviation in an entry.

The AMS supplies the `mrabbrev.bib` file containing the standard abbreviations for many mathematical journals. Find it at `ams.org`, under `Reference Tools`, click on `MR Serials Abbreviations for BibTeX`. Based on this file, you can make your own `abbrev.bib` file containing entries for all the journals you reference with whatever abbreviations you find easiest to remember.

If you use this scheme, the command you use to specify the `bib` files may look like

```
\bibliography{abbrev,... }
```

Section 15.2.1 explains the `\bibliography` command.

15.2 Using BIBTEX

In Section 15.1, you learned how to create database files. The sample `bib` files are `template.bib` and `secondarticleb.bib` in the `samples` folder (see page 5). In this section, you learn how to use BIBTEX to process these files to create a bibliography. We illustrate the process of working with BIBTEX with the `secondarticleb` sample article.

We use the `amsplain` style. To obtain all six examples of different styles shown on pages 422–423, just change `amsplain` to the appropriate style name in your document and typeset it.

One BIBTEX style behaves differently. The `apacite` style of the American Psychological Association requires that the preamble of your document include the line

```
\usepackage{apalike}
```

in addition to using the style file. The package can also be modified by a large number of options.

15.2.1 Sample files

Type the following two lines to replace the `thebibliography` environment in the `secondarticle.tex` sample document:

```
\bibliographystyle{amsplain}
\bibliography{secondarticleb}
```

Save the new sample article as `secondarticleb.tex`. The first line specifies the `bst` file, `amsplain.bst`, which is part of the AMS distribution (see Section 9.6). The second line specifies the database files used, `secondarticleb.bib`; in this case there is only one.

The contents of the `secondarticleb.bib` bibliographic database file are as follows:

```
@BOOK{gM68,
   author = "George A. Menuhin",
   title = "Universal Algebra",
   publisher = "D.~Van Nostrand",
   address = "Princeton",
   year = 1968,
   }
@BOOK{fR82,
   author = "Ferenc R. Richardson",
   title = "General Lattice Theory",
   edition = "Expanded and Revised",
   language = "Russian",
   publisher = "Mir",
   address = "Moscow",
   year = 1982,
   }
@ARTICLE{eM57,
   author = "Ernest T. Moynahan",
   title = "On a Problem of {M. Stone}",
   journal = "Acta Math. Acad. Sci. Hungar.",
   pages = "455-460",
   volume = 8,
   year = 1957,
   }

@ARTICLE{eM57a,
   author = "Ernest T. Moynahan",
   title = "Ideals and Congruence Relations in
      Lattices.~\textup{II}",
   journal = "Magyar Tud. Akad. Mat. Fiz. Oszt. K{\"{o}}zl.",
   language = "Hungarian",
   pages = "417-434",
   volume = 7,
   year = 1957,
   }

@PHDTHESIS{sF90,
   author = "Soo-Key Foo",
   title = "Lattice Constructions",
```

```
school = "University of Winnebago",
address = "Winnebago, MN",
year = 1990,
month = dec,
}
```

Type `secondarticleb.bib` or copy it from the `samples` folder to your `work` folder.

15.2.2 *Setup*

Before you start BIBTEX, make sure that everything is set up properly as described in this section.

To list database entries in the bibliography, use the `\cite` command. Refer to Section 8.5.1 for details on how to use citations. If you want to have a reference listed in the bibliography without a citation in the text, then use the `\nocite` command. For example,

```
\cite{pK57}
```

includes the reference in the bibliography and cites the entry with label `pK57`, whereas

```
\nocite{pK57}
```

includes the reference in the bibliography but does not cite the entry. In either case, one of the `bib` files specified in the argument of the `\bibliography` command must contain an entry with the label `pK57`. The `\nocite{*}` command includes *all* the entries from the bibliographic databases you've specified.

Your document must specify the bibliography style and must name the `bib` files to be used. For instance, the `secondarticleb.tex` sample article contains the lines

```
\bibliographystyle{amsplain}
\bibliography{secondarticleb}
```

The `\bibliographystyle` command specifies `amsplain.bst` as the style and the `\bibliography` command specifies the database file `secondarticleb.bib`. To use several database files, separate them with commas, as in

```
\bibliography{abbrev,gg,lattice,secondarticleb}
```

where

- `abbrev.bib` contains custom abbreviations
- `gg.bib` contains personal articles
- `lattice.bib` contains lattice theory articles by other authors
- `secondarticleb.bib` contains additional references needed

It is important to make sure that the `bst` file, the `bib` file(s), and the LaTeX document(s) are in folders where BibTeX can find them. If you are just starting out, you can simply copy all of them into one folder. Later, you may want to look for a more permanent solution by keeping the files `abbrev.bib` and `lattice.bib` in one "central" location, while placing `secondarticleb.bib` in the same folder as its corresponding LaTeX document.

15.2.3 *Four steps of BibTeXing*

The following steps produce a typeset bibliography in your LaTeX document. We use the `secondarticleb.tex` sample article as an example.

Step 1 Check that BibTeX, your LaTeX document, and the `bib` files are placed in the appropriate folders.

Step 2 Typeset `secondarticleb.tex` to get a fresh `aux` file. This step is illustrated in Figure 15.1.

Step 3 Run BibTeX on the `secondarticleb.aux` file in one of the following three ways:

- by invoking it with the argument `secondarticleb`
- by starting the application and then opening `secondarticleb.aux`
- by running it by choosing it as a menu option of your editor or GUI front end or by clicking on an icon

If BibTeX cannot find a crucial file, for example, the `bst` file, it stops. The reason it stopped is shown in the `log` window and also written to a `blg` (bibliography log) file, `secondarticleb.blg`. Correct the error(s) and go back to step 2. A successful run creates a `bbl` (bibliography) file, `secondarticleb.bbl`, in addition to `secondarticleb.blg`. This step is illustrated in Figure 15.2.

Step 4 Typeset the LaTeX document `secondarticleb.tex` *twice*.

15.2.4 *BibTeX files*

BibTeX uses and creates a number of files when it is run. To illustrate this process, complete the four steps using `secondarticleb.tex`.

Step 1 Start fresh by deleting the `aux`, `blg`, and `bbl` files, if they are present.

Step 2 Typeset the article `secondarticleb.tex` to get an `aux` file (see Figure 15.1). Notice that the `log` file contains warnings about missing references and a number of other lines not relevant to the current discussion. The lines in the `aux` file containing bibliographic information are

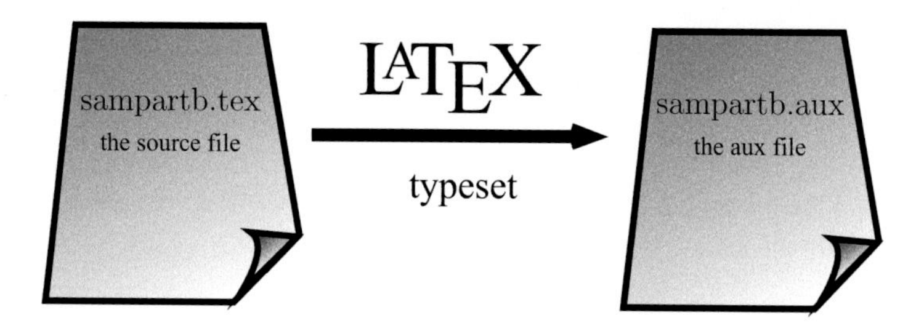

Figure 15.1: Using BIBTEX, step 2.

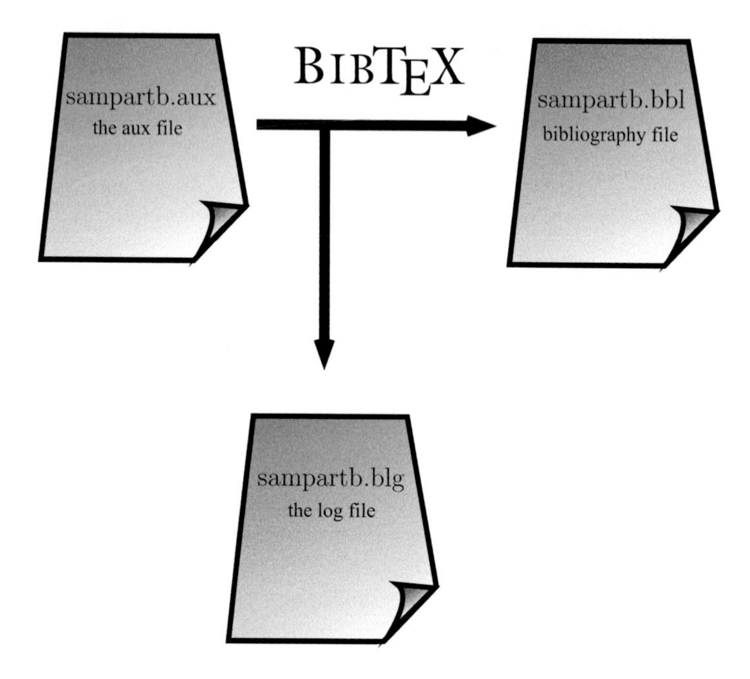

Figure 15.2: Using BIBTEX, step 3.

```
\citation{fR82}
\citation{gM68}
\citation{eM57}
\citation{sF90}
\citation{eM57a}
\bibstyle{amsplain}
\bibdata{secondarticleb}
```

Each `\citation` command in this file corresponds to a `\cite` or `\nocite` command in the article. The lines

```
\bibliographystyle{amsplain}
\bibliography{secondarticleb}
```

in `secondarticleb.tex` are written as

```
\bibstyle{amsplain}
\bibdata{secondarticleb}
```

in the `secondarticleb.aux` file.

Step 3 Now run BIBTEX on the `secondarticleb.aux` file (see Figure 15.2). How we do this, depends on the LATEX installation you have.

BIBTEX generates two new files: `secondarticleb.blg` and `secondarticleb.bbl`. Look at `secondarticleb.blg`:

```
This is BibTeX, C Version 0.99c
The top-level auxiliary file: secondarticleb.aux
The style: amsplain.bst
Database file #1: secondarticleb.bib
```

On some systems, this file may be much longer than the one I show here. At present, this `blg` file does not contain much important information. If there were any warnings or errors, they would be listed in this file.

The `secondarticleb.bbl` file, in which BIBTEX created a `thebibliography` environment (see Section 8.5.1) is more interesting:

```
\providecommand{\bysame}{\leavevmode%
\hbox to3em {\hrulefill}\thinspace}
\begin{thebibliography}{1}

\bibitem{sF90}
Soo-Key Foo, \emph{Lattice constructions},
Ph.D. thesis, University of Winnebago,
Winnebago, MN, December 1990.

\bibitem{gM68}
George~A. Menuhin, \emph{Universal algebra},
D.~Van Nostrand, Princeton, 1968.

\bibitem{eM57a}
Ernest~T. Moynahan, \emph{Ideals and congruence
relations in lattices.~\textup{II}},
Magyar Tud. Akad. Mat. Fiz. Oszt. K{\"{o}}zl.
\textbf{7} (1957), 417-434 (Hungarian).

\bibitem{eM57}
```

```
\bysame, \emph{On a problem of {M. Stone}}, Acta
Math. Acad. Sci. Hungar. \textbf{8} (1957),
455-460.

\bibitem{fR82}
Ferenc~R. Richardson, \emph{General lattice theory},
expanded and revised ed., Mir, Moscow,
1982 (Russian).

\end{thebibliography}
```

Observe that the nonbreakable spaces (ties) and the `\bysame` command have been provided in the author fields.

Step 4 Now typeset `secondarticleb.tex` again. The typeset version now has a REFERENCES section, constructed from the `bbl` file, but the new `log` file has warnings about missing entries. The new `aux` file contains five interesting new lines:

```
\bibcite{sF90}{1}
\bibcite{gM68}{2}
\bibcite{eM57a}{3}
\bibcite{eM57}{4}
\bibcite{fR82}{5}
```

These lines identify the cross-reference label `sF90` (see the first line shown—the symbol designates Foo's thesis in `secondarticleb.bib`) with the number `1`, and so on. Now typeset `secondarticleb.tex` again, and all the citations are correctly placed in the typeset article.

Observe:

1. The crucial step 3, running the BibTeX application, gives different messages and obeys different rules from LaTeX—see Section 15.2.5.
2. The `secondarticleb.bbl` file was created by BibTeX. It is not changed by running LaTeX.

15.2.5 BibTeX rules and messages

Rule ■ **BibTeX and %**
You cannot comment out a field with an %.

For example, the entry

```
@ARTICLE{eM57,
   author = "Ernest T. Moynahan",
   title = "On a Problem of {M. Stone}",
   journal = "Acta Math. Acad. Sci. Hungar.",
   % pages = "455-460",
   volume = 8,
   year = 1957,
   }
```

causes BIBTEX to generate the message

Recall that BIBTEX ignores field names it cannot recognize. So changing the field name pages, for example to pages-comment, does not give a message. However, doing so removes a required field, so you get the warning message

```
Warning--missing pages in eM57
```

Rule ■ BIBTEX field names
Do not abbreviate field names.

For instance, if you abbreviate volume to vol, as in

```
@ARTICLE{eM57,
   author = "Ernest T. Moynahan",
   title = "On a Problem of {M. Stone}",
   journal = "Acta Math. Acad. Sci. Hungar.",
   pages = "455-460",
   vol = 8,
   year = 1957,
   }
```

the vol field is simply ignored. This entry is typeset as

3. Ernest T. Moynahan, *On a problem of M. Stone*, Acta Math. Acad. Sci. Hungar. (1957), 455–460.

instead of

3. Ernest T. Moynahan, *On a problem of M. Stone*, Acta Math. Acad. Sci. Hungar. **8** (1957), 455–460.

Rule ■ BIBTEX field terminations
Make sure that every field of an entry, except possibly the last, is terminated with a comma.

If you drop a comma before a field, you get a message such as

```
I was expecting a ',' or a ')'
                line 6 of file secondarticleb.bib
 :
 :     year = 1968,
(ay have been on previous line)
I'm skipping whatever remains of this entry
Warning--missing year in gM68+
```

Rule ■ BIBTEX field value terminations
Make sure that the field value is properly terminated.

You should be careful not to drop a double quote or brace. If you drop the closing quote on line 11 of the bib file,

```
title = "General Lattice Theory
```

you get the message

```
I was expecting a ',' or a '}'
                   line 12 of file secondarticleb.bib
 :     edition = "+\\
 :                 Expanded and Revised",
I'm skipping whatever remains of this entry
Warning--missing publisher in fR82
Warning--missing year in fR82

I was expecting a ',' or a '}'
                   line 12 of file secondarticleb.bib
 :     edition = "
 :                 Expanded and Revised",
I'm skipping whatever remains of this entry
Warning--missing publisher in fR82
Warning--missing year in fR82
```

If, instead, you drop the opening double quote in the same line, you get the message

```
Warning--string name "general" is undefined
--line 11 of file secondarticleb.bib
I was expecting a ',' or a '}'
```

```
                    line 11 of file secondarticleb.bib
 :    title =  general
 :                      Lattice Theory",
I'm skipping whatever remains of this entry
Warning--missing title in fR82
Warning--missing publisher in fR82
Warning--missing year in fR82
(There was 1 message)+
```

BIBTEX assumed that `general` was an abbreviation, since it was not preceded by a ".

The obvious conclusion is that you have to be very careful about typing your bibliographic entries for BIBTEX. If you have access to special tools for maintaining your bibliographic data, use them. Otherwise, refer to the `template.bib` file that contains templates of often-used entry types.

15.2.6 *Submitting an article*

If you submit an article to a journal that provides you with a BIBTEX style file, then you can submit the article and the BIBTEX database file, pared down of course. If this is not the case, create the `bbl` file with `amsplain.bst` and copy and paste the content into the `thebibliography` environment in the article. Then the journal's editor can edit the bibliography.

15.3 *Concluding comments*

There is a lot more to BIBTEX than what has been covered in this chapter. For example, BIBTEX's algorithm to alphabetize names is fairly complicated. Some names create additional difficulties. Where should John von Neumann be placed, under the "v"-s or the "N"-s? It depends on the style. How do we handle names where the first word is the family name, as in Ho Chi Minh or Grätzer György? Again, it depends on the style.

Oren Patashnik's *BIBTEXing* [62] has many helpful hints. It includes a clever hack to order entries correctly even when the style does not do so. Chapter 13 of *The LATEX Companion,* 2nd edition [56] has a long discussion of BIBTEX. It also contains a long list of styles.

There are many tools to make BIBTEXing easier. *BibDesk* for the Mac is an excellent graphical BIBTEX-bibliography manager. For Windows, there is *BibTexMng*. For UNIX, there is *pybibliographer* and if you are an Emacs user, there is *Ebib*. Written in Java, so available on most platforms, is *JBibtexManager*. There are many BIBTEX databases. The largest one may be "The Collection of Computer Science Bibliographies" with more than two million references.

You can easily build your own mathematical databases with MathSciNet from the AMS. Do a search. When the result page comes up, go to the pull down menu next to `Batch Download` and select `Citations (BibTeX)`. Now you can check mark the items you want by clicking on the little squares and then click on `Retrieve Marked` next to the pull down menu or click on `Retrieve First 50`. For the latter to work well, before your search, click on the `Preferences` button and click on the circle next to 50, so you get at most 50 items per result page. Then `Retrieve First 50` retrieves them all.

Finally, after many years of development, the AMS released `amsrefs`, the kid brother of BibTeX, at its annual meeting in January 2002. The presentation was made by Michael Downes, who designed and coded the package. I was very excited to hear his lecture—bibliographic management was the last block needed to complete the rebuilding of LaTeX. It turned out that `amsrefs` is not simply a BibTeX replacement. It has a number of very important new features.

1. You can, with `amsrefs`, enter the marked up bibliographic entries into the document. This means that the document class of the journal publishing the paper can format your bibliography.

2. `amsrefs` is a LaTeX package. Therefore, you do not have to learn (another) esoteric language to control the formatting of your bibliography. Developing a format for a journal is very easy.

3. The bibliographic data files are also LaTeX files, so you can print them within LaTeX, making it easy to maintain them.

After Michael Downes passed away, David Jones took over the project, and released version 2.0 in June of 2004.

Unfortunately, unlike the BibTeX and its `bbl` file, `amsrefs` directly creates the typeset file. So if the journal you want to submit your article to does not have an `amsrefs` style file—and today only the AMS journals have them—then you have to manually convert the `amsrefs` entries into the format the journal would accept. There is no option to set in `amsrefs` to produce a LaTeX source file for the bibliographic entries. Therefore, unless you know that you intend to submit to an AMS journal—and you know that it will accept your article for publication—you should not use `amsrefs`.

There is a second obstacle. There is no BibDesk or BibTexMng for `amsrefs`. But I believe that if `amsrefs` overcomes the first obstacle, then the second obstacle would resolve itself fast.

Philipp Lehman's `biblatex` package shares some of Michael Downes' goals. It works with (some) BibTeX databases and uses LaTeX to format the bibliography.

CHAPTER

MakeIndex

Pehong Chen's *MakeIndex* application, described in Pehong Chen and Michael A. Harrison's *Index preparation and processing* [10], helps LaTeX users create long indexes. For short indexes, you can easily do without it (see Section 8.5.2).

In Section 16.1, we show you by an example how to prepare an article for indexing. We introduce formally the index commands in Section 16.2. In Section 16.3, we describe how LaTeX and *MakeIndex* process the index entries. The rules are stated in Section 16.4.

Multiple indexes are almost as easy as single indexes. They are described in Section 16.5. We conclude with glossaries in Section 16.6.

Indexing is a difficult task. For an extensive discussion on how to create a useful index, consult *The Chicago Manual of Style,* 15th edition [11].

16.1 Preparing the document

LaTeX provides the `theindex` environment (see Section 8.5.2). Within this environment, it provides the `\item`, `\subitem`, and `\subsubitem` commands to typeset entries, subentries, and subsubentries, respectively, and the `\indexspace` command for adding vertical space between alphabetical blocks, see Figure 16.1 for an example.

G. Grätzer, *More Math Into* LaTeX, DOI 10.1007/978-3-319-23796-1_16

The makeidx package provides the \index command for specifying the index entry at a particular point in the document, which becomes a page reference for the entry in the typeset index.

Making an index entry with *MakeIndex* is easy. You simply place the index commands in your source file, and then let LaTeX and *MakeIndex* do the work of gathering the entries and the page numbers for the entries, sorting them, and formatting the typeset index.

There are three steps:

1. In the preamble of your LaTeX document, include the line

   ```
   \makeindex
   ```

   If you do not use an AMS document class, include the two lines

   ```
   \usepackage{makeidx}
   \makeindex
   ```

2. Type the line

   ```
   \printindex
   ```

   at the point in your document where you want the index to appear, usually as part of the back matter (see Section 8.5).

3. Mark all entries in your document with \index commands.

We illustrate this procedure with the firstarticlei.tex article, which modifies the article firstarticle.tex by inserting a number of index entries (both these files are in the samples folder; see page 5).

We now add a dozen \index commands to firstarticle.tex.

Command 1

Retype the line

```
\begin{theorem}
```

to read

```
\begin{theorem}\index{Main Theorem}
```

Commands 2 and 3

Type the commands

```
\index{pistar@$\Pi^{*}$ construction}%
\index{Main Theorem!exposition|(}%
```

after the line

```
\section{The $\Pi^{*}$ construction}\label{S:P*}
```

Command 4

Retype the line

```
See also Ernest~T. Moynahan~\cite{eM57a}.
```

as follows:

```
See also Ernest~T.
\index{Moynahan, Ernest~T.}%
Moynahan~\cite{eM57a}.
```

Commands 5 to 7

Type the three index items

```
\index{lattice|textbf}%
\index{lattice!distributive}%
\index{lattice!distributive!complete}%
```

before the line

```
\begin{theorem}\label{T:P*}
```

Command 8

Type

```
\index{Main Theorem!exposition|)}
```

after the line

```
hence $\Theta = \iota$.
```

Command 9

Retype the line

```
\bibitem{sF90}
```

as follows:

```
\bibitem{sF90}\index{Foo, Soo-Key}%
```

Command 10

Retype the line

```
\bibitem{gM68}
```

as follows:

```
\bibitem{gM68}\index{Menuhin, George~A.}%
```

Command 11

Retype the line

```
\bibitem{eM57}
```

as follows:

```
\bibitem{eM57}\index{Moynahan, Ernest~T.}%
```

Command 12

Retype the line

```
\bibitem{eM57a}
```

as follows:

```
\bibitem{eM57a}\index{Moynahan, Ernest~T.}%
```

These `\index` commands produce the index for the `firstarticlei.tex` article shown in Figure 16.1. Notice that although you typed 12 index commands, only 11 entries appear in the index. The last two entries for Moynahan (commands 11 and 12) occur on the same typeset page, so only one page number shows up in the index.

INDEX

Foo, Soo-Key, 2

lattice, **1**
  distributive, 1
    complete, 1

Main Theorem, 1
  exposition, 1–2
Menuhin, George A., 2
Moynahan, Ernest T., 1, 2

Π^* construction, 1

Figure 16.1: A simple index.

The `showidx` package (see Section 10.3) lists all the index items of a page in a top corner on the margin. The top of the first page of the typeset `firstarticlei.tex` is shown in Figure 16.2.

A CONSTRUCTION OF COMPLETE-SIMPLE DISTRIBUTIVE LATTICES

GEORGE A. MENUHIN

ABSTRACT. In this note, we prove that there exist *complete-simple distributive lattices,* that is, complete distributive lattices with only two complete congruences.

Main Theorem
pistar@$"Pi^*$
construction
Main
Theorem!exposition—(
Moynahan, Ernest~T.
lattice—textbf
lattice!distributive
lattice!distributive!complete

Figure 16.2: Using `showidx`.

16.2 Index commands

There are a few major forms of `\index` commands. They are discussed in this section, illustrated by the commands shown in Section 16.1.

Simple \index ***commands***

The index entry

Foo, Soo-Key, 2

was created by command 9,

`\index{Foo, Soo-Key}`

This entry is an example of the simplest form of an index command:

`\index{`*`entry`*`}`

The entry

lattice, **2**

was created as command 5,

`\index{lattice|textbf}`

Ignore, for the time being, the `|textbf` part. This entry has a subentry,

lattice, **2**
    distributive, 2

which was created by command 6,

`\index{lattice!distributive}`

There is also a subsubentry,

lattice, **2**
  distributive, 2
    complete, 2

which was created by command 7,

```
\index{lattice!distributive!complete}
```

The form of the \index command for subentries is

```
\index{entry!subentry}
```

and for subsubentries it is

```
\index{entry!subentry!subsubentry}
```

Modifiers

Command 5

```
\index{lattice|textbf}
```

produces a bold page number in the entry lattice.

The command whose name follows the symbol | (in this case, the command name is textbf) is applied to the page number. For instance, if you want a large bold page number, then define the command \LargeBold as

```
\newcommand{\LargeBoldB}[1]{\textbf{\Large #1}}
```

and type the \index command as

```
\index{entry|LargeBold}
```

You can also modify \index commands to indicate *page ranges:*

Main Theorem, 1
  exposition, 1–2

The latter index entry has a page range. It was created with commands 3 and 8:

```
\index{Main Theorem!exposition|(}
\index{Main Theorem!exposition|)}
```

Separate an entry from its modifier with |, open the page range with (, and close it with).

Modifiers can also be combined. The index commands

```
\index{Main Theorem!exposition|(textbf}
\index{Main Theorem!exposition|)textbf}
```

produce a bold page range.

Sorting control

Observe the `\index` command

```
\index{pistar@$\Pi^{*}$ construction}
```

This produces the entry

Π^* construction, 1

To place this entry in the correct place in the index, use a *sort key*. The general form of an `\index` command with a sort key is

`\index{`*sortkey*`@`*entry*`}`

In this example, the sortkey is `pistar`. When the entries are sorted, the *sortkey* is used to sort the entry. A few typical examples follow:

Example 1 An `\index` command for G.I. Žitomirskiĭ,

```
\index{Zitomirskii@\v{Z}itomirski\u{\i}, G.I.}
```

sorts Žitomirskiĭ with the Z entries.

If you used the command

```
\index{\v{Z}itomirski\u{\i}, G.I.}
```

Žitomirskiĭ would be sorted with the v's.

Example 2 An `\index` command for the Őrmester lemma,

```
\index{Ormester@\H{O}rmester lemma}
```

would sort Őrmester lemma with the O entries.

If you used the command

```
\index{\H{O}rmester lemma}
```

Őrmester lemma would be sorted with the H's.

Example 3 An `\index` command for *truncated* lattice,

```
\index{truncated lattice@\emph{truncated} lattice}
```

sorts *truncated* lattice with the t entries.

If you use the command

```
\index{\emph{truncated} lattice}
```

this would sort *truncated* lattice with the e's.

Example 4 We want to place the symbol Truncat f, typed as `\Trunc f` (see Section 14.1.6) in the index, sorted as Trunc.

```
\index{$\Trunc f$}
```

would place Truncat f near the beginning of the index, sorted with the $ symbol. If you use the command

```
\index{Trunc@$\Trunc f$}
```

this would sort Truncat f with the T's.

Sorting control and subentries

If you want to place a subentry under an entry with a sort key, you must include the sort key part of the entry as well:

```
\index{sortkey@entry!subentry}
```

For instance,

```
\index{Zitomirskii@\v{Z}itomirski\u{\i}, G.I.!education}
```

You can also use a sort key for subentries (and subsubentries), such as

```
\index{lattice!weakly distributive@
       \emph{weakly} distributive}
```

or, a more complicated example,

```
\index{Zitomirskii@\v{Z}itomirski\u{\i}, G.I.!elementary
education@\textbf{elementary} education}
```

Special characters

Since the `!`, `@`, and `|` characters have special meanings within an `\index` command, you need to *quote* those characters if you want them to appear as themselves. *MakeIndex* uses the double quote character (`"`) for this purpose: `"!`, `"@`, and `"|`.

Because this usage makes the double quote a special character itself, it also has to be quoted if you need to use it in an `\index` command: `""`.

Example 1 To produce the entry Start here!, type the `\index` command as

```
\index{Start here"!}
```

Example 2 To produce the entry @ symbol, type the `\index` command as

```
\index{"@ symbol}
```

Example 3 To produce the entry $|A|$, type the `\index` command as

```
\index{"|A"|@$"|A"|$}
```

Cross-references

It is easy to make a cross-reference to another index entry. For instance, to list distributive lattice by cross-referencing it to lattice, distributive, the command is

```
\index{distributive lattice|seeonly{lattice,
                                    distributive}}
```

which produces the entry

distributive lattice, *see* lattice, distributive

For non-AMS document classes, `seeonly` should be `see`.
A command of this form can be placed anywhere in the document.

Tip Put all cross-referencing `\index` commands in one place in your document, so they are easy to keep track of.

Placement of `\index` ***commands***

The principle is simple.

Rule ■ Placement of `\index` commands
An `\index` command should:

1. Reference the correct page
2. Not introduce unwanted space into the typeset document

For example, you should avoid placing `\index` commands as shown here:

```
Let $L$ be a distributive lattice
\index{lattice}
\index{distributive lattice}
that is strongly complete.
```

This placement may result in unwanted extra space following the word lattice:

Let L be a distributive lattice that is strongly complete.

Note the placement of the `\index` commands in Section 16.1. In each case I have placed them as close to the referenced item as I could. If you place an index entry on a separate line, use `%` to comment out unwanted spaces including the end-of-line character (see Section 3.5.1), as in

```
Let $L$ be a distributive lattice
\index{lattice}%
\index{distributive lattice}%
that is strongly complete.
```

Read also Section 17.5 on page breaks and index entries.

Listing the forms of the* `\index` *command

We have discussed the following forms:

```
\index{entry}
\index{entry!subentry}
\index{entry!subentry!subsubentry}
\index{entry|modifier}
\index{entry|open/close modifier}
\index{sortkey@entry}
\index{sortkey@entry!subentry}
\index{sortkey@entry!subsortkey@subentry}
```

Of course, more combinations are possible; the following may be the longest form:

```
\index{sortkey@entry!subsortkey@subentry
!subsubsortkey@subsubentry|open/close modifier}
```

16.3 Processing the index entries

Once you are satisfied with the `\index` commands, the index is ready to be created.

Step 1 Typeset `firstarticlei.tex` (see Figure 16.3).

Step 2 Run the *MakeIndex* application on `firstarticlei.idx` (see Figure 16.4).

Step 3 Typeset `firstarticlei.tex` again.

You find the index on page 3 of the typeset document.

Let us look at this process in detail. In step 1 (see Figure 16.3), LaTeX creates the `firstarticlei.idx` file:

```
\indexentry{Main Theorem}{1}
\indexentry{pistar@$\Pi^{*}$ construction}{1}
\indexentry{Main Theorem!exposition|(}{1}
\indexentry{Moynahan, Ernest~T.}{1}
\indexentry{lattice|textbf}{1}
\indexentry{lattice!distributive}{1}
\indexentry{lattice!distributive!complete}{1}
\indexentry{Main Theorem!exposition|)}{2}
\indexentry{Foo, Soo-Key}{2}
```

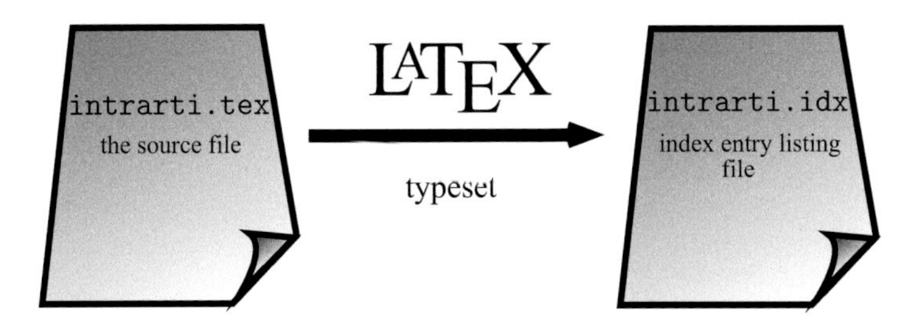

Figure 16.3: Using *MakeIndex,* step 1.

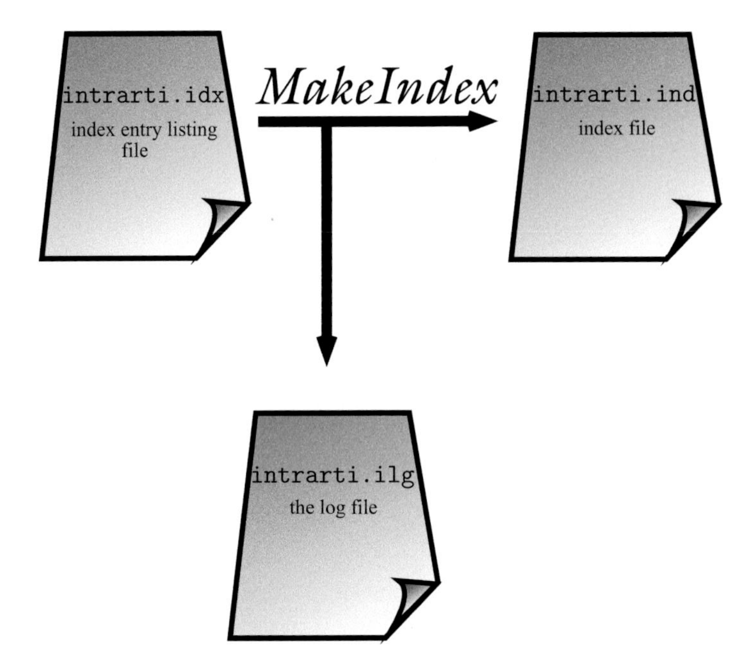

Figure 16.4: Using *MakeIndex,* step 2.

```
\indexentry{Menuhin, George~A.}{2}
\indexentry{Moynahan, Ernest~T.}{2}
\indexentry{Moynahan, Ernest~T.}{2}
```

In step 2 (see Figure 16.4), *MakeIndex* processes `firstarticlei.idx` and creates the index file `firstarticlei.ind`, which contains a `theindex` environment with all the index entries:

```
\begin{theindex}

  \item Foo, Soo-Key, 2

  \indexspace

  \item lattice, \textbf{1}
    \subitem distributive, 1
      \subsubitem complete, 1

  \indexspace

  \item Main Theorem, 1
    \subitem exposition, 1--2
  \item Menuhin, George~A., 2
  \item Moynahan, Ernest~T., 1, 2

  \indexspace

  \item $\Pi^{*}$ construction, 1

\end{theindex}
```

The `\printindex` command reads `firstarticlei.ind` during the next typesetting cycle.

MakeIndex also produces the index `log` file `firstarticlei.ilg`:

```
This is makeindex, version 2.14 [02-Oct-2002]
(kpathsea + Thai support).
Scanning input file firstarticlei.idx....done
(12 entries accepted, 0 rejected).
Sorting entries....done (43 comparisons).
Generating output file firstarticlei.ind....done
(22 lines written, 0 warnings).
Output written in firstarticlei.ind.
Transcript written in firstarticlei.ilg.
```

It is important to understand that in step 1, LaTeX does not process the index entries, it simply writes the arguments of the `\index` commands in the source file to the `idx` file as arguments of `\indexentry` commands verbatim (that is, with no change). *MakeIndex* then processes the `idx` file by removing the double quote marks for the special characters, sorting the entries, and collating the page numbers. The resulting `ind` file is a normal LaTeX source file (you can edit it, if necessary) that is included in the original document by the `\printindex` command the next time you run LaTeX.

In Step 2, we run the *MakeIndex* application on `firstarticlei.idx`. How we do this depends on the LaTeX installation you have. In UNIX installations, you type

```
MakeIndex firstarticlei
```

```
\usepackage[original]{imakeidx}
```

Also note the very useful `\indexprologue` command. It provides text (the argument of the command) to be typeset between the index header and the entries.

The `imakeidx` package has lots of nice features; see the excellent documentation.

16.6 Glossary

Using the glossary commands is very similar to using the corresponding index commands.

Instead of the

```
index           makeindex
```

commands, use the

```
glossary        makeglossary
```

commands, respectively. Glossary entries are written in the `glo` file, which corresponds to the `idx` file. LaTeX gives you no further assistance in making a glossary file. There is no `\printglossary` command, `theglossary` environment, or *MakeGlossary* application. There is, however, Thomas Henlich's `makeglos` package (see Section D.1).

16.7 Concluding comments

Style files are available in a very limited form also for *MakeIndex*. Google

```
makeindex style
```

for an up-to-date listing.

Indexing is a very complex process, so it is not surprising that there are many index packages available. CTAN lists a number of them in the `index` directory. The best known is `xindy`, described in detail in *The LaTeX Companion,* 2nd edition [56].

CHAPTER

17

Books in LaTeX

Since the introduction of LaTeX, the visual quality of articles published in mathematical journals has improved dramatically. Unfortunately, the same cannot be said of books published using LaTeX. A record number of very ugly books have appeared.

It is easy to understand why. While `amsart` has been designed to produce high-quality printed output, the standard book document classes do not produce attractive books without additional work.

LaTeX provides the `book` and the `amsbook` document classes to serve as foundations for well-designed books. Better quality books have to use document classes designed by professionals. We briefly discuss logical and visual design in Section 2.3.

So this chapter is not about how to produce a finished book using LaTeX. Our goal is much more modest, how to prepare a book manuscript for your publisher. In Section 17.1 we describe the book document classes `book` and `amsbook`. The table of contents and lists of figures and tables are discussed in Section 17.2.

Typesetting a book involves dozens of files. Section 17.3.3 gives some tips on how to organize them. Section 17.4 covers logical design. Section 17.5 deals with the final preparation of your edited manuscript for your publisher. Finally, Section 17.6 suggests a few more things to do if you typeset your book yourself.

G. Grätzer, *More Math Into* LaTeX, DOI 10.1007/978-3-319-23796-1_17

17.1 Book document classes

In this section, we briefly discuss the way in which book and amsbook, the two standard book document classes, differ from the corresponding article document classes.

17.1.1 Sectioning

Book document classes have chapters, invoked with the \chapter command and *parts,* invoked with \part. The \part command is generally used to group chapters in longer documents, for instance in this book. Parts have no effect on the numbering of chapters, sections, and so on, so Chapter 1 of Part I is not numbered as I.1 but as 1.

Both \chapter and \part take a title as an argument, but the \chapter command also has an optional argument:

```
\chapter[short_title]{title}
```

The optional *short_title* argument is used in the running head. You may need to protect any fragile commands in *title* and *short_title* with the \protect command (see Section 3.3.3).

Here is the whole hierarchy:

```
\part

\chapter
   \section
      \subsection
         \subsubsection
            \paragraph
               \subparagraph
```

Book document classes, as a rule, do not number subsubsections or any of the sectioning divisions below that level.

Equations in chapters

By default, equations are numbered from 1 within chapters. So in Chapter 1 as well as in Chapter 3, the equations are numbered (1), (2), and so forth. If you have the

```
\numberwithin{equation}{chapter}
```

command in the preamble, then equations in Chapter 2 are numbered as (2.1), (2.2), and so on.

17.1.2 Division of the body

The book document classes formalize the division of the body into three parts.

Front matter The material that appears in the front of the document, including the title pages (normally four), table of contents, preface, introduction, and so on. LaTeX numbers these pages using roman numerals. The front matter is introduced with the `\frontmatter` command.

Main matter The main part of the book, including the appendices if any. Page numbering starts from 1 using arabic numerals. The main matter is introduced with the `\mainmatter` command.

Back matter Material that appears in the back of the book, including the bibliography, index, and various other sections, such as the colophon, afterword, and so on. The back matter is introduced with the `\backmatter` command.

For the `book` document class—and the document classes built on it—in the front and back matter, the `\chapter` command does not produce a chapter number but the title is listed in the table of contents. So you can start your introduction with

```
\chapter{Introduction}
```

Within such a chapter, you should use the `*`-ed forms of the sectioning commands `\section`, `\subsection`, and so on, otherwise you have sections with numbers such as 0.1.

In the main matter, the `\appendix` command marks the beginning of the appendices. Each subsequent chapter becomes a new appendix. For example,

```
\appendix
\chapter{A proof of the Main Theorem}\label{A:Mainproof}
```

produces an appendix with the given title.

Note that appendices may be labeled and cross-referenced. In Appendix A, sections are numbered A.1, A.2, and so on, subsections in A.1 are numbered A.1.1, A.1.2, and so on. The precise form these numbers take depends, of course, on the document class, packages, and user-specific changes (see Section 14.5.1).

See Section 17.3.1 for a detailed example.

The `\chapter` and `\chapter*` commands always produce a title listed in the table of contents for the `amsbook` document class, and the document classes built on it.

The following two questions are frequently asked:

My book has only one appendix. How can I get it to be called just "Appendix", not "Appendix A"?

The single appendix in my book is being labeled "Appendix A". How can I change this to just "Appendix"? This appendix has a title, so the answer to the preceding question doesn't apply. These questions are answered in the author FAQ of the AMS, go to

```
http://www.ams.org/authors/author-faq.html
```

17.1.3 Document class options

The options and defaults for the book document classes are the same as those of other document classes (see Sections 9.5 and 10.1.2) with a few exceptions.

Two-sided printing

Options: `twoside` *default*
`oneside`

The `twoside` option formats the output for printing on both sides of a page.

Titlepage

Options: `titlepage` *default*
`notitlepage`

The `titlepage` option creates a separate title page. The `notitlepage` option creates no separate pages.

Chapter start

Options: `openright` *default*
`openany`

A chapter always starts on a new page. The `book` document class—and the document classes built on it—uses the option `openright` to start each chapter on an odd page, while the option `openany` starts each chapter on the first available new page. If you use the default option, end each chapter with the command

```
\cleardoublepage
```

Then if a chapter ends on an odd page, a blank page is added with no header or page number. The `\cleardoublepage` command is correctly coded if you use `amsbook`. Otherwise, use the package `cleardoublepage.sty` (in the `samples` folder).

The `amsbook` document class—and the document classes built on it—automatically clears to a right-hand page and leaves a totally blank page if needed.

17.1.4 Title pages

The `book` document class supports the commands: `\title`, `\author`, `\date`, and `\maketitle` (see Section 10.1.1). The `amsbook` document class supports the same commands as `amsart` (see Section 9.2), except for `\date`.

You can design your own title page within the `titlepage` environment, which does not require the use of the `\maketitle` command. Title pages for books, of course, should be created by a book designer for the publisher.

17.2 Tables of contents, lists of tables and figures

A long document, as a rule, has a table of contents. It may also include a list of figures and a list of tables.

17.2.1 Tables of contents

What goes into the table of contents?

For the amsbook document class—and the document classes built on it—all titles, not the short titles, of the sectioning commands, whether *-ed or not, subject only to the value of the tocdepth counter, as described in the last subsection of Section 14.5.1. For instance, if tocdepth is set to 2, the default, then the titles of chapters, sections, and subsections are included in the table of contents, and subsubsections are excluded.

This leaves us with the problem, what do we do if the title is too long? You cannot break the line with \\, because this would the appear in table of contents. The AMS coded the following solution: enter the line break in the form

```
\except{toc}{\linebreak}
```

For the book document class—and the document classes built on it—the title or optional argument of the sectioning commands, subject to the value of the tocdepth counter, with the following exceptions:

- In Section 8.4.1 we discuss the *-ed versions of sectioning commands. They are excluded from the table of contents.
- If the sectioning command has a short title, then it is the short title that is utilized. The example in Section 9.2 shows why this is important. If you have \\ in the title, you must have a short title without it, otherwise the linebreak would show up in the running head and the table of contents.

When you typeset your document with a table of contents, LaTeX creates a file with the toc extension. The next time the document is typeset, the toc file is typeset too and included in your typeset document at the point where the command

```
\tableofcontents
```

appears in the source file, normally in the front matter. If your source file is named myart.tex, the toc file is named myart.toc. This file lists all the sectioning units as well as their titles and page numbers.

If you already have a toc file, the \tableofcontents command typesets a table of contents using the previously created toc file and creates a new toc file.

LaTeX adds a line to the table of contents, formatted like a section title, if you include the command

```
\addcontentsline{toc}{section}{text_to_be_added}
```

in your source file. There are three arguments:

1. The first argument informs LaTeX that a line, the third argument, should be added to the `toc` file.
2. The second argument specifies how the line should be formatted in the table of contents. In our example, the second argument is `section`, so the line is formatted as a section title in the table of contents. The second argument must be the name of a sectioning command.
3. The third argument is the text to be added.

You can add an unformatted line to the table of contents with the command

```
\addtocontents{toc}{text_to_be_added}
```

Such a command can also be used to add vertical spaces into the table of contents. For instance, if you want to add some vertical space before a part, you should insert the following line before the sectioning command for the part:

```
\addtocontents{toc}{\protect\vspace{10pt}}
```

Tip If you have a `\addcontentsline` or `\addtocontents` command in a file that is `\include`-ed, then place it as a first line of this file.

The `toc` file is easy to read. The following are typical lines from the table of contents file for a document using the `book` document class:

```
\contentsline{section}{\numberline {5-4.}Top matter}{119}
\contentsline{subsection}{\numberline {5-4.1.}
Article info}{119}
\contentsline {subsection}{\numberline {5-4.2.}
Author info}{121}
```

Section 14.5.1 explains how you can specify which levels of sectioning appear in the table of contents. Section 2.3 of *The LaTeX Companion,* 2nd edition [56] lists the style parameters for the table of contents. It also shows you how to define new `toc`-like files and use multiple tables of contents in a single document, for instance, adding a mini table of contents for each chapter.

Tip You may have to typeset the document three times to create the table of contents and set the numbering of the rest of the document right.

1. The first typesetting creates the `toc` file.
2. The second inserts the table of contents with the old page numbers into the typeset document, re-records in the `aux` file the page numbers, which may have changed as a result of the insertion, and cross-references in the `aux` file, and generates a new `toc` file with the correct page numbers.
3. The third typesetting uses these new `aux` and `toc` files to typeset the document correctly and creates a new `toc` file.

Fragile commands in a movable argument, such as a section (short) title, must be `\protect`-ed (see Section 3.3.3). Here is a simple example using the table of contents. If the document contains the `\section` command

```
\section{The function \( f(x^{2}) \)}
```

the section title is stored in the `toc` file as

```
\contentsline {section}{\numberline
{1}The function\relax $ f(x^{2}) \relax \GenericError { }
{LaTeX Error: Bad math environment delimiter}{Your
command was ignored.\MessageBreak Type I <command>
<return> to replace it with another command,\MessageBreak
or <return> to continue without it.}}{1}
```

and the `log` file contains the message

```
! LaTeX Error: Bad math environment delimiter.

 ...
l.1 ...continue without it.}}{1}
```

Error messages usually refer to a line in the source file, but in this case the error message refers to a line in the `toc` file.

The correct form for this section title is

```
\section{The function \protect\( f(x^{2}) \protect\)}
```

or, even simpler,

```
\verb+\section{The function $f(x^{2})$}
```

Note that this example is merely an illustration of unprotected fragile commands in movable arguments. As a rule, avoid using formulas in (sectioning) titles.

17.2.2 Lists of tables and figures

If you place a `\listoftables` command in the document, LaTeX stores information for the list of tables in a `lot` file. The list of tables is inserted into the body of your document at the point where the command appears, normally in the front matter, following the table of contents.

A list of figures, similar to a list of tables, can be compiled with the command `\listoffigures`. This command creates an auxiliary file with the extension `lof`.

An optional argument of the `\caption` commands in your tables and figures can replace the argument in the list of tables and figures. Typically, the optional argument is used to specify a shorter caption for the list of tables or list of figures. There are other uses. For instance, you may notice that, as a rule, captions should be terminated by periods. If in the list of tables or list of figures, your book style fills the space between the text and the page number with dots, the extra period looks bad. This problem goes away if you use the following form of the `\caption` command:

```
\caption[title]{title.}
```

There are analogs of the table of contents commands for use with tables and figures. The command

```
\addtocontents{lot}{line_to_add}
```

adds a line to the list of tables or to the list of figures with the first argument `lof`.

17.2.3 Exercises

For exercises, amsbook provides the `xcb` environment. It is used for a series of exercises at the end of a section or chapter. The argument of the environment specifies the phrase (such as `Exercises`) to begin the list:

```
\begin{xcb}{Exercises}
\begin{enumerate}
\item A finite lattice $L$ is modular if{f} it does not
contain a pentagon.\label{E:pent}
\item Can the numbers of covering pairs in\label{E:incr}
Exercise~\ref{E:pent} be increased?\label{E:incr}+
\end{enumerate}
\end{xcb}
```

which typesets as

Exercises

(1) A finite lattice L is modular iff it does not contain a pentagon.
(2) Can the numbers of covering pairs in Exercise 1 be increased?

17.3 Organizing the files for a book

An article is typically one `tex` file and maybe some PDF files for the illustrations. On the other hand, a book, like this one, is composed of hundreds of files. In this section, I describe how the files for a book like this may be organized.

There are three commands that help with the organization:
`\include` `\includeonly` `\graphicspath`
We discuss these commands in this section.

17.3.1 The folders and the master document

All the files for this book are in a folder `MiL5` and in this folder the most important document is `MiL5.tex`, the *master document*.

The master document, `MiL5.tex` reads, in a somewhat simplified form, as follows:

```
%MiL5 master document
\documentclass[leqno]{book}
\usepackage{MiL5}
\usepackage{makeidx}
\makeindex
\usepackage{cleardoublepage}
\includeonly{
%frontmatter,
%intro,
%Chapter1,% terminology
...
Chapter17,% books
%appA,% Math symbol tables
...
}
\begin{document}
\frontmatter
\include{frontmatter}
\tableofcontents
\listoftables\listoffigures
\include{intro}%Intro
\mainmatter
\include{Chapter1}%Mission Impossible
...
\include{Chapter17}% Books

\appendix+\\
\include{AppendixA}%Math symbol tables+\\
... +\\
```

```
\backmatter+\\
\printindex+\\
\end{document}+\\[8pt]
```

Some parts of the master file deserve comment, for example, the third line,

```
\usepackage{MiL5}
```

loads the command file

```
MiL5.sty
```

which contains all the commands defined for the book and the code for the book style. Since the book style is based on `book.cls`, in line 5, we load the `makeidx` package and print the index with `\printindex` (see Section 16.1).

Line 7 states

```
\usepackage{cleardoublepage}
```

This creates blank pages after chapters that end on an odd page number (see Section 17.1.3). For the three `...matter` commands, see Section 17.1.2.

17.3.2 Inclusion and selective inclusion

This book is pieced together by the `\include` commands in the master document. For example,

```
\include{Chapter17}
```

inserts the contents of the file `Chapter17.tex`, starting on a new page, as though its contents had been typed at that place in the document. The master document for this book has a lot of `\include` commands...

Rule ■ File termination

Terminate every file you `\include` with an `\endinput` command.

If you terminate an `\include`-ed file with `\end{document}`, LaTeX gives a warning such as:

```
(\end occurred when \iftrue on line 6 was incomplete)
(\end occurred when \ifnum on line 6 was incomplete)
```

If you use `\include` commands in the master file, as in the example in Section 17.3.1, then you can use the `\includeonly` command for selective inclusion. The lines of the `\includeonly` command parallel the `\include` commands. Block comment all the lines of the argument of the `\includeonly` command, and uncomment the chapter you are working on. In the example above, I am working on this chapter.

The argument of the `\includeonly` command is a list of files separated by commas. If you want to typeset the whole book, uncomment all the lines.

17.3.3 Organizing your files

The MiL5 folder, containing the files of this book, contains MiL5.tex, the master document, the command file MiL5.sty, and all the tex files listed in the master document, that is, the chapters, the frontmatter, the introduction, the appendices, and of course, all the auxiliary files that LaTeX creates.

This book contains about 300 illustrations in a subfolder Graphics of the folder MiL5.tex. We have to tell LaTeX to look for the illustrations in this folder. We do this with the command

```
\graphicspath{{Graphics/}}
```

in the preamble. If you have two folders, Illustr1 and Illustr2 for illustrations, the \includegraphics command takes the form

```
\graphicspath{{Illustr1/}{Illustr2/}}
```

Even if you have more than one folder for the illustrations, you must make sure that each graphics file has a unique name.

We place the \graphicspath command in MiL5.sty.

In the above commands, / is appropriate for Mac and UNIX computers. For a Windows computer, use \ instead.

If you submit a dvi file, you cannot use the \graphicspath command.

17.4 Logical design

The discussion of logical and visual design in Section 2.3 applies to books even more than to articles. Since books are long and complex documents, errors in the logical design are much harder to correct.

Let us review some common sense rules.

Rule 1 ■ Stick with the sectioning commands provided by the document class. Define the non-standard structures you wish to use as environments.

Here is an example which is obviously bad:

```
\vspace{18pt}

\noindent \textbf{Theorem 1.1.}
\textit{This is bad.}

\vspace{18pt}
```

And a good way to achieve the same result:

```
\begin{theorem}\label{T:Goodtheorem}
This is a good theorem.
\end{theorem}
```

The bad example creates a number of difficulties.

- You have to number the theorems yourself. Adding, deleting, and rearranging theorems becomes difficult and updating cross-references is even harder.
- It is difficult to keep such constructs consistent.
- If the publisher decides to increase the white space before and after the theorems to 20 points, finding and changing all the appropriate commands becomes a tedious and error prone task.

Rule 2 ■ Define frequently used constructs as commands.

Rather than

```
\textbf{Warning! Do not exceed this amount!}
```

define

```
\newcommand{\important}[1]{\textbf{#1}}
```

and type your warnings as

```
\important{Warning! Do not exceed this amount!}
```

You or your editor can then change all the warnings to a different style with ease.

Rule 3 ■ Avoid text style commands.

If you use small caps for acronyms, do not type

```
\textsc{ibm}
```

but rather define

```
\newcommand{\ibm}{\textsc{ibm}}
```

and then

```
\ibm
```

or more generally

```
\newcommand{\acronym}[1]{\textsc{#1}}
```

and then

```
\verb+\acronym{ibm}+
```

Rule 4 ■ Avoid white space commands.

Occasionally, you may feel that there should be some white space separating two paragraphs, so you do the following:

```
paragraph 1

\medskip

paragraph 2
```

It would be better to define a new command, say `\separate`, as

```
\newcommand{\separate}{\medskip}
```

and type the previous example as

```
paragraph 1

\separate

paragraph 2
```

Now such white space can be adjusted throughout the entire document by simply redefining one command. Note that redefining `\medskip` itself may have unintended side effects:

- Many environments depend on LaTeX's definition of `\medskip`.
- You may have used `\medskip` in other situations as well.

Here is a short list of commands should not redefine:

```
\bigskip   \hfil    \hspace    \parskip     \vfill    \vspace
\break     \hfill   \kern      \smallskip   \vglue
\eject     \hglue   \medskip   \vfil        \vskip
```

17.5 Final preparations for the publisher

Throughout this book, there are a number of "don'ts". Most are practices you should avoid while writing articles. When writing a book, it is even more important not to violate these rules.

When the editors, including the copy editor, are finished with your manuscript and you have the document class designed for the book, then you can start on the final preparations.

Step 1 ■ Eliminate all TeX commands.

TeX commands, that is, Plain TeX commands that are not part of LaTeX (not listed as LaTeX commands in the index of this book) may interfere with LaTeX in unexpected ways. They may also cause problems with the style file that is created for your book.

Step 2 ■ Collect all your custom commands and environments together in one place, preferably in a separate command file (see Section 14.3).

Step 3 ■ Make sure that custom commands for notations and custom environments for structures are used consistently throughout your document.

This book uses the command `\doc` for document names, so `firstarticle` is typed as `\doc{firstarticle}`. Of course, `\texttt{firstarticle}` gives the same result,but if you intermix `\doc{firstarticle}` and `\texttt{firstarticle}` commands, you lose the ability to easily change the way document names are displayed.

Step 4 ■ Watch out for vertical white space adding up.

- Do not directly follow one displayed math environment with another. Multiple adjacent lines of displayed mathematics should all be in the same environment.
- If your style file uses interparagraph spacing, avoid beginning paragraphs with displayed math.

For instance,

```
\[
   x=y
\]
```

```
\[
  x=z
\]
```

is wrong. Use, instead, an `align` or `gather` environment.

Step 5 ■ If possible, do not place "tall" mathematical formulas inline. All formulas that might change the interline spacing, as a rule, should be displayed.

You can find examples on pages 138 and 368. Here is one more example, double hat accents used inline: $\hat{\hat{A}}$.

Step 6 ■ Read the `log` file.

- Watch for line-too-wide warnings (see Section 1.4).
- Check for font substitutions (see Section 3.6.7).

Adobe Acrobat Professional has a preflight utility. It will check whether the PDF version of your typeset document has all the fonts it requires.

Step 7 ■ Do not assume that gray boxes or color illustrations appear when published exactly the way that they look on your monitor or printer.

Color work requires calibration of monitors and printers. It is often best left to the experts at the publisher.

Step 8 ■ Do not assume that the application that created your PDF files (see Section 8.4.3) can create high-quality PDF files.

Many applications can create PDF files or convert files to PDF format. Very few do it right. Ask your publisher what applications they recommend.

Font substitutions can also cause problems:

- A font that was used in typesetting the document may not be the font you intended. Missing fonts are substituted and the substitute fonts are rarely satisfactory.
- A special trap: Your publisher may have more, or maybe fewer, fonts than you do! As a result, the font substitutions on your publisher's system may be different from those on yours. Make sure that the fonts you use are not substituted.

17.6 If you create the PDF file for your book

Many publishers take your manuscript, prepared as described in Section 17.5, and guide it through the final steps for printing. Some books, however, are prepared by the authors for printing using a custom document class for books and submitted to the publisher as PDF files. If your book falls into this category, there are a few more things you should do before you create the final PDF file for your book.

Adjust the pages

Make sure that you are satisfied with the way the document is broken into pages by LaTeX and with the placement of the `figure` and `table` environments (see Section 8.4.3). If necessary, you should make last-minute changes to adjust page breaks. You may find the `\enlargethispage` command (see Section 3.7.3) very helpful at this stage. Just be sure to apply it on both facing pages.

To ensure that

- Page numbers in the index are correct
- `\pageref` references (see Section 8.4.2) are correct
- Marginal comments (see Section 3.9.4) are properly placed
- Tables and figures are properly placed

insert page breaks where necessary. Where pages break, add the three commands `\linebreak`, `\pagebreak`, and `\noindent`. Here is an example. The bottom of page 3 and the top of page 4 of my book *General Lattice Theory* [28] are shown in Figure 17.1.

In other words, lattice theory singles out a special type of poset for detailed investigation. To make such a definition worthwhile, it must be shown that this class of posets is a very useful class, that there are many such posets in various branches of mathematics (analysis, topology, logic, algebra, geometry, and so on), and that a general study of these posets will lead to a better understanding of the behavior of the examples. This was done in the first edition of G. Birkhoff's

---

4 I. First Concepts

Lattice Theory [1940]. As we go along, we shall see many examples, most of them in the exercises. For a general survey of lattices in mathematics, see G. Birkhoff [1967] and H. H. Crapo and G.-C. Rota [1970].

Figure 17.1: A page break.

Now let us assume that we have to manually do this page break because some index items attached to this paragraph generate incorrect page numbers. The paragraph split by the page break is

```
In other words, lattice theory singles out a special type
of poset for detailed investigation. To make such a
definition worthwhile, it must be shown that this class
of posets is a very useful class, that there are many
such posets in various branches of mathematics (analysis,
topology, logic, algebra, geometry, and so on), and that
a general study of these posets will lead to a better
understanding of the behavior of the examples.
This was done in the first edition of  G.~Birkhoff's
\emph{Lattice Theory} \cite{gB40}. As we go along,
we shall see many examples, most of them in the
exercises. For a general survey of lattices in
mathematics, see G.~Birkhoff \cite{gB67} and H.~H.~Crapo
and G.-C.~Rota \cite{CR70}.
```

When typesetting this paragraph, LaTeX inserts a page break following

```
This was done in the first edition of G.~Birkhoff's+.
```

So we edit four lines as follows:

```
understanding of the behavior of the examples.
This was done in the first edition of  G.~Birkhoff's
\linebreak

\pagebreak

\noindent \emph{Lattice Theory} \cite{gB40}. As we go
along, we shall see many examples, most of them in the
```

This change does not affect the appearance of the typeset page, but now pages 3 and 4 are separated by a `\pagebreak`. Make sure that any `\index` or `\label` commands are moved to the appropriate half of the paragraph. Now all index commands generate the correct page numbers.

Of course, if the page break is between paragraphs, only the `\pagebreak` command is needed. If the break occurs in the middle of a word, use `\-\linebreak` to add a hyphen.

This method works about 95 percent of the time. Occasionally, you have to drop either the `\linebreak` or the `\pagebreak` command.

Check for missing fonts and other defects

Open the PDF file of your book in Adobe Reader (or even better, in Adobe Acrobat Pro). Under `File`, go to `Properties...` and click on the `Fonts` tab. You will find a long list of fonts. Each one should be marked `Embedded Subset`. If all your fonts are embedded, you are in good shape.

Adobe Acrobat Pro has an excellent set of utilities to check whether your PDF file is ready for printing. You find them under `Preflight`. Adobe Acrobat Pro will correct all the mistakes it finds in the file and presents a detailed report.

Other adjustments

- Move the `figure` and `table` environments (see Section 8.4.3) physically close to where they appear in the typeset version, and change the optional argument of the `figure` and `table` environments to `!h`.
- Balance the white space on each page as necessary.
- Generate the index only after the page breaks are fixed.

Polish the auxiliary files

- Typeset the document one last time and then place the `\nofiles` command in the preamble (see Section C.2.4) to make sure that the auxiliary files are not overwritten.
- Normally, you should not have to edit the table of contents (`toc`) file or the `lot` and `lof` files (see Section 17.2) and your style file should take care of the formatting. Sometimes, however, an unfortunate page break makes editing necessary. In an appropriate place, you may want to add to the text the command

  ```
  \addtocontents{toc}{\pagebreak}
  ```

  to avoid such edits.
- Create the index (`ind`) file from the new `aux` file, as described in Section 16.3. A lot of help is available in *The Chicago Manual of Style,* 16th edition [11]; it has a section on bad breaks, remedies, and *Continued* lines in the index. Break the `ind` file into pages. To minimize bad breaks, use the `\enlargethispage` command where necessary (see Section 3.7.3). Add any *Continued* entries.

  Many book document classes, including `book`, have two problems with the Index.

  (i) There is no Index entry in the Table of Contents.

  (ii) The first page of the Index is numbered.

  These are easy to correct. Add to the beginning of the `ind` file the command

  ```
  \thispagestyle{empty}
  ```

  and precede the `\printindex` with

  ```
  \addtocontents{toc}{Index}
  ```

APPENDIX

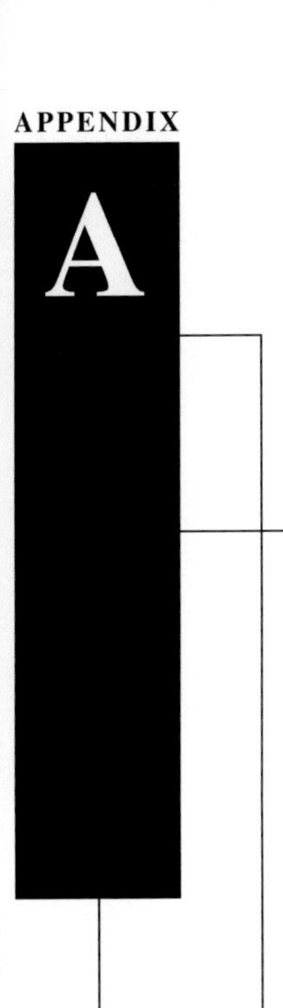

Math symbol tables

A.1 Hebrew and Greek letters

Hebrew letters

Type	Typeset
`\aleph`	ℵ
`\beth`	ℶ
`\daleth`	ℸ
`\gimel`	ℷ

G. Grätzer, *More Math Into* LaTeX, DOI 10.1007/978-3-319-23796-1

Greek letters

Lowercase

Type	Typeset	Type	Typeset	Type	Typeset
`\alpha`	α	`\iota`	ι	`\sigma`	σ
`\beta`	β	`\kappa`	κ	`\tau`	τ
`\gamma`	γ	`\lambda`	λ	`\upsilon`	υ
`\delta`	δ	`\mu`	μ	`\phi`	ϕ
`\epsilon`	ϵ	`\nu`	ν	`\chi`	χ
`\zeta`	ζ	`\xi`	ξ	`\psi`	ψ
`\eta`	η	`\pi`	π	`\omega`	ω
`\theta`	θ	`\rho`	ρ		
`\varepsilon`	ε	`\varpi`	ϖ	`\varsigma`	ς
`\vartheta`	ϑ	`\varrho`	ϱ	`\varphi`	φ
	`\digamma`	$\digamma$	`\varkappa`	$\varkappa$	

Uppercase

Type	Typeset	Type	Typeset	Type	Typeset
`\Gamma`	Γ	`\Xi`	Ξ	`\Phi`	Φ
`\Delta`	Δ	`\Pi`	Π	`\Psi`	Ψ
`\Theta`	Θ	`\Sigma`	Σ	`\Omega`	Ω
`\Lambda`	Λ	`\Upsilon`	Υ		
`\varGamma`	$\varGamma$	`\varXi`	$\varXi$	`\varPhi`	$\varPhi$
`\varDelta`	$\varDelta$	`\varPi`	$\varPi$	`\varPsi`	$\varPsi$
`\varTheta`	$\varTheta$	`\varSigma`	$\varSigma$	`\varOmega`	$\varOmega$
`\varLambda`	$\varLambda$	`\varUpsilon`	$\varUpsilon$		

A.2 Binary relations

Type	Typeset	Type	Typeset
<	$<$	>	$>$
=	$=$	:	$:$
\in	$\in$	\ni or \owns	$\ni$
\leq or \le	$\leq$	\geq or \ge	$\geq$
\ll	$\ll$	\gg	$\gg$
\prec	$\prec$	\succ	$\succ$
\preceq	$\preceq$	\succeq	$\succeq$
\sim	$\sim$	\approx	$\approx$
\simeq	$\simeq$	\cong	$\cong$
\equiv	$\equiv$	\doteq	$\doteq$
\subset	$\subset$	\supset	$\supset$
\subseteq	$\subseteq$	\supseteq	$\supseteq$
\sqsubseteq	$\sqsubseteq$	\sqsupseteq	$\sqsupseteq$
\smile	$\smile$	\frown	$\frown$
\perp	$\perp$	\models	$\models$
\mid	$\mid$	\parallel	$\parallel$
\vdash	$\vdash$	\dashv	$\dashv$
\propto	$\propto$	\asymp	$\asymp$
\bowtie	$\bowtie$		
\sqsubset	$\sqsubset$	\sqsupset	$\sqsupset$
\Join	$\Join$		

Note the \colon command used in $f\colon x \to x^2$, typed as

```
f \colon x \to x^2
```

More binary relations

Type	Typeset	Type	Typeset
\leqq	$\leqq$	\geqq	$\geqq$
\leqslant	$\leqslant$	\geqslant	$\geqslant$
\eqslantless	$\eqslantless$	\eqslantgtr	$\eqslantgtr$
\lesssim	$\lesssim$	\gtrsim	$\gtrsim$
\lessapprox	$\lessapprox$	\gtrapprox	$\gtrapprox$
\approxeq	$\approxeq$		
\lessdot	$\lessdot$	\gtrdot	$\gtrdot$
\lll	$\lll$	\ggg	$\ggg$
\lessgtr	$\lessgtr$	\gtrless	$\gtrless$
\lesseqgtr	$\lesseqgtr$	\gtreqless	$\gtreqless$
\lesseqqgtr	$\lesseqqgtr$	\gtreqqless	$\gtreqqless$
\doteqdot	$\doteqdot$	\eqcirc	$\eqcirc$
\circeq	$\circeq$	\triangleq	$\triangleq$
\risingdotseq	$\risingdotseq$	\fallingdotseq	$\fallingdotseq$
\backsim	$\backsim$	\thicksim	$\thicksim$
\backsimeq	$\backsimeq$	\thickapprox	$\thickapprox$
\preccurlyeq	$\preccurlyeq$	\succcurlyeq	$\succcurlyeq$
\curlyeqprec	$\curlyeqprec$	\curlyeqsucc	$\curlyeqsucc$
\precsim	$\precsim$	\succsim	$\succsim$
\precapprox	$\precapprox$	\succapprox	$\succapprox$
\subseteqq	$\subseteqq$	\supseteqq	$\supseteqq$
\Subset	$\Subset$	\Supset	$\Supset$
\vartriangleleft	$\vartriangleleft$	\vartriangleright	$\vartriangleright$
\trianglelefteq	$\trianglelefteq$	\trianglerighteq	$\trianglerighteq$
\vDash	$\vDash$	\Vdash	$\Vdash$
\Vvdash	$\Vvdash$		
\smallsmile	$\smallsmile$	\smallfrown	$\smallfrown$
\shortmid	$\shortmid$	\shortparallel	$\shortparallel$
\bumpeq	$\bumpeq$	\Bumpeq	$\Bumpeq$
\between	$\between$	\pitchfork	$\pitchfork$
\varpropto	$\varpropto$	\backepsilon	$\backepsilon$
\blacktriangleleft	$\blacktriangleleft$	\blacktriangleright	$\blacktriangleright$
\therefore	$\therefore$	\because	$\because$

Negated binary relations

Type	Typeset	Type	Typeset
`\neq` or `\ne`	$\neq$	`\notin`	$\notin$
`\nless`	$\nless$	`\ngtr`	$\ngtr$
`\nleq`	$\nleq$	`\ngeq`	$\ngeq$
`\nleqslant`	$\nleqslant$	`\ngeqslant`	$\ngeqslant$
`\nleqq`	$\nleqq$	`\ngeqq`	$\ngeqq$
`\lneq`	$\lneq$	`\gneq`	$\gneq$
`\lneqq`	$\lneqq$	`\gneqq`	$\gneqq$
`\lvertneqq`	$\lvertneqq$	`\gvertneqq`	$\gvertneqq$
`\lnsim`	$\lnsim$	`\gnsim`	$\gnsim$
`\lnapprox`	$\lnapprox$	`\gnapprox`	$\gnapprox$
`\nprec`	$\nprec$	`\nsucc`	$\nsucc$
`\npreceq`	$\npreceq$	`\nsucceq`	$\nsucceq$
`\precneqq`	$\precneqq$	`\succneqq`	$\succneqq$
`\precnsim`	$\precnsim$	`\succnsim`	$\succnsim$
`\precnapprox`	$\precnapprox$	`\succnapprox`	$\succnapprox$
`\nsim`	$\nsim$	`\ncong`	$\ncong$
`\nshortmid`	$\nshortmid$	`\nshortparallel`	$\nshortparallel$
`\nmid`	$\nmid$	`\nparallel`	$\nparallel$
`\nvdash`	$\nvdash$	`\nvDash`	$\nvDash$
`\nVdash`	$\nVdash$	`\nVDash`	$\nVDash$
`\ntriangleleft`	$\ntriangleleft$	`\ntriangleright`	$\ntriangleright$
`\ntrianglelefteq`	$\ntrianglelefteq$	`\ntrianglerighteq`	$\ntrianglerighteq$
`\nsubseteq`	$\nsubseteq$	`\nsupseteq`	$\nsupseteq$
`\nsubseteqq`	$\nsubseteqq$	`\nsupseteqq`	$\nsupseteqq$
`\subsetneq`	$\subsetneq$	`\supsetneq`	$\supsetneq$
`\varsubsetneq`	$\varsubsetneq$	`\varsupsetneq`	$\varsupsetneq$
`\subsetneqq`	$\subsetneqq$	`\supsetneqq`	$\supsetneqq$
`\varsubsetneqq`	$\varsubsetneqq$	`\varsupsetneqq`	$\varsupsetneqq$

A.3 Binary operations

Type	Typeset	Type	Typeset
+	$+$	-	$-$
\pm	$\pm$	\mp	$\mp$
\times	$\times$	\cdot	$\cdot$
\circ	$\circ$	\bigcirc	$\bigcirc$
\div	$\div$	\bmod	mod
\cap	$\cap$	\cup	$\cup$
\sqcap	$\sqcap$	\sqcup	$\sqcup$
\wedge or \land	$\wedge$	\vee or \lor	$\vee$
\triangleleft	$\triangleleft$	\triangleright	$\triangleright$
\bigtriangleup	$\bigtriangleup$	\bigtriangledown	$\bigtriangledown$
\oplus	$\oplus$	\ominus	$\ominus$
\otimes	$\otimes$	\oslash	$\oslash$
\odot	$\odot$	\bullet	$\bullet$
\dagger	$\dagger$	\ddagger	$\ddagger$
\setminus	$\setminus$	\smallsetminus	$\smallsetminus$
\wr	$\wr$	\amalg	$\amalg$
\ast	$\ast$	\star	$\star$
\diamond	$\diamond$		
\lhd	$\lhd$	\rhd	$\rhd$
\unlhd	$\unlhd$	\unrhd	$\unrhd$
\dotplus	$\dotplus$	\centerdot	$\centerdot$
\ltimes	$\ltimes$	\rtimes	$\rtimes$
\leftthreetimes	$\leftthreetimes$	\rightthreetimes	$\rightthreetimes$
\circleddash	$\circleddash$	\uplus	$\uplus$
\barwedge	$\barwedge$	\doublebarwedge	$\doublebarwedge$
\curlywedge	$\curlywedge$	\curlyvee	$\curlyvee$
\veebar	$\veebar$	\intercal	$\intercal$
\doublecap or \Cap	$\doublecap$	\doublecup or \Cup	$\doublecup$
\circledast	$\circledast$	\circledcirc	$\circledcirc$
\boxminus	$\boxminus$	\boxtimes	$\boxtimes$
\boxdot	$\boxdot$	\boxplus	$\boxplus$
\divideontimes	$\divideontimes$	\vartriangle	$\vartriangle$
\And	$\And$		

A.4 Arrows

Type	Typeset	Type	Typeset
\leftarrow	$\leftarrow$	\rightarrow or \to	$\rightarrow$
\longleftarrow	$\longleftarrow$	\longrightarrow	$\longrightarrow$
\Leftarrow	$\Leftarrow$	\Rightarrow	$\Rightarrow$
\Longleftarrow	$\Longleftarrow$	\Longrightarrow	$\Longrightarrow$
\leftrightarrow	$\leftrightarrow$	\longleftrightarrow	$\longleftrightarrow$
\Leftrightarrow	$\Leftrightarrow$	\Longleftrightarrow	$\Longleftrightarrow$
\uparrow	$\uparrow$	\downarrow	$\downarrow$
\Uparrow	$\Uparrow$	\Downarrow	$\Downarrow$
\updownarrow	$\updownarrow$	\Updownarrow	$\Updownarrow$
\nearrow	$\nearrow$	\searrow	$\searrow$
\swarrow	$\swarrow$	\nwarrow	$\nwarrow$
\iff	$\iff$	\mapstochar	$\mapstochar$
\mapsto	$\mapsto$	\longmapsto	$\longmapsto$
\hookleftarrow	$\hookleftarrow$	\hookrightarrow	$\hookrightarrow$
\leftharpoonup	$\leftharpoonup$	\rightharpoonup	$\rightharpoonup$
\leftharpoondown	$\leftharpoondown$	\rightharpoondown	$\rightharpoondown$
\leadsto	$\leadsto$		
\leftleftarrows	$\leftleftarrows$	\rightrightarrows	$\rightrightarrows$
\leftrightarrows	$\leftrightarrows$	\rightleftarrows	$\rightleftarrows$
\Lleftarrow	$\Lleftarrow$	\Rrightarrow	$\Rrightarrow$
\twoheadleftarrow	$\twoheadleftarrow$	\twoheadrightarrow	$\twoheadrightarrow$
\leftarrowtail	$\leftarrowtail$	\rightarrowtail	$\rightarrowtail$
\looparrowleft	$\looparrowleft$	\looparrowright	$\looparrowright$
\upuparrows	$\upuparrows$	\downdownarrows	$\downdownarrows$
\upharpoonleft	$\upharpoonleft$	\upharpoonright	$\upharpoonright$
\downharpoonleft	$\downharpoonleft$	\downharpoonright	$\downharpoonright$
\leftrightsquigarrow	$\leftrightsquigarrow$	\rightsquigarrow	$\rightsquigarrow$
\multimap	$\multimap$		
\nleftarrow	$\nleftarrow$	\nrightarrow	$\nrightarrow$
\nLeftarrow	$\nLeftarrow$	\nRightarrow	$\nRightarrow$
\nleftrightarrow	$\nleftrightarrow$	\nLeftrightarrow	$\nLeftrightarrow$
\dashleftarrow	$\dashleftarrow$	\dashrightarrow	$\dashrightarrow$
\curvearrowleft	$\curvearrowleft$	\curvearrowright	$\curvearrowright$
\circlearrowleft	$\circlearrowleft$	\circlearrowright	$\circlearrowright$
\leftrightharpoons	$\leftrightharpoons$	\rightleftharpoons	$\rightleftharpoons$
\Lsh	$\Lsh$	\Rsh	$\Rsh$

A.5 Miscellaneous symbols

Type	Typeset	Type	Typeset
\hbar	$\hbar$	\ell	ℓ
\imath	$\imath$	\jmath	$\jmath$
\wp	$\wp$	\partial	∂
\Im	$\Im$	\Re	$\Re$
\infty	∞	\prime	$\prime$
\emptyset	$\emptyset$	\varnothing	$\varnothing$
\forall	$\forall$	\exists	$\exists$
\smallint	$\smallint$	\triangle	$\triangle$
\top	$\top$	\bot	$\bot$
\P	$\P$	\S	$\S$
\dag	$\dag$	\ddag	$\ddag$
\flat	$\flat$	\natural	$\natural$
\sharp	$\sharp$	\angle	$\angle$
\clubsuit	$\clubsuit$	\diamondsuit	$\diamondsuit$
\heartsuit	$\heartsuit$	\spadesuit	$\spadesuit$
\surd	$\surd$	\nabla	∇
\pounds	$\pounds$	\neg or \lnot	$\neg$
\Box	$\Box$	\Diamond	$\Diamond$
\mho	$\mho$		
\hslash	$\hslash$	\complement	$\complement$
\backprime	$\backprime$	\nexists	$\nexists$
\Bbbk	$\Bbbk$		
\diagup	$\diagup$	\diagdown	$\diagdown$
\blacktriangle	$\blacktriangle$	\blacktriangledown	$\blacktriangledown$
\triangledown	$\triangledown$	\eth	$\eth$
\square	$\square$	\blacksquare	$\blacksquare$
\lozenge	$\lozenge$	\blacklozenge	$\blacklozenge$
\measuredangle	$\measuredangle$	\sphericalangle	$\sphericalangle$
\circledS	$\circledS$	\bigstar	$\bigstar$
\Finv	$\Finv$	\Game	$\Game$

A.6 Delimiters

<table>
<tr><th>Name</th><th>Type</th><th>Typeset</th></tr>
<tr><td>left parenthesis</td><td>(</td><td>$($</td></tr>
<tr><td>right parenthesis</td><td>)</td><td>$)$</td></tr>
<tr><td>left bracket</td><td>[or \lbrack</td><td>$[$</td></tr>
<tr><td>right bracket</td><td>] or \rbrack</td><td>$]$</td></tr>
<tr><td>left brace</td><td>\{ or \lbrace</td><td>$\{$</td></tr>
<tr><td>right brace</td><td>\} or \rbrace</td><td>$\}$</td></tr>
<tr><td>backslash</td><td>\backslash</td><td>$\backslash$</td></tr>
<tr><td>forward slash</td><td>/</td><td>$/$</td></tr>
<tr><td>left angle bracket</td><td>\langle</td><td>$\langle$</td></tr>
<tr><td>right angle bracket</td><td>\rangle</td><td>$\rangle$</td></tr>
<tr><td>vertical line</td><td>| or \vert</td><td>$\vert$</td></tr>
<tr><td>double vertical line</td><td>\| or \Vert</td><td>$\Vert$</td></tr>
<tr><td>left floor</td><td>\lfloor</td><td>$\lfloor$</td></tr>
<tr><td>right floor</td><td>\rfloor</td><td>$\rfloor$</td></tr>
<tr><td>left ceiling</td><td>\lceil</td><td>$\lceil$</td></tr>
<tr><td>right ceiling</td><td>\rceil</td><td>$\rceil$</td></tr>
<tr><td>upward</td><td>\uparrow</td><td>$\uparrow$</td></tr>
<tr><td>double upward</td><td>\Uparrow</td><td>$\Uparrow$</td></tr>
<tr><td>downward</td><td>\downarrow</td><td>$\downarrow$</td></tr>
<tr><td>double downward</td><td>\Downarrow</td><td>$\Downarrow$</td></tr>
<tr><td>up-and-down</td><td>\updownarrow</td><td>$\updownarrow$</td></tr>
<tr><td>double up-and-down</td><td>\Updownarrow</td><td>$\Updownarrow$</td></tr>
<tr><td>upper-left corner</td><td>\ulcorner</td><td>$\ulcorner$</td></tr>
<tr><td>upper-right corner</td><td>\urcorner</td><td>$\urcorner$</td></tr>
<tr><td>lower-left corner</td><td>\llcorner</td><td>$\llcorner$</td></tr>
<tr><td>lower-right corner</td><td>\lrcorner</td><td>$\lrcorner$</td></tr>
</table>

A.7 Operators

"Pure" operators, with no limits

Type	Typeset	Type	Typeset	Type	Typeset	Type	Typeset
`\arccos`	$\arccos$	`\cot`	$\cot$	`\hom`	$\hom$	`\sin`	$\sin$
`\arcsin`	$\arcsin$	`\coth`	$\coth$	`\ker`	$\ker$	`\sinh`	$\sinh$
`\arctan`	$\arctan$	`\csc`	$\csc$	`\lg`	$\lg$	`\tan`	$\tan$
`\arg`	$\arg$	`\deg`	$\deg$	`\ln`	$\ln$	`\tanh`	$\tanh$
`\cos`	$\cos$	`\dim`	$\dim$	`\log`	$\log$		
`\cosh`	$\cosh$	`\exp`	$\exp$	`\sec`	$\sec$		

Operators with limits

Type	Typeset	Type	Typeset
`\det`	$\det$	`\limsup`	$\limsup$
`\gcd`	$\gcd$	`\max`	$\max$
`\inf`	$\inf$	`\min`	$\min$
`\lim`	$\lim$	`\Pr`	$\Pr$
`\liminf`	$\liminf$	`\sup`	$\sup$
`\injlim`	$\injlim$	`\projlim`	$\projlim$
`\varliminf`	$\varliminf$	`\varlimsup`	$\varlimsup$
`\varinjlim`	$\varinjlim$	`\varprojlim`	$\varprojlim$

A.7.1 Large operators

Type	Inline	Displayed
`\int_{a}^{b}`	$\int_{a}^{b}$	$\displaystyle\int_{a}^{b}$
`\oint_{a}^{b}`	$\oint_{a}^{b}$	$\displaystyle\oint_{a}^{b}$
`\iint_{a}^{b}`	$\iint_{a}^{b}$	$\displaystyle\iint_{a}^{b}$
`\iiint_{a}^{b}`	$\iiint_{a}^{b}$	$\displaystyle\iiint_{a}^{b}$
`\iiiint_{a}^{b}`	$\iiiint_{a}^{b}$	$\displaystyle\iiiint_{a}^{b}$
`\idotsint_{a}^{b}`	$\idotsint_{a}^{b}$	$\displaystyle\idotsint_{a}^{b}$
`\prod_{i=1}^{n}`	$\prod_{i=1}^{n}$	$\displaystyle\prod_{i=1}^{n}$
`\coprod_{i=1}^{n}`	$\coprod_{i=1}^{n}$	$\displaystyle\coprod_{i=1}^{n}$
`\bigcap_{i=1}^{n}`	$\bigcap_{i=1}^{n}$	$\displaystyle\bigcap_{i=1}^{n}$
`\bigcup_{i=1}^{n}`	$\bigcup_{i=1}^{n}$	$\displaystyle\bigcup_{i=1}^{n}$
`\bigwedge_{i=1}^{n}`	$\bigwedge_{i=1}^{n}$	$\displaystyle\bigwedge_{i=1}^{n}$
`\bigvee_{i=1}^{n}`	$\bigvee_{i=1}^{n}$	$\displaystyle\bigvee_{i=1}^{n}$
`\bigsqcup_{i=1}^{n}`	$\bigsqcup_{i=1}^{n}$	$\displaystyle\bigsqcup_{i=1}^{n}$
`\biguplus_{i=1}^{n}`	$\biguplus_{i=1}^{n}$	$\displaystyle\biguplus_{i=1}^{n}$
`\bigotimes_{i=1}^{n}`	$\bigotimes_{i=1}^{n}$	$\displaystyle\bigotimes_{i=1}^{n}$
`\bigoplus_{i=1}^{n}`	$\bigoplus_{i=1}^{n}$	$\displaystyle\bigoplus_{i=1}^{n}$
`\bigodot_{i=1}^{n}`	$\bigodot_{i=1}^{n}$	$\displaystyle\bigodot_{i=1}^{n}$
`\sum_{i=1}^{n}`	$\sum_{i=1}^{n}$	$\displaystyle\sum_{i=1}^{n}$

A.8 Math accents and fonts

Math accents

		amsxtra	
Type	Typeset	Type	Typeset
\acute{a}	$\acute{a}$		
\bar{a}	$\bar{a}$		
\breve{a}	$\breve{a}$	\spbreve	$\breve{}$
\check{a}	$\check{a}$	\spcheck	$\vee$
\dot{a}	$\dot{a}$	\spdot	$\cdot$
\ddot{a}	$\ddot{a}$	\spddot	$\cdot\cdot$
\dddot{a}	$\dddot{a}$	\spdddot	$\cdots$
\ddddot{a}	$\ddddot{a}$		
\grave{a}	$\grave{a}$		
\hat{a}	$\hat{a}$		
\widehat{a}	$\widehat{a}$	\sphat	$\wedge$
\mathring{a}	$\mathring{a}$		
\tilde{a}	$\tilde{a}$		
\widetilde{a}	$\widetilde{a}$	\sptilde	$\sim$
\vec{a}	$\vec{a}$		

Math fonts

Type	Typeset
LaTeX	
\mathbf{A}	$\mathbf{A}$
\mathcal{A}	$\mathcal{A}$
\mathit{A}	A
\mathnormal{A}	$\mathnormal{A}$
\mathrm{A}	A
\mathsf{A}	A
\mathtt{A}	$\mathtt{A}$
\boldsymbol{\alpha}	$\boldsymbol{\alpha}$
\mathbb{A}	$\mathbb{A}$
\mathfrak{A}	$\mathfrak{A}$
\mathscr{a}	$\mathscr{A}$

And thousand more from STIX (see Section 6.2)!
\mathscr requires the eucal package with the mathscr option

A.9 Math spacing commands

Name	Width	Short	Long
1 mu (math unit)	ı	`\mspace{1mu}`	
thinspace	␣	`\,`	`\thinspace`
medspace	␣	`\:`	`\medspace`
thickspace	␣	`\;`	`\thickspace`
interword space	␣	`\␣`	
1 em	␣		`\quad`
2 em	␣		`\qquad`
Negative space			
1 mu	ı		`\mspace{-1mu}`
thinspace	␣	`\!`	`\negthinspace`
medspace	␣		`\negmedspace`
thickspace	␣		`\negthickspace`

APPENDIX

Text symbol tables

B.1 Some European characters

Name	Type	Typeset	Type	Typeset
a-ring	`\aa`	å	`\AA`	Å
aesc	`\ae`	æ	`\AE`	Æ
ethel	`\oe`	œ	`\OE`	Œ
eszett	`\ss`	ß	`\SS`	SS
inverted question mark	`?‘`	¿		
inverted exclamation mark	`!‘`	¡		
slashed L	`\l`	ł	`\L`	Ł
slashed O	`\o`	ø	`\O`	Ø

G. Grätzer, *More Math Into* LaTeX, DOI 10.1007/978-3-319-23796-1

B.2 Text accents

Name	Type	Typeset	Name	Type	Typeset
acute	`\'{o}`	ó	macron	`\={o}`	ō
breve	`\u{o}`	ŏ	overdot	`\.{g}`	ġ
caron/haček	`\v{o}`	ǒ	ring	`\r{u}`	ů
cedilla	`\c{c}`	ç	tie	`\t{oo}`	o͡o
circumflex	`\^{o}`	ô	tilde	`\~{n}`	ñ
dieresis/umlaut	`\"{u}`	ü	underdot	`\d{m}`	ṃ
double acute	`\H{o}`	ő	underbar	`\b{o}`	o̲
grave	`\`{o}`	ò			
dotless i	`\i`	ı	dotless j	`\j`	ȷ
	`\'{\i}`	í		`\v{\j}`	ǰ

B.3 Text font commands

B.3.1 Text font family commands

Command with Argument	Command Declaration	Switches to the font family
`\textnormal{...}`	`{\normalfont ...}`	document
`\emph{...}`	`{\em ...}`	*emphasis*
`\textrm{...}`	`{\rmfamily ...}`	roman
`\textsf{...}`	`{\sffamily ...}`	sans serif
`\texttt{...}`	`{\ttfamily ...}`	`typewriter style`
`\textup{...}`	`{\upshape ...}`	upright shape
`\textit{...}`	`{\itshape ...}`	*italic shape*
`\textsl{...}`	`{\slshape ...}`	*slanted shape*
`\textsc{...}`	`{\scshape ...}`	SMALL CAPITALS
`\textbf{...}`	`{\bfseries ...}`	**bold**
`\textmd{...}`	`{\mdseries ...}`	normal weight and width

B.3.2 Text font size changes

Command	LaTeX sample text	AMS sample text
`\Tiny`	[not available]	sample text
`\tiny`	sample text	sample text
`\SMALL` or `\scriptsize`	sample text	sample text
`\Small` or `\footnotesize`	sample text	sample text
`\small`	sample text	sample text
`\normalsize`	sample text	sample text
`\large`	sample text	sample text
`\Large`	sample text	sample text
`\LARGE`	sample text	sample text
`\huge`	sample text	sample text
`\Huge`	sample text	sample text

B.4 Additional text symbols

Name	Type	Typeset
ampersand	\&	&
asterisk bullet	\textasteriskcentered	∗
backslash	\textbackslash	\
bar (caesura)	\textbar	|
brace left	\{	{
brace right	\}	}
bullet	\textbullet	•
circled a	\textcircled{a}	ⓐ
circumflex	\textasciicircum	^
copyright	\copyright	©
dagger	\dag	†
double dagger (diesis)	\ddag	‡
dollar	\$	$
double quotation left	\textquotedblleft or ``	“
double quotation right	\textquotedblright or ''	”
em dash	\textemdash or ---	—
en dash	\textendash or --	–
exclamation down	\textexclamdown or !`	¡
greater than	\textgreater	>
less than	\textless	<
lowline	_	_
midpoint	\textperiodcentered	·
octothorp	\#	#
percent	\%	%
pilcrow (paragraph)	\P	¶
question down	\textquestiondown or ?`	¿
registered trademark	\textregistered	®
section	\S	§

Additional text symbols, *continued*

Name	Type	Typeset
single quote left	\textquoteleft or ‘	‘
single quote right	\textquoteright or ’	’
sterling	\pounds	£
superscript	^{a}	ª
tilde	\textasciitilde	~
trademark	\texttrademark	™
visible space	\textvisiblespace	␣

For the \textsubscript command, see Section 10.3.

B.5 Additional text symbols with T1 encoding

An accent

Name	Type	Typeset
Ogonek	\k{e}	ę

European characters

Name	Type	Typeset	Type	Typeset
Eth	\dh	ð	\DH	Ð
Dyet	\dj	đ	\DJ	Đ
Eng	\ng	ŋ	\NG	Ŋ
Thorn	\th	þ	\TH	Þ

Quotation marks

Name	Type	Typeset	Type	Typeset
Single Guillemet	\guilsinglleft	‹	\guilsinglright	›
Double Guillemet	\guillemotleft	«	\guillemotright	»
Single Quotation	\quotesinglbase	‚	\textquoteright	’
Double Quotation	\quotedblbase	„	\textquotedbl	"

B.6 Text spacing commands

Name	Width	Short command	Long command
Positive Space			
Normal	varies	␣	
Intersentence	varies	\@.␣	
Interword	varies	\␣	
Italic Corr.	varies	\/␣	
Tie	varies	~	
Thinspace	␣	\,	\thinspace
Medspace	␣	\:	\medspace
Thickspace	␣	\;	\thickspace
1 em	␣		\quad
2 em	␣		\qquad
Negative Space			
Thinspace	␣	\!	\negthinspace
Medspace	␣		\negmedspace
Thickspace	␣		\negthickspace

APPENDIX

C

Some background

In this book we define LaTeX as the foundation TeX, the work platform LaTeX, and the superstructure AMS packages rolled into one. While you do not need to know anything about LaTeX's detailed structure and history to use it, such knowledge may help you understand how and why LaTeX works the way it does.

In Section C.1, we present a short history of LaTeX, where it has come from and where it is going. Section C.2 provides a description of how LaTeX works. In Section C.3 the various prompts are defined and Section C.4 discusses the separation of visual and logical design elements.

C.1 A short history

C.1.1 TeX

Donald E. Knuth's multivolume work, *The Art of Computer Programming* [47], caused its author a great deal of frustration because it was very difficult to keep the volumes typographically uniform. To solve this problem, Knuth decided to create his own typesetting language. The result is described in *The* TeX*book* [48].

G. Grätzer, *More Math Into* LaTeX, DOI 10.1007/978-3-319-23796-1

A mathematical typesetting language takes care of the multitude of details that are so important in mathematical typesetting, including

- Spacing formulas properly
- Breaking text into pleasingly typeset lines and paragraphs
- Hyphenating words where necessary
- Providing hundreds of symbols for typesetting mathematics

LaTeX does all this and more on almost any computer: Windows computer, Mac, UNIX, workstation, or mainframe. You can write your document on a Windows computer and e-mail it to a coworker who makes corrections on a Mac. The final manuscript might be sent to a publisher who uses a UNIX computer to prepare the document for printing.

Knuth realized that typesetting is only half the solution to the manuscript production problem. You also need a style designer—a specialist who determines what fonts to use, how large a vertical space to put before and after a theorem, and numerous other design issues.

C.1.2 LaTeX 2.09 *and* AMS-TeX

Knuth also realized that typesetting a complex document in TeX requires a very knowledgeable user. So TeX was designed as a platform on which *convenient work environments*—macro packages—could be built, more suitable for the average user to work with. It is somewhat unfortunate that *two* such platforms were made available to the mathematical community in the early 1980s, AMS-TeX and LaTeX.

AMS-TeX was written by Michael D. Spivak for the American Mathematical Society, whereas LaTeX was developed by Leslie Lamport. The strengths of the two systems were somewhat complementary. AMS-TeX provided many features needed by mathematical articles, including

- Sophisticated math typesetting capabilities
- Extensive options for formatting multiline formulas
- Flexible bibliographic references

LaTeX also provided many features, including

- The use of logical units to separate the logical and the visual design of an article
- Automatic numbering and cross-referencing
- Bibliographic databases

Both AMS-TeX and LaTeX became very popular, causing a split in the mathematical community as some chose one system over the other.

C.1.3 LaTeX 3

When Lamport decided not to develop LaTeX any further, the LaTeX 3 *team*[1] took over with the aim of actively supporting, maintaining, and updating LaTeX.

The goals for LaTeX 3 are very ambitious. LaTeX 3 will

- Provide high-quality typesetting for a wide variety of document types and typographic requirements
- Support direct formatting commands for editors and designers, which are essential to the fine-tuning of document layout and page design
- Process complex structured documents and support a document syntax that allows automatic translation of documents conforming to the international document-type definition standard SGML (Standard Generalized Markup Language, ISO 8879)
- Provide a common foundation for a number of incompatible LaTeX variants that have been developed, including the old LaTeX 2.09, LaTeX with the New Font Selection Scheme, and AMS-LaTeX

See two articles by Frank Mittelbach and Chris Rowley, LaTeX *2.09* → LaTeX *3* [57], 1992, and *The* LaTeX *3 Project* [59], 1994, for a statement of goals. Go to `The LaTeX3 project` at
http://www.latex-project.org/latex3.html
for more up-to-date articles and reports.

A number of LaTeX 3 projects have already been completed and are part of LaTeX, including:

The New Font Selection Scheme LaTeX uses Knuth's Computer Modern fonts. The New Font Selection Scheme, NFSS, of Frank Mittelbach and Rainer Schöpf, written in 1989, allows the *independent changing* of font attributes and the integration of new font families into LaTeX. With the proliferation of PostScript fonts and printers, more and more users want to use PostScript fonts in their LaTeX documents.

New and improved environments Frank Mittelbach wrote a new multicolumn environment and Rainer Schöpf improved the `verbatim` and `comment` environments. There have also been several improvements made to the `tabular` and `array` environments. The extremely important `graphicx` package by David Carlisle and Sebastian Rahtz was released.

[1] A talented group of mathematicians and programmers, Frank Mittelbach, Chris Rowley, and Rainer Schöpf. The group has since expanded with the addition of Johannes Braams, David Carlisle, Michael Downes, Denys Duchier, Robin Fairbairns, Alan Jeffrey, and Martin Schröder; many volunteers have also contributed to the project. The current LaTeX 3 project team personnel are: Frank Mittelbach, Rainer Schöpf, Chris Rowley, David Carlisle, Johannes Braams, Robin Fairbairns, Morten Høgholm, Thomas Lotze, Javier Bezos, Will Robertson, Joseph Wright, and Bruno Le Floch.

The first interim solution

In 1990, the AMS released AMS-LaTeX , version 1.0—see Rainer Schöpf's *Foreword* to this book for a personal account. This release contained

- AMS-TeX recoded to work with LaTeX
- The NFSS styles for proclamations
- The new `verbatim` environment

AMS-LaTeX , version 1.0, is a LaTeX *dialect*. It was incompatible with the then current LaTeX—version 2.09.

While the LaTeX 3 team wanted to unify the mathematical community, this first attempt by the AMS split it even further apart. Many AMS-TeX users simply refused to switch. Even today, 17 years later, many mathematicians cling to AMS-TeX. Even the LaTeX community was split into users of the old LaTeX, those whose LaTeX incorporated the NFSS, and AMS-LaTeX users.

The second interim solution

When it became obvious that the goals of LaTeX 3 could not be fulfilled any time soon, the LaTeX 3 team decided to issue a new version of LaTeX, version 2e (also called LaTeXe) in June of 1994. This version replaced LaTeX 2.09, see the two Mittelbach and Rowley articles cited above. This interim release accomplished some of LaTeX 3's goals, including the projects listed previously. Since then, LaTeXe (called LaTeX today) has become accepted as the standard LaTeX.

In February of 1995, the AMS released version 1.2 of AMS-LaTeX (which I call the AMS packages in this book) built on top of the new LaTeX. Michael Downes was the project leader.

The changes in AMS-LaTeX were substantial. The `align` environment, for example, was completely rewritten by David M. Jones. The recoded AMS-TeX had now become a LaTeX package, `amsmath`.

It is extremely important to note that while AMS-LaTeX 1.0 and 1.1 were monolithic structures, versions 1.2 and 2.0 (see Section C.1.4) are just collections of packages that fit nicely into the LaTeX model. You can use one AMS package or all, by themselves or mixed with other LaTeX packages. This book was typeset using the LaTeX document class (`book`) and the AMS packages, version 2.13, along with a number of other LaTeX (non-AMS) packages.

C.1.4 More recent developments

Since 1996, changes to LaTeX have been minor. A few new symbols have been added. Much work has been done on character encoding and LM (Latin Modern) fonts by Bogusław Jackowski and Janusz M. Nowacki to extend LaTeX to languages other than American English (see Appendixes E and F).

In 1999, the American Mathematical Society released version 2.0 of the AMS packages and in 2004, version 2.2. About the same time, a consortium (made up of the AMS, Blue Sky Research, and Y&Y) released free PostScript versions of the CM and AMS fonts. These PostScript fonts are now part of any LaTeX distribution.

Interestingly, there are still those who argue that the AMS packages are not part and parcel of LaTeX and typesetting math. In life, almost everything is a compromise, in software design, even more so. Using the AMS packages to typeset math is an exception. It costs you nothing—if you do not need their features for a document, then you don't have to use them. You need not sacrifice anything in order to have the power of the AMS packages available when you need them. This is why, in this book, by LaTeX we mean LaTeX with the AMS packages.

C.2 How LaTeX works

In this section, I present a very simplified overview of the inner workings of LaTeX.

C.2.1 The layers

TeX and LaTeX consist of many layers. These include:

`virtex` TeX's core, containing about 350 primitive commands such as

```
input    accent    hsize
```

`virtex` can also read *format files,* which are precompiled sets of commands. LaTeX is nothing more than `virtex` reading in a large set of commands, built layer upon layer.

`plain.tex` The most basic layer built on `virtex`. It adds about 600 commands to `virtex`. When you invoke the `TeX` command, `virtex` loads the `plain` format, which is the default. The core TeX commands combined with the commands defined by the `plain` format are called Plain TeX.

Plain TeX is described in detail in Appendix B of Knuth's *The TeXbook* [48]. You can also read `plain.tex`, a text file in the LaTeX distribution. Plain TeX is powerful enough that you could do all your work in it. This approach is advocated by many, including Michael Doob in his book, TeX *Starting from* 1 [12].

`virtex` cannot build (compile) format files. For that you need another version of TeX called `initex`, which loads the most basic information a format needs, such as the hyphenation tables and `plain.tex`, and creates a format file.

LaTeX

LaTeX is a format file containing a compiled set of commands written by Leslie Lamport and others. It provides tools for logical document design, automatic numbering and cross-referencing, tables of contents, and many other features. The new LaTeX we are using is under the control of the LaTeX3 group.

Document classes

The document class forms the next layer. You may choose

- `amsart`, `amsbook`, or `amsproc`, provided by the AMS
- `article`, `book`, `letter`, `proc`, `report`, or `slides`, the legacy classes
- or any one of a large (and growing) number of other document classes provided by publishers of books and journals, universities, and other interested parties

Packages

The next layer is made up of the packages loaded by the document. You can use standard LaTeX packages, AMS packages, or any of hundreds of other packages in the LaTeX universe, mixed together as necessary. Any package may require other packages, or may automatically load other packages.

Documents

At the top of this hierarchy sit your documents, with their custom commands and environments, utilizing all the power derived from the layers below.

C.2.2 *Typesetting*

When typesetting, LaTeX uses two basic types of files, the source files and the font metric files.

A font metric file is designed to hold the information for a font of a given size and style. Each LaTeX font metric file, called a `tfm` file, contains the size of each character, the kerning (the space placed between two adjacent characters), the length of the italic correction, the size of the interword space, and so on. A typical `tfm` file is `cmr10.tfm`, which is the LaTeX font metric file for the font `cmr` (CM roman) at 10-point size.

LaTeX reads the source file one line at a time. It converts the characters of each line into a *token sequence*. A token is either a character—together with an indication of what role the character plays—or a command. The argument of a command is the token following it unless a group enclosed in braces follows it, in which case the contents of the group becomes the argument.[2] An example of this behavior can be seen when you specify an exponent. LaTeX looks for the next token as the exponent unless a group enclosed in braces follows the ^ symbol. This explains why `$2^3$` and `$2^\alpha$`

[2]Delimited commands work somewhat differently (see Section 14.1.9).

work, but `$2^\mathfrak{m}$` does not. Indeed, 3 and `\alpha` each become a single token but `\mathfrak{m}` becomes more than one, four, in fact. Of course, if you *always* use braces, as in

```
$2^{3}$, $2^{\alpha}$, $2^{\mathfrak{m}}$
```

then you never have to think about tokens to type such expressions.

After tokenizing the text, LaTeX hyphenates it and attempts to split the paragraph into lines of the required width. The measurements of the characters—also called glyphs—are absolute, as are the distances between characters—called kerning. The spaces, interword space, intersentence space, and so on, are made of *glue* or rubber length (see Section 14.5.2). Glue has three parameters:

- the length of the space
- stretchability, the amount by which it can be made longer
- shrinkability, the amount by which it can be made shorter

LaTeX stretches and shrinks glue to form lines of equal length.

LaTeX employs a formula to measure how much stretching and shrinking is necessary in a line. The result is called badness. A badness of 0 is perfect, while a badness of 10,000 is very bad. Lines that are too wide are reported with messages such as

```
Overfull \hbox (5.61168pt too wide) in paragraph
                at lines 49--57
```

The badness of a line that is stretched too much is reported as follows:

```
Underfull \hbox (badness 1189) in paragraph
                at lines 93--93
```

Once enough paragraphs are put together, LaTeX composes a page from the typeset paragraphs using vertical glue. A short page generates a warning message such as

```
Underfull \vbox (badness 10000) has
occurred while \output is active
```

The typeset file is stored as a `dvi` (Device Independent) file or a PDF file.

C.2.3 Viewing and printing

Viewing and printing LaTeX's typeset output are not really part of LaTeX proper, but they are obviously an important part of your work environment. The printer driver prints the `dvi` and PDF files, and the video driver lets you view them on your monitor.

C.2.4 LaTeX's files

Auxiliary files

LaTeX is a *one-pass compiler,* that is, it reads the source file once only for typesetting. As a result, LaTeX must use auxiliary files to store information it generates during a run. For each typesetting run, LaTeX uses the auxiliary files compiled during the *previous* typesetting run. This mechanism explains why you have to typeset twice or more (see Section 17.2) to make sure that changes you have made to the source files are reflected in the typeset document. Such an auxiliary file has the same base name as the source file, the extension indicates its type.

The most important auxiliary file, the `aux` file, contains a great deal of information about the document, most importantly, the data needed for symbolic referencing. Here are two typical entries:

```
\newlabel{struct}{{5}{2}}
\bibcite{eM57a}{4}
```

The first entry indicates that a new symbolic reference was introduced on page 2 of the typeset document in Section 5 using the command

```
\label{struct}
```

The command `\ref{struct}` produces 5, while `\pageref{struct}` yields 2.

The second entry indicates that the bibliographic entry with label `eM57a` has been assigned the number 4, so `\cite{eM57a}` produces [4].

There is an `aux` file for the source file being processed, and another one for each file included in the main file by an `\include` command.

No auxiliary file is written if the `\nofiles` command is given. The message

```
No auxiliary output files.
```

in the `log` file reminds you that `\nofiles` is in effect.

The `log` file contains all the information shown in the `log` window during the typesetting. The `dvi` file contains the typeset version of the source file.

There are five auxiliary files that store information for special tasks. They are written only if that special task is invoked by a command and there is no `\nofiles` command. The additional auxiliary files are

`glo` Contains the glossary entries produced by `\glossary` commands. A new file is written only if there is a

```
\makeglossary
```

command in the source file (see Section 16.6).

`lof` Contains the entries used to compile a list of figures. A new file is written only if there is a

```
\listoffigures
```

command in the source file (see Section 8.4.3).

`lot` Contains the entries used to compile a list of tables. A new file is written only if there is a

```
\listoftables
```

command in the source file (see Section 8.4.3).

`toc` Contains the entries used to compile a table of contents. A new file is written only if there is a

```
\tableofcontents
```

command in the source file (see Section 17.2).

For information about the auxiliary files created by BibTeX and *MakeIndex,* see Sections 15.2.3 and 16.3, respectively. Some classes and packages create additional auxiliary files (see Section 11.2.3 for an example).

Versions

A complete LaTeX distribution consists of hundreds of files, all of which interact in some way. Since most of these files have had many revisions, you should make sure that they are all up-to-date and compatible with each other. You can check the version numbers and dates by reading the first few lines of each file in a text editor or by checking the dates and version numbers that are shown on the list created by the command `\listfiles`, which I discuss later in this section.

LaTeX has been updated every year. While writing this book, I used the version of LaTeX that was issued on May 5, 2014.

When you typeset a LaTeX document, LaTeX prints its release date in the `log` file with a line such as

```
LaTeX2e <2014/05/01>
```

If you use a LaTeX feature that was introduced recently, you can put a command such as the following into the preamble of your source file:

```
\NeedsTeXFormat{LaTeX2e}[2008/12/01]
```

This command specifies the date of the oldest version of LaTeX that may be used to typeset your file. If someone attempts to typeset your file with an older version, LaTeX generates a warning.

The AMS math package amsmath is at version 2.13, the document classes at version 2.26, and the AMSFonts set is at version 2.2d. See Section D.1 for more information on obtaining updated versions.

If you include the `\listfiles` command in the preamble of your document, then the `log` file contains a detailed listing of all the files used in the typesetting of your document. Here are the first few (truncated) lines from such a listing:

```
*File List*
    book.cls    1999/01/07 v1.4a Standard LaTeX document class
   leqno.clo    1998/08/17 v1.1c Standard LaTeX option
                    (left equation numbers)
    bk10.clo    2007/10/19 v1.4h Standard LaTeX file (size option)
    MiL5.sty    2014/12/15 Commands for MiL5
 amsmath.sty    2013/01/14 v2.14 AMS math features
 amstext.sty    2000/06/29 v2.01
  amsgen.sty    1999/11/30 v2.0
  amsbsy.sty    1999/11/29 v1.2d
  amsopn.sty    1999/12/14 v2.01 operator names
  amsthm.sty    2004/08/06 v2.20
verbatim.sty    2003/08/22 v1.5q LaTeX2e package for
                   verbatim enhancements
 amsxtra.sty    1999/11/15 v1.2c
   eucal.sty    2009/06/22 v3.00 Euler Script fonts
 amssymb.sty    2013/01/14 v3.01 AMS font symbols
amsfonts.sty    2013/01/14 v3.01 Basic AMSFonts support
 omxcmex.fd    1999/05/25 v2.5h Standard LaTeX
                    font definitions
latexsym.sty    1998/08/17 v2.2e Standard LaTeX package
                    (lasy symbols)
   amscd.sty    1999/11/29 v1.2d
   alltt.sty    1997/06/16 v2.0g defines alltt environment
  xspace.sty    2009/10/20 v1.13 Space after command
                    names (DPC,MH)
graphicx.sty    2014/04/25 v1.0g Enhanced LaTeX Graphics
                    (DPC,SPQR)
  keyval.sty    2014/05/08 v1.15 key=value parser (DPC)
graphics.sty    2009/02/05 v1.0o Standard LaTeX Graphics
                    (DPC,SPQR)
    trig.sty    1999/03/16 v1.09 sin cos tan (DPC)
```

This list looks quite up-to-date (in fact, it is completely up-to-date). To confirm this, open the file `alltt.sty` in the latest LaTeX distribution. You find the lines

```
\ProvidesPackage{alltt}
           [1997/06/16 v2.0g defines alltt environment]
```

that explain the date found in the listing.

C.3 Interactive LaTeX

If LaTeX cannot carry out your instructions, it displays a *prompt* and possibly an error message in the `log` window.

- The `**` prompt means that LaTeX needs to know the name of a source file to typeset. This usually means that you misspelled a file name, you are trying to typeset a document that is not located in LaTeX's current folder, or that there is a space in the name of your source file.
- The `?` prompt indicates that LaTeX has found an error in your source file, and wants you to decide what to do next. You can try to continue typesetting the file by pressing
  - Return
  - `q` to typeset in quiet mode, not stopping for errors. Depending on the nature of the error, LaTeX may either recover or generate more error messages
  - `x` to stop typesetting your file
  - `h` to get advice on how to correct the error
- If you have misspelled the name of a package in a `\usepackage` command, or if LaTeX cannot find a file, it displays a message similar to the following:

```
! LaTeX Error: File 'misspelled.sty' not found.
Type X to quit or <RETURN> to proceed,
or enter new name. (Default extension: sty)

Enter file name:
```

  You can either type the correct name of the file at the prompt, or type `x` to quit LaTeX.
- The `*` prompt signifies that LaTeX is in *interactive mode* and is waiting for instructions. To get such a prompt, comment out the line

```
\end{document}
```

  in a source file, then typeset the file. Interactive instructions, such as `\show` and `\showthe` (see Section 14.1.8) may be given at the `*` prompt. To exit, type

```
\end{document}
```

  at the `*` prompt, and press Return.
- If you get the `*` prompt and no error message, type `\stop` and press Return.

C.4 Separating form and content

In Section 2.3, we discuss logical and visual design and how LaTeX allows you to concentrate on the logical design and takes care of the visual design.

LaTeX uses four tools to separate the logical and visual design of a document:

1. **Commands** Information is given to LaTeX in the arguments of commands. For instance, title page information is given in this form. The final organization and appearance of the title page is completely up to the document class and its options.

   A more subtle example is the use of a command for distinguishing a term or notation. For instance, you may want to use an `\env` command for environment names. You may define `\env` as follows:

   ```
   \newcommand{\env}[1]{\texttt{#1}}
   ```

   This gives you a command that typesets all environment names in typewriter style (see Section 3.6.2). Logically, you have decided that an environment name should be marked up. Visually, you may change your decision any time. By changing the definition to

   ```
   \newcommand{\env}[1]{\textbf{#1}}
   ```

   all environment names are typeset in bold (see Section 3.6.5).

   The following example is taken from `secondarticleccom.tex` (see Section 9.3 and the `samples` folder). This article defines the construct $D^{\langle 2 \rangle}$ with the command

   ```
   \newcommand{\Dsq}{D^{\langle 2 \rangle}}
   ```

   If a referee or coauthor suggests a different notation, editing this *one line* changes the notation throughout the entire article.

2. **Environments** Important logical structures are placed within environments. For example, list items are typed within a list environment (see Section 4.2) and formatted accordingly. If you later decide to change the type of the list, you can do so by simply changing the name of the environment.

3. **Proclamations** You can change the style or numbering scheme of any proclamation at any time by changing that proclamation's definition in the preamble. See the typeset `secondarticle` article on pages 272–275 for examples of proclamations typeset with different styles.

4. **Numbering and cross-referencing** Theorems, lemmas, definitions, sections, and equations are logical units that can be freely moved around. LaTeX automatically recalculates the numbers and cross-references.

You write articles to communicate your ideas. The closer you get to a separation of logical and visual design, the more you are able to concentrate on that goal. Of course, you can never quite reach this ideal. For instance, a `line too wide` warning (see Sections 1.4 and 3.7.1) is a problem of visual design. When a journal changes the document class in an article you submitted, unless the new document class retains the same fonts and line width of the document class you used, new `line too wide` problems arise. LaTeX is successful in automatically solving visual design problems well over 95% of the time. That is getting fairly close to the ideal.

APPENDIX

LaTeX *and the Internet*

While LaTeX is pretty stable, the rest of the world around us is changing very fast and the Internet plays an ever larger role in our lives. This appendix deals with the Internet as a useful source of LaTeX information.

The Internet is clearly the main repository of all matters LaTeX, and the Comprehensive TeX Archive Network (CTAN) is the preeminent collection of TeX-related material. Section D.1 discusses how and where to find the LaTeX distribution, AMS and LaTeX packages, and the sample files for this book on CTAN.

Various international TeX user groups (especially TUG, the TeX Users Group) and the American Mathematical Society play a significant role in supporting LaTeX. I discuss some of the major user groups in Section D.2.

Finally, you find a great deal of useful information on the Internet concerning LaTeX. I provide some pointers in Section D.3.

D.1 Obtaining files from the Internet

Say you are interested in using Piet van Oostrum's `fancyhdr` package mentioned in Section 8.6. Chances are you can go ahead and use it, your LaTeX installation already has it. In this age of gigantic hard disks, your LaTeX installation places pretty much

G. Grätzer, *More Math Into* LaTeX, DOI 10.1007/978-3-319-23796-1

everything on your computer. But what if your version of fancyhdr needs updating or you need a new package. How you go about getting it?

We discuss below the proper way of doing this, with an FTP client or a Web browser. But maybe the simplest approach is to google fancyhdr. The first line of the first entry of the complete list of 82,100 responses is

```
The TeX Catalogue OnLine, Entry for fancyhdr, Ctan Edition
```

Clicking on it takes you to a page describing the package. You can get the package by clicking on Download. It is this simple.

In general, there are two types of Internet sites from which you can download files:

- FTP sites (using the file transfer protocol)
- Web sites (using the HTTP protocol)

To access them, use a *client* application on your computer to connect to a *server* on another machine. Most *Web browsers*, which are designed to connect to Web sites, also handle FTP transfers.

All operating systems include a browser and an FTP client as part of the system.

The Comprehensive TeX Archive Network

The Comprehensive TeX Archive Network (CTAN) is the preeminent collection of TeX-related material on the Internet. There are three main CTAN hosts:

- U.S.
  - FTP address: ftp://tug.ctan.org/
  - Web address: http://www.ctan.org/
- U.K.
  - FTP address: ftp://ftp.tex.ac.uk/
  - Web address: http://www.tex.ac.uk/
- Germany
  - FTP address: ftp://ftp.dante.de/
  - Web address: http://www.dante.de/

If you go to a CTAN site, at the very root you find README.structure, a very important file. It describes the bottom of the archive tree.

- biblio Systems for maintaining and presenting bibliographies within documents typeset using LaTeX

- `digests` Collections of TEX mailing list digests, TEX-related 'electronic magazines', and indexes, etc., of printed publications
- `dviware` Printer drivers and previewers, etc., for DVI files
- `fonts` Fonts written in Metafont, and support for using fonts from other sources (e.g., those in Adobe Type 1 format)
- `graphics` Systems and TEX macros for producing graphics
- `help` FAQs and similar direct assistance, the catalogue
- `indexing` Systems for maintaining and presenting indexes of documents typeset using TEX.
- `info` Manuals and extended how-to information, errata for TEX-related publications, collections of project (e.g., LATEX and NTS) documents, etc.
- `language` Support for various languages
- `macros` TEX macros. Several directories have significant sub-trees.
- `obsolete` Material which is now obsolete, including all of LATEX 2.09
- `support` TEX support environments and the like
- `systems` TEX systems. Organized by operating environment
- `tds` The TEX Directory Structure standard
- `usergrps` Information supplied by TEX User Groups
- `web` Literate Programming tools and systems

All of these have many subdirectories, for instance, `info` has the `examples` subdirectory that contains the sample files for this book. This is a rather new subdirectory, older sample files are in `info` proper.

So if you are interested in BIBTEX, you go to `biblio/`, and so on. The explanations are clear. All matters LATEX are in `macros/latex/`, which has a number of subdirectories, including

- `contrib`—Contributed LATEX macros
- `unpacked`—Unpacked copy of the LATEX sources
- `required`—Packages "required" of a LATEX distribution

There are many *full mirrors*, exact duplicates, of CTAN and many *partial mirrors*. At the root of CTAN you find the `README.mirrors` file listing them all. To reduce network load, you should try to use a mirror located near you.

Many CTAN sites now have easy search access with Web browsers. For instance, point your browser to

http://tug.ctan.org/search.html

In the search field, type fancyhdr, and you get a long list of links. Click on

macros/latex/contrib/fancyhdr.zip

and you are done. If you type gratzer, you get the links to the help files of my various books—in info/ and info/examples/.

The AMS *packages*

Chances are that you received the AMS packages with your LaTeX distribution. If you did not, or if you want to update them, go to a CTAN site:

- /tex-archive/fonts/amsfonts/latex/
- /tex-archive/macros/latex/required/amslatex/

or to the AMS site:

http://www.ams.org/tex/amslatex.html

The sample files

The sample files for this book, introduced in Section 1.1.2 on page 5, live on CTAN in the directory

/info/examples/Math_into_LaTeX-5

You can go to /info/examples/ and download it, or you can search for the directory name Math_into_LaTeX-5. If you forget these, just search for gratzer.

You can also find the *Mission Impossible* (Part I) on CTAN:

/info/Math_into_LaTeX-4/Mission_Impossible.pdf

D.2 The TeX Users Group

The TeX Users Group (TUG) does a tremendous job of supporting and promoting TeX, by publishing a journal, *TUGboat*, three times a year and organizing an annual international conference. TUG also helps support the LaTeX 3 team in maintaining LaTeX and developing LaTeX 3.

Consider joining TUG if you have an interest in LaTeX. TUG's contact information is:

PO Box 2311
Portland, OR 97208–23110

Telephone: (503) 223-9994
E-mail: office@tug.org
Web page: `http://www.tug.org/`

If you are a member, you receive every year a brand new TeX Live DVD, which contains everything you need to install LaTeX.

The American Mathematical Society

The AMS provides excellent technical advice for using the AMS packages and AMS-Fonts. You can reach the AMS technical staff by e-mail at `tech-support@ams.org`, or by telephone at (800) 321-4267 or (401) 455-4080. You can also find a great deal of helpful TeX information on the AMS Web site in the `Author Resource Center`.

D.3 Some useful sources of LaTeX information

You may find useful the Frequently Asked Questions (FAQ) documents maintained on CTAN; search `FAQ`. The U.K. TeX Users Group maintains its own FAQ list at

`http://www.tex.ac.uk/cgi-bin/texfaq2html?introduction=yes`

The AMS FAQ is at

`http://www.ams.org/authors/author-faq.html`

You can also ask most TeX-related questions in the Usenet newsgroup `comp.text.tex`.

APPENDIX

PostScript fonts

In the late 1990s, as we mentioned in Section C.1.4, a consortium (the AMS, Blue Sky Research, and Y&Y) released a free PostScript version of the CM and AMS fonts, so everyone could switch to PostScript fonts, a tremendous advance for LaTeX users.

The Computer Modern fonts were originally "hardwired" into LaTeX. Many users liked LaTeX but disliked the Computer Modern font, and with the spread of personal computers and PostScript laser printers, it was imperative that more PostScript fonts be integrated into LaTeX. In Section E.1, I describe how easy it is to use standard PostScript fonts, such as Times. In Section E.2, I show you how to replace the CM and AMS fonts in a LaTeX document with the Lucida Bright fonts.

"PostScript font" is the terminology that lay people, like me, use. The proper terminology is *Adobe Type 1 format font.* PostScript has provision for a wide range of fonts including Type 3 and Type 1 (as well as Type 42 and Type 5, and so on). The Type 3 font category is very general and includes bitmap fonts, grayscaled fonts, and so on. Type 1 fonts are tightly constrained *outline* fonts, which can be accurately rendered at almost any resolution, and have a special purpose code that deals only with Type 1 fonts.

G. Grätzer, *More Math Into* LaTeX, DOI 10.1007/978-3-319-23796-1

E.1 The Times font and MathTime

In this section, we step through the process of incorporating the Adobe Times font into a LaTeX document to replace the Computer Modern text fonts, and, optionally, of using the *MathTime Pro 2* math fonts to replace the Computer Modern math fonts. To do so, we use the PSNFSS packages (see Section 10.3).

A document class specifies three standard font families (see Section 3.6.2):

- A roman (or serif) font family
- A sans serif font family
- A typewriter style font family

The `times` package in the PSNFSS distribution makes Times the roman font family, Helvetica the sans serif font family, and Courier the typewriter style font family.

Setting up Times

First, install the Adobe Times, Helvetica, and Courier PostScript fonts and their TeX font metric files.

Now typeset the `psfonts.ins` file—in the PSNFSS distribution. This produces `sty` files for the standard PostScript fonts. The Times style file is called `times.sty`. If you do not already have it, copy it into a folder LaTeX can access.

To use the `times` package, you must have the *font definition* (`fd`) files for the fonts specified. By checking the `times.sty` file, you see that you need three files for the three fonts: Times, Helvetica, and Courier. In the `times` package these are named `ptm`, `phv`, and `pcr`, respectively. The three file names, each comprising three characters, are the font names in the naming scheme devised by Karl Berry. In `ptm`, `p` stands for the foundry's name (in this case, Adobe), `tm` stands for Times, `hv` for Helvetica, and `cr` for Courier. The corresponding font definition files are named `ot1ptm.fd`, `ot1phv.fd`, and `ot1pcr.fd`, respectively. `OT1` designates the old TeX font encoding scheme, which is not discussed here. You can get these files from CTAN (see Section D.1). If you do not already have it, copy it into a folder LaTeX can access.

Using Times

In the preamble of your document, type

```
\usepackage{times}
```

after the `\documentclass` line. Then Times becomes the roman, Helvetica the sans serif, and Courier the typewriter style document font family. That is all there is to it.

Using the `times` package changes the document font family throughout your document. To switch to Times only occasionally, type

```
{\fontfamily{ptm}\selectfont phrase}
```

The text preceding and following this construct is not affected. For example,

```
{\fontfamily{ptm}\selectfont
This text is typeset in the Times font.}
```

typesets as

This text is typeset in the Times font.

Similarly,

```
\fontfamily{ptm}\selectfont
This text is typeset in the Times font.
\normalfont
```

also typesets the same phrase in Times. Recall that the `\normalfont` command restores the document font family (see Section 3.6.2).

Setting up *MathTime*

Looking at a mathematical article typeset with the Times text font, you may find that the Computer Modern math symbols look too thin. To more closely match Times and other PostScript fonts, Michael Spivak modified the CM math symbols, calling these modified fonts *MathTime Pro 2*. You can purchase these fonts from Personal TeX,

```
http://store.pctexstore.com/
```

Install the *MathTime Pro 2* PostScript fonts and the TEX font metric files. If you do not already have them, copy from PSNFSS the files

```
mathtime.ins   mathtime.dtx  mtfonts.fdd
```

into a folder LATEX can access.

Typeset `mathtime.ins` to produce the necessary `fd` files and the `mathtime.sty` file.

Using *MathTime*

If you want to use Times as the document font family and *MathTime* as the default math font, specify

```
\usepackage[LY1]{fontenc}            %specify font encoding
\usepackage[LY1,mtbold]{mathtime}  %switch math fonts
\usepackage{times}                   %switch text fonts
```

in the preamble of your document.

The `mathtime` package has many options. See its documentation for more information; typeset `mathtime.dtx` to get it.

E.2 Lucida Bright fonts

Another alternative to Computer Modern fonts is *Lucida Bright* for both text and math fonts. You can purchase the Lucida Bright fonts from TUG.

Copy the files

```
lucidabr.ins, lucidabr.dtx,
lucidabr.fdd,lucidabr.yy
```

into your TeX input folder. Typeset `lucidabr.yy`, producing the `lucidabr.sty` file and a large number of `fd` files.

Now add the lines

```
\usepackage[LY1]{fontenc}     %specify font encoding
\usepackage[LY1]{lucidabr}    %switch text and math fonts
```

in the preamble of your document. The `lucidabr` package has many options. See its documentation—typeset `lucidabr.dtx` to get it.

E.3 More PostScript fonts

You can obtain PostScript fonts from a wide variety of sources. There are many free PostScript fonts on CTAN. Table E.1 is a short list of the more prominent commercial vendors.

See also the Web page at `http://www.microsoft.com/typography/` for a lot of useful information and links.

Foundry	URL
Adobe	`www.adobe.com/type/`
Agfa/Monotype	`www.agfamonotype.com/`
Berthold	`www.bertholdtypes.com/`
Bitstream	`www.bitstream.com/`
Emigre	`www.emigre.com/`
Hoefler	`www.typography.com/`
ITC	`www.itcfonts.com/`
Linotype	`www.linotype.com/`
Monotype	`www.fonts.com/`
Scriptorium	`www.fontcraft.com/`
Vintage	`www.vintagetype.com/`

Table E.1: Some type foundries on the Internet.

APPENDIX

F

LaTeX *localized*

If the language in which you write articles is not American English and/or your keyboard is not the standard American keyboard, you may find it annoying and sometimes difficult to use standard LaTeX. The annoyance may start with finding out how to type `~` for a nonbreakable space, to LaTeX's inability to properly hyphenate `Gr\"{a}tzer`, and LaTeX's inability to use a different alphabet.

Many of the improvements to LaTeX in recent years have been to localize LaTeX, that is, to adapt LaTeX for use with languages other than American English and keyboards other than standard American keyboards. The `babel`, `fontenc`, `inputenc` packages are the major players, along with new font-encoding schemes, including the T1 encoding. You find these packages as part of the LaTeX distribution (see Section 10.3).

The `babel` package is described in detail in Johannes Braams, *Babel, a multilingual package for use with LaTeX's standard document classes* [7] and in Chapter 9 of *The* LaTeX *Companion,* 2nd edition [56].

G. Grätzer, *More Math Into* LaTeX, DOI 10.1007/978-3-319-23796-1

If you are interested in using a localized LaTeX, you should turn to the TeX user group for that linguistic group to find out what is available. You should also consult the `babel` user guide.

At a minimum, a supported language has translated redefinable names (see Table 14.1), and a localized variant of the `\today` command. Two very advanced language adaptations are German and French.

We first illustrate the use of the `babel` package with the German language, which gives you a rich set of features, including

- Allows you to type `"a` for `\"{a}`
- Introduces `"s` for sharp s (eszett)
- Introduces `"ck` for a `ck` that becomes k-k when hyphenated

Type the following test file: (`german.tex` in the `samples` folder):

```
\documentclass{article}
\usepackage[german]{babel}
\usepackage[T1]{fontenc}

\begin{document}
\section{H"ullenoperatoren}
Es sei $P$ eine teilweise geordnete Menge. Wir sagen,
dass in $P$ ein \emph{H"ullenoperator} $\lambda$
erkl"art ist, wenn sich jedem $a \in  P$ ein eindeutig
bestimmtes $\lambda(a) \in P$ zuordnen l"a"st, so  dass
die folgenden Bedingungen erf"ullt sind.
\end{document}
```

And here it is typeset:

1 Hüllenoperatoren

Es sei P eine teilweise geordnete Menge. Wir sagen, dass in P ein *Hüllenoperator* λ erklärt ist, wenn sich jedem $a \in P$ ein eindeutig bestimmtes $\lambda(a) \in P$ zuordnen läßt, so dass die folgenden Bedingungen erfüllt sind.

The second example uses the following options for the packages:

```
\usepackage[T2A]{fontenc}
\usepackage[koi8-u]{inputenc}
\usepackage[ukrainian]{babel}
```

The encoding `koi8-u` is appropriate for Ukrainian.

And here is the typeset Ukrainian sample file:

Поняття теорії ігор

Віктор Анякін

31 липня 2006 р.

Логічною основою теорії ігор є формалізація трьох понять, які входять в її визначення і є фундаментальними для всієї теорії:

- Конфлікт,
- Прийняття рішення в конфлікті,
- Оптимальність прийнятого рішення.

Ці поняття розглядаються в теорії ігор у найширшому сенсі. Їх формалізації відповідають змістовним уявленням про відповідні об'єкти.

Змістовно, конфліктом можна вважати всяке явище, відносно якого можна казати про його учасників, про їхні дії, про результати явищ, до яких призводять ці дії, про сторони, які так чи інакше зацікавлені в таких наслідках, і про сутність цієї зацікавленості.

Якщо назвати учасників конфлікту *коаліціями дії* (позначивши їхню множину як $\Re_D$, можливі дії кожної із коаліції дії — її *стратегіями* (множина всіх стратегій коаліції дії K позначається як S), результати конфлікту — *ситуаціями* (множина всіх ситуацій позначається як S; вважається, що кожна ситуація складається внаслідок вибору кожної із коаліцій дії деякої своєї стратегії, так, що $S \subset \prod\limits_{K \in \Re} S_K$), зацікавлені сторони — *коаліціями інтересів* (їх множина — $\Re_I$) і, нарешті, говорити про можливі переваги для кожної коаліції інтересів K однієї ситуації s' перед іншою s'' (цей факт позначається як $s' \underset{K}{\prec} s''$), то конфлікт в цілому може бути описаний як система

$$\Gamma = \langle \Re_D, \{S_K\}_{K \in \Re_D}, S, \Re_I, \{\underset{K}{\prec}\}_{K \in \Re_I} \rangle$$

Така система, яка представляє конфлікт, називається *грою*. Конкретизації складових, які задають гру, призводять до різноманітних класів ігор.

APPENDIX

LaTeX on the iPad

A few years back, personal computing was desktop-centric. For many tasks, for instance, for back up and for updating the operating system, you had to connect your smartphone and tablet with a computer. Tim Cook (Apple's CEO as I am writing this book) coined the term "Post PC revolution" to describe the trend that a tablet is no longer a younger brother of a PC, but an equal partner; in fact, for many users, it can be the only computer they will ever need.

But can you use it for your LaTeX documents? Isn't the iPad designed only for e-mail, to read news, and enjoy entertainment? Certainly. While it has a fast CPU, it has an even more powerful graphics chip so viewing videos and complex Web pages is quick. The operating system is designed to make performing these basic tasks very easy and intuitive. iOS masks the complexities of the underlying computer.

Nevertheless, underneath this easy-to-use interface there is a Mac. Get a little familiar with the iPad as a computer, and you can work with your LaTeX documents pretty well.

There are good reasons why the iPad is the only tablet I'll discuss. Today, the iPad is clearly the dominant tablet of more than a hundred on the market and the iPad is the only tablet with a decent market share that is in an *ecosystem*: the iPad is just one device under iCloud along with the iPhone, the Mac desktops, and the Mac notebooks.

G. Grätzer, *More Math Into* LaTeX, DOI 10.1007/978-3-319-23796-1

I work on a LaTeX document on my iMac, and when I am away from home, I continue my work on my MacBook Air or iPad; there is no interruption, all the devices are fully synchronized.

In Section G.1, we discuss the iPad file system, sandboxing, file transfers, printing, and text editing. We discuss where are the files to be LaTeXed and where the LaTeX process takes place in Section G.2. Finally, in Section G.3, we introduce two LaTeX implementations for the iPad: Texpad and TeX Writer.

This appendix is based on my articles in the Notices of the Amer. Math. Soc. **60** (2013), pp. 332–334 and 434–439. You can find these two articles, `NoticesV.pdf` and `NoticesVI.pdf`, in the `samples` folder for some more detail.

G.1 The iPad as a computer

To work on a LaTeX document, you sit in front of your computer, in the complex folder hierarchy you find `document.tex`, double clicks it to start the LaTeX implementation, edit the document, typeset it. Then you print `document.pdf`, proofread it, and then you go back to editing...

How do you do these steps on an iPad? On the iPad, there is only a rectangular array of apps, see Figure G.1. No documents are visible. There may be folders containing more apps, but no folder in a folder. There are no Library folders, no Download folder. And no File menu containing the Print command!

I have `document.tex` on my desktop, but how do I transfer it to the iPad? I would plug in my thumb drive to facilitate the transfer, but the iPad has no USB port.

G.1.1 File system

As we pointed out, the iPad starts up displaying a rectangular array of icons and folders for apps, as in see Figure G.1. There are no icons for documents. There is no familiar Desktop for documents and folders. No Applications folder. The screen is always occupied by a single window; the file system, as we know it from desktop computers, is gone.

In its place is an app-centric starting point. Touch the icon of an app and you are in business. When the app opens, you get access to the documents of the app.

For security reasons, the apps are sandboxed, limiting an app's access to files, preferences, network resources, hardware, and so on. Ars Technica's John Siracusa described the goal of sandboxing as follows: "Running an application inside a sandbox is meant to minimize the damage that could be caused if that application is compromised by a piece of malware. A sandboxed application voluntarily surrenders the ability to do many things that a normal process run by the same user could do. For example, a normal application run by a user has the ability to delete every single file owned by that user. Obviously, a well-behaved application will not do this. But if an application becomes compromised, it can be coerced into doing something destructive."

Figure G.1: A rectangular array of apps

Of course, the iPad is a computer, and it has a File System, we just do not see it. But it is important to visualize it. To help us along, we will use an app.

G.1.2 FileApp

If you search the iPad's App Store for "file" apps, there are more than 1,000 of them. Many of them could be used to help us understand the iPad file system. I choose FileApp by DigiDNA (Figures G.2 and G.3).

To get started, plug the iPad into a desktop computer, download and start the application iMazing on the computer; download and start FileApp on the iPad. On the left panel of iMazing, click on Apps, then on FileApp. Anything you drag into the right pane of DiskAid is copied to FileApp. So much for file transfer. To see the file structure of the various iPad apps, click on their names.

Of course, for file transfers I should also mention the ubiquitous Dropbox. Download it for the iPad, sign in (as you did for your computer Dropbox); that's it.

Figure G.2: iMazing

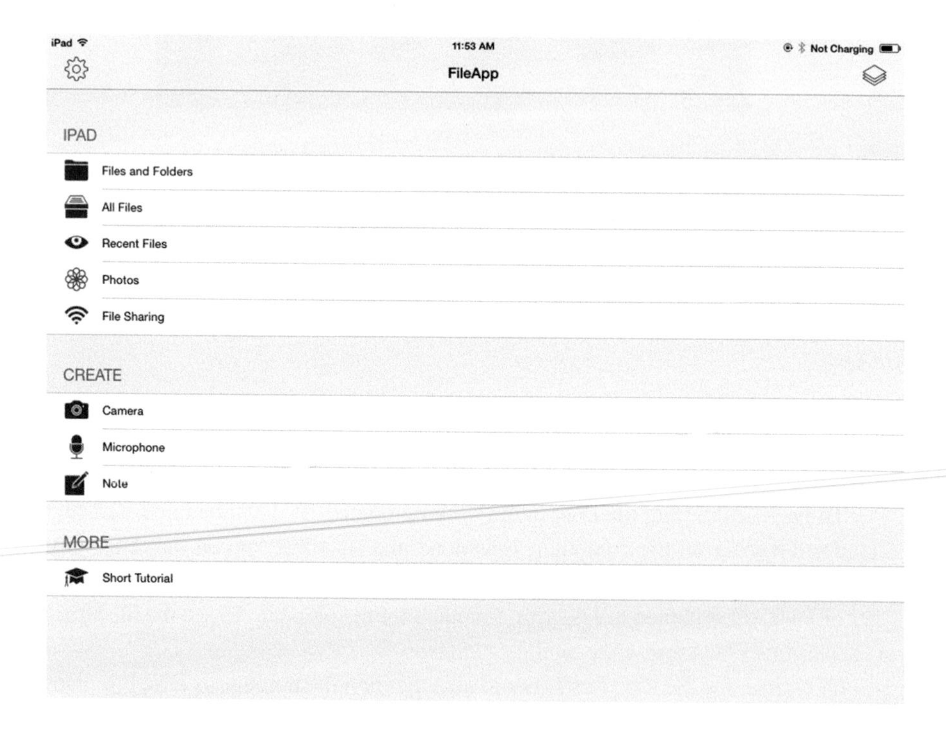

Figure G.3: FileApp

G.1.3 Printing

When I first wanted to print from my iPhone, I realized that there is no print command. However, lots of apps would do the job. In fact, searching for "print" in the App Store, I discovered over 600 apps; many of them print, utilizing my desktop computer.

Typical of these apps is PrintDirect (EuroSmartz) and Printer Pro (Readdle Productivity). They can use any printer connected with your desktop computer. They wirelessly connect to your computer and print with its help.

If so many apps can help me out with printing, how come iOS does not? Read the comments about iOS printing; I was not the only one confused.

However, if the iPad is the poster child of the Post PC Revolution, its native printing solution cannot involve desktop computers. Apple introduced the appropriate technology; they named it AirPrint. The idea is simple: the iPad collaborates with the printer. Of course, for this you need a wireless printer that is AirPrint aware. Apple lists all the AirPrint aware printers:

`http://support.apple.com/kb/ht4356`

as of this writing, about 2,000. If you are lucky and have one of these printers, test it. Open an e-mail and touch the Action icon (here it is the Reply icon); this offers you the options: Reply, Forward, and Print. Touch Print. Printer Options appears, and you can choose how many copies and on which printer. (Lots of apps provide more choices, such as page range.) Choose the printer and print.

For a second test, open a Web page in Safari. There is only one difference: the action icon is a curved arrow in a rectangle.

As a third test, open the Drudge Report. It has the familiar Action icon; we are in business. Finally, open the Politico app, read the news and look for an action icon. There is none. So to use AirPrint, you need an AirPrint aware printer and an AirPrint aware app! For the time being, these are limiting restrictions.

G.1.4 Text editors

Many of us edit LaTeX documents in text editors more sophisticated than the text editor that comes with the LaTeX implementation. There are so many text editors, well over 200..., see the table at

`http://brettterpstra.com/ios-text-editors/`

Keeping the iPad horizontal, the keyboard gobbles up too much real estate. Keeping it vertical, the keyboard is less intrusive, but the keys are smaller. If you want to do serious work on the iPad, use a keyboard.

The iOS's touch text editing is nice, but it lacks a feature crucial for text editing: moving the cursor a character ahead or back. (Of course, keyboards have cursor keys!) Text editors offer a variety of solutions, for instance, finger swiping.

I will discuss briefly a very sophisticated text editor: Textastics. If you want Syntax Highlighting, Search and Replace, and Text Expander, this a good choice. In Figure G.4, you see me editing a document.

You can see the cursor navigation wheel (which appears with a two finger tap—finger swipe also moves the cursor). It comes with an excellent user manual. Textastics also has a Mac version. And if you spend time shaping it to your liking, then you would like the same tamed editor for all your work.

G.2 Files

The LaTeX files, of course, can always be composed in the app. You can obtain your existing files in two ways:

1. Using iTunes. To transfer files—one at a time—to your app from your computer using iTunes, connect your iPad to your computer and start iTunes by double clicking on its icon. Under Devices, we selected the iPad from the left side of the iTunes window (see Figure G.5). At the top of the iTunes window, next to Summary and Info, select Apps (see Figure G.6). The lower part of the window now has File Sharing; see Figure G.7. On the left, you see a listing of the apps available for file transfer. Select the app; the files already in the app are then listed in the right pane. Click on the add button and a file browser appears. Choose the file you want to transfer.

2. Via Dropbox. I assume that you have Dropbox. For an introduction, go to `dropbox.com`. In the app, you sign in to Dropbox. Now the app can see the contents of your Dropbox, or some part of it (at the Dropbox server) as long as you have an Internet connection.

3. With FileApp. See the discussion in Section G.1.2 (Figures G.2 and G.3).

G.3 Two LaTeX implementations for the iPad

We now discuss two LaTeX implementations.

G.3.1 Texpad

There are three ways Texpad can typeset.

A. On your iPad. The app places a LaTeX distribution on the iPad and you typeset with it. However, a complete LaTeX distribution is about 4 GB! No app can be this big. So you only get a small LaTeX distribution.

B. On your computer via Dropbox. This is the most powerful option. You have all the packages and fonts on your computer available to you. An app (such as AutomaTeX by Jonathan Weisberg) monitors if there is any change in the LaTeX file in Dropbox. If there is, the file is retypeset and the pdf is made available to you via the Dropbox.

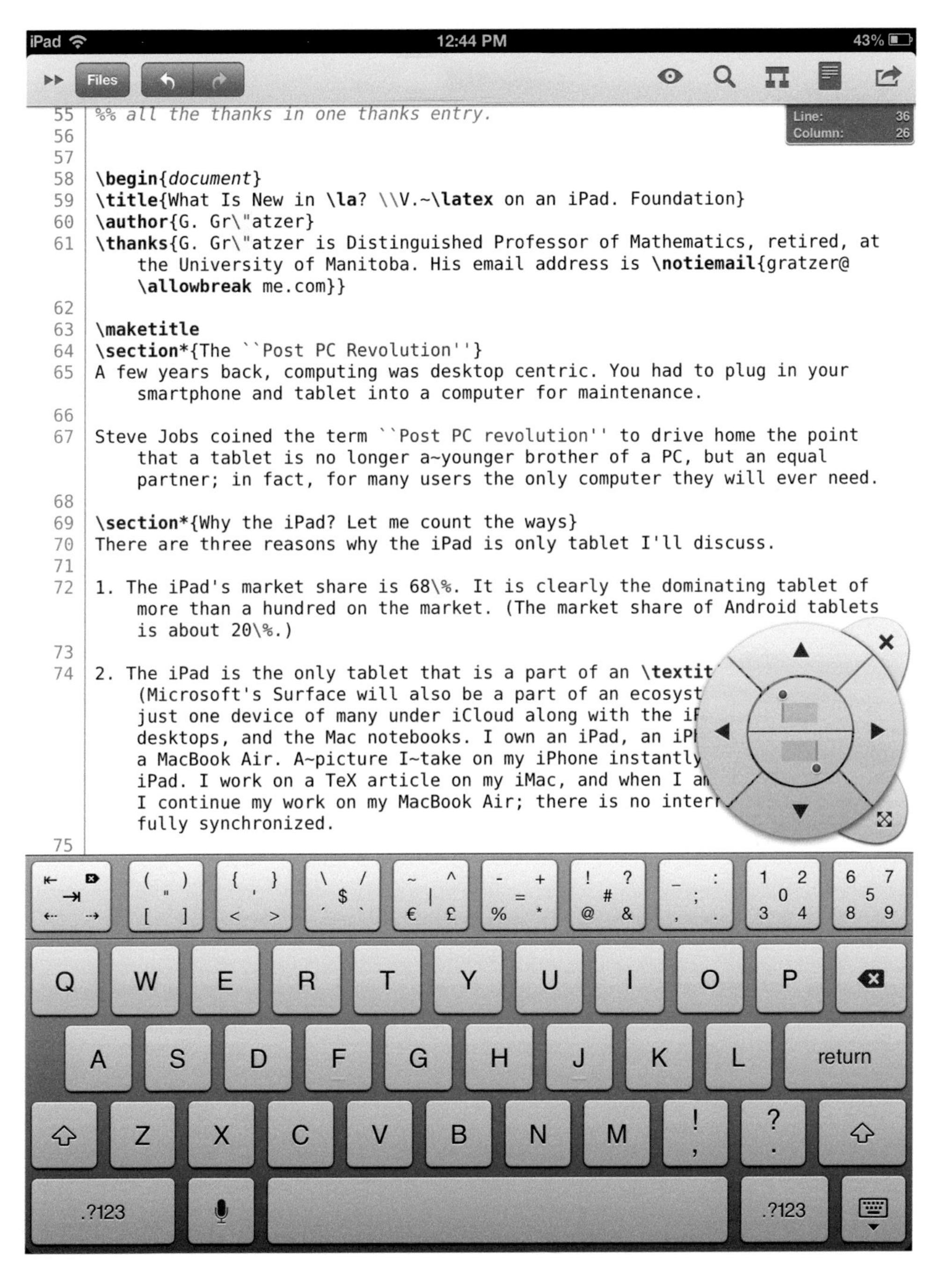

Figure G.4: Editing with Textastics

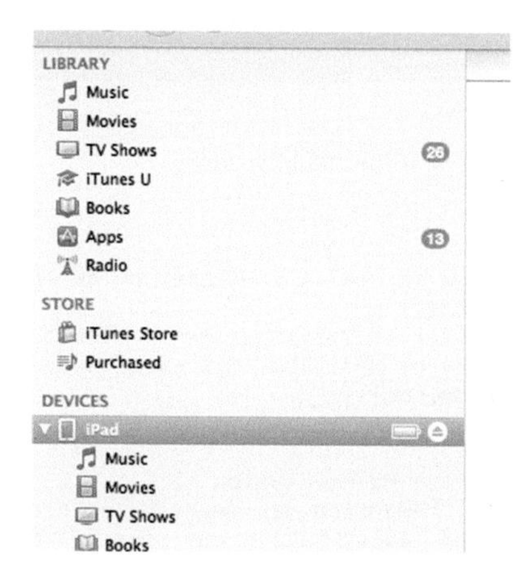

Figure G.5: Under Devices, we selected the iPad

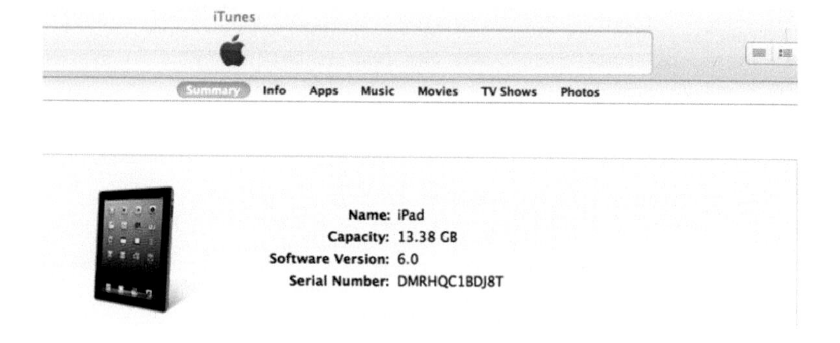

Figure G.6: Choose Apps

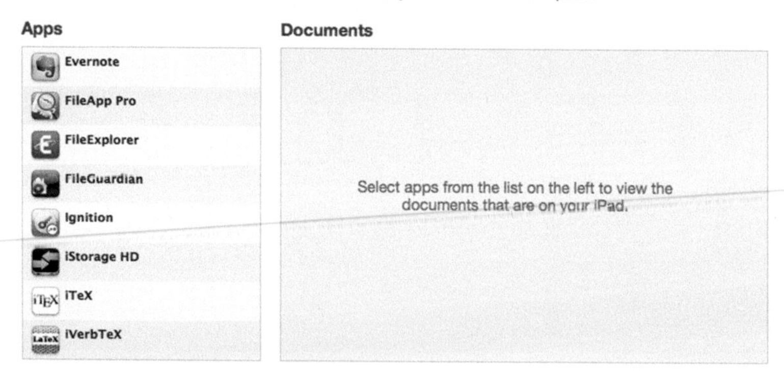

Figure G.7: Select app

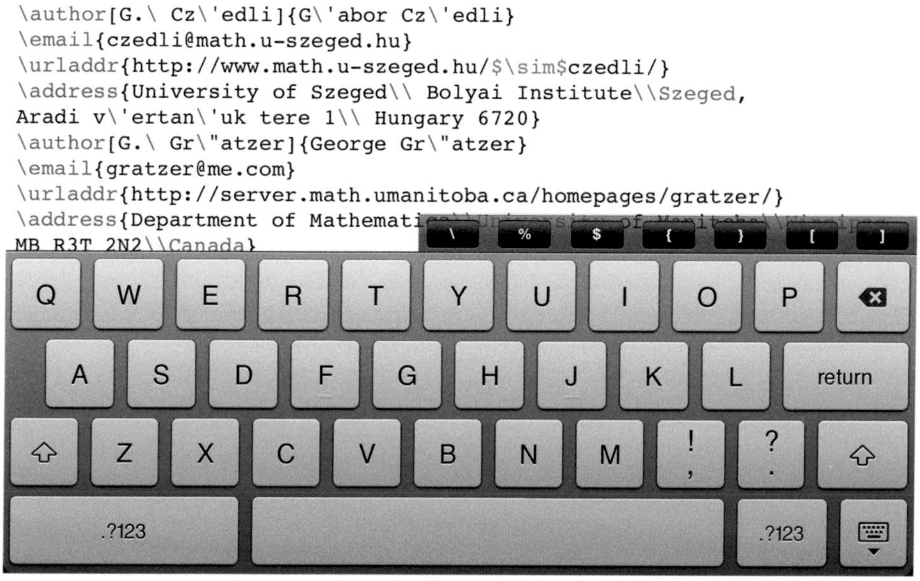

```
\author[G.\ Cz\'edli]{G\'abor Cz\'edli}
\email{czedli@math.u-szeged.hu}
\urladdr{http://www.math.u-szeged.hu/$\sim$czedli/}
\address{University of Szeged\\ Bolyai Institute\\Szeged,
Aradi v\'ertan\'uk tere 1\\ Hungary 6720}
\author[G.\ Gr\"atzer]{George Gr\"atzer}
\email{gratzer@me.com}
\urladdr{http://server.math.umanitoba.ca/homepages/gratzer/}
\address{Department of Mathemati
MB R3T 2N2\\Canada}
```

\ % $ { } [ ]

Q W E R T Y U I O P
A S D F G H J K L return
Z X C V B N M ! , ? .
.?123 .?123

Figure G.8: Editing with soft keyboard

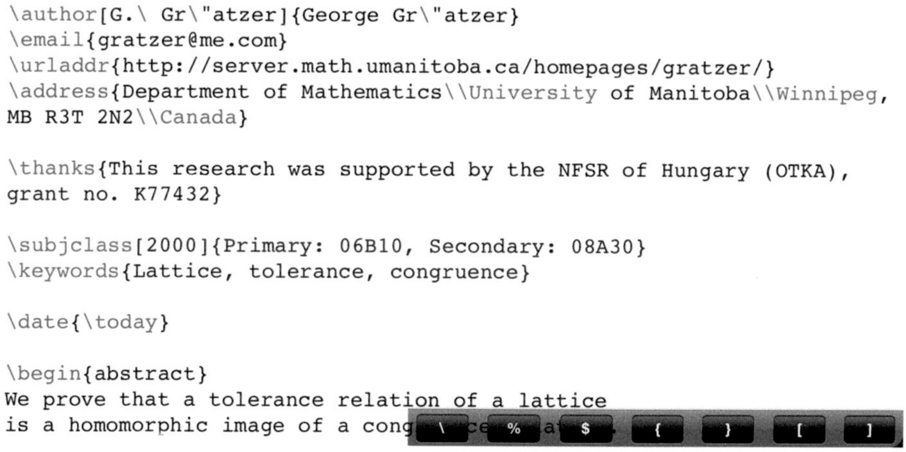

```
\author[G.\ Gr\"atzer]{George Gr\"atzer}
\email{gratzer@me.com}
\urladdr{http://server.math.umanitoba.ca/homepages/gratzer/}
\address{Department of Mathematics\\University of Manitoba\\Winnipeg,
MB R3T 2N2\\Canada}

\thanks{This research was supported by the NFSR of Hungary (OTKA),
grant no. K77432}

\subjclass[2000]{Primary: 06B10, Secondary: 08A30}
\keywords{Lattice, tolerance, congruence}

\date{\today}

\begin{abstract}
We prove that a tolerance relation of a lattice
is a homomorphic image of a cong
```

\ % $ { } [ ]

Figure G.9: Editing with Bluetooth keyboard

C. In the Cloud. This option provides you with a remote server, the Cloud; you connect to it with Wi-Fi. The server has a full LaTeX implementation, so you miss only the special fonts. And, of course, you must have Wi-Fi to use it. So you can polish up your lecture on the airplane on the way to a meeting.

Texpad has some interesting features, including:

- Autocompletion of all common commands and autofilling `\cite`-s and `\ref`-s.
- Replacement of the LaTeX console with a list of errors and warnings linked to the source.
- Global search, outline view, and syntax highlight.

Step 1. To get started with Texpad, go to the iPad App Store and install Texpad. Sign up for Dropbox with the same e-mail address and password as for your computer's Dropbox.

Step 2. Now open Texpad. Figure G.10 shows Texpad at the first startup. The Help button gets the help file.

Step 3. Touch Off to turn Dropbox On. (If you have Dropbox installed and connected, it's even simpler, you just have to Allow the connection.) Your File Storage now gives two options: iPad and Dropbox (see Figure G.11). It is important to understand that your LaTeX files will live in the Dropbox (in the Cloud, at the Dropbox server) or locally on your iPad.

Step 4. The Dropbox files are now available to you by touching Dropbox under File Storage, see Figure G.11.

- First, create a folder for the LaTeX files to be transferred. Navigate to iPad file storage. Touch the + in the bottom right, and choose Folder. Name the folder.
- Second, navigate to the Dropbox file system view and to the folder containing the file you want to copy. Touch Edit. Select the file to transfer. At the bottom center, touch Copy. Navigate to the folder into which you want to copy the file and touch Copy.

Step 5. Typesetting will take place either on the iPad or in the Cloud. Go to the folder of a LaTeX file, touch the file (on the iPad or in the Dropbox), and typeset it on the iPad (touch Local Typeset) or in the Cloud, that is, at Valletta's server (touch Cloud Typeset).

Step 6. Try to visualize what is happening.

- If you typeset on the iPad and the file is on the iPad, it just typesets locally; that is it.
- If you typeset on the iPad and the file is in Dropbox, the file is transferred to the iPad, typeset, and the resulting pdf is sent back to the Dropbox; nothing is kept at the iPad.
- If you typeset in the Cloud and the file is in Dropbox, the file is transferred to the Cloud, typeset, and the resulting pdf is sent back to the Dropbox; nothing is kept in the Cloud.

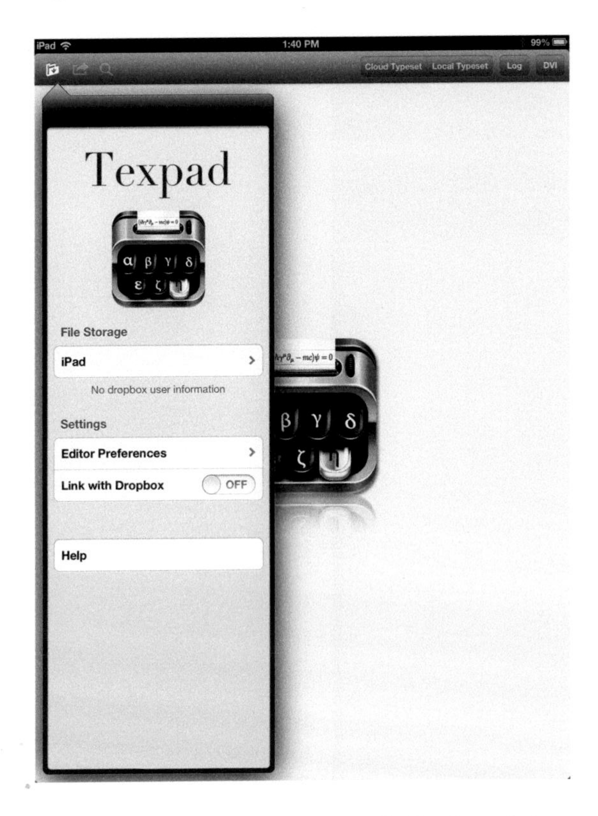

Figure G.10: Texpad first start up

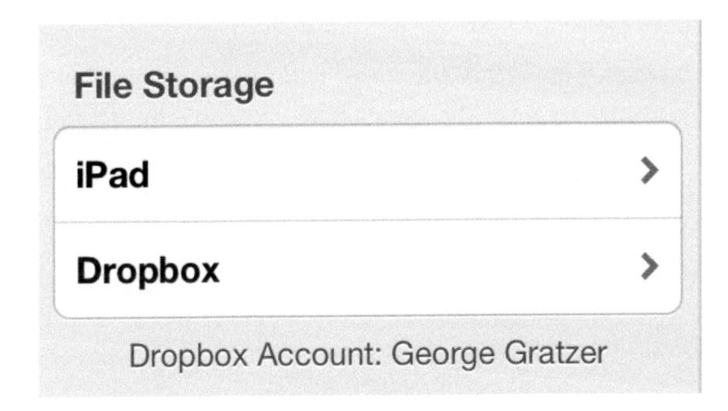

Figure G.11: Expanded File Storage

- If you typeset in the Cloud and the file is on the iPad, the file is transferred to the Cloud, typeset, and the resulting pdf is sent back to the iPad.

Step 7. Once you touch a LaTeX file, you are ready to edit it. Cursor control is very important. You do it with a two finger swipe. Of course, this is not so important if you use a Bluetooth keyboard; it has cursor keys.

Figure G.12: Organizer window

Step 8. You edited and typeset your LaTeX file. You want to get to another file. Touch the organize button (the folder icon on the upper left). You get the Organizer window (see Figure G.12). Touch the button in the upper left of the window, you get back to Dropbox, eventually, to the expanded File Storage of Figure 7.

These eight steps should be enough to get you started. Read the detailed Help file for some more information. It is available as a help file and also at

`https://www.texpadapp.com/support/ios`

G.3.2 *TeX Writer*

You get your files via Dropbox, typeset on your iPad. Documentation: `readme.pdf` is no quick start, but it is useful for understanding how TeX Writer works and how to customize it. TeX Writer was the first to typeset on the iPad. It could only typeset TeX files. Now it has LaTeX and the AMS packages on board.

Step 1. When you start up TeX Writer, first link to Dropbox. In TeX Writer, you get a display showing the source file `readme.tex`; see Figure G.13. Pressing the

More icon (right pointing arrow), you get more icons, to read the pdf version or Air Printing `readme.pdf`. On the left is the Organize icon; touching it, you get a file listing: `readme.tex` and `readme.pdf`. At the bottom is New File; touch it to compose one.

Step 2. So you are perplexed about what to do next, you ran out of icons. You have to know that TeX Writer accesses the Dropbox in a special way. When you connect to Dropbox from TeX Writer, it creates a new folder `App` in Dropbox. In the folder `App` it creates the subfolder `TeX Writer`. In this subfolder you find `readme.tex`. Anything you put in the `TeX Writer` subfolder is visible in the file listing window on the iPad; anything not in this subfolder is not visible to TeX Writer.

Step 3. TeX Writer gets your files from this subfolder in Dropbox. Place a folder in there with the files of your current project. These will be available to you on your iPad. Moreover, these files are fully synchronized, so the editing changes you make on your iPad show up in Dropbox.

Step 4. LaTeXing, you spend most of your time editing. TeX Writer's editor has some interesting features. Excellent cursor control. Touch `begin{}`, type in the name of the environment, and the environment is placed in your document. You also have undo, redo, search, and so on.

When typing, you retain the editing functions you get at the start, and in addition, you get an extra row of LaTeX specific keys. You do not get them with a Bluetooth keyboard; however, the keyboard can have many of these keys you need for typing LaTeX. Nice feature: the Log viewer links to error lines.

G.4 Conclusion

Jason Snell was interviewing Craig Federighi, Apple senior vice president of software engineering (and two more executives of Apple), for MacWorld. Snell writes:

"When I walked into Apple's offices for my conversation with the three executives, they noticed that I had brought a phone, a tablet, and a laptop, and had ultimately selected my MacBook Air as my tool of choice for the interview.

'You had a bunch of tools,' Federighi said, pointing at my bag. 'And you pulled out the one that felt right for the job that you were doing. It wasn't because it had more computing power… You pulled it out because it was the most natural device to accomplish a task.' "

I'm not suggesting that you write all your document on an iPad. I do suggest, however, that you can LaTeX with ease, say on a trip, correcting a document or adding a slide to your presentation. Use your iPad to LaTeX when appropriate.

LaTeXing on an iPad requires some compromises, for instance, you cannot use nonstandard fonts. Nevertheless, when not at your desk, the iPad will be nearly as functional as your MacBook Air, and it is so much easier to carry around…

And the best is yet to come: the larger iPad will make working on the iPad easier.

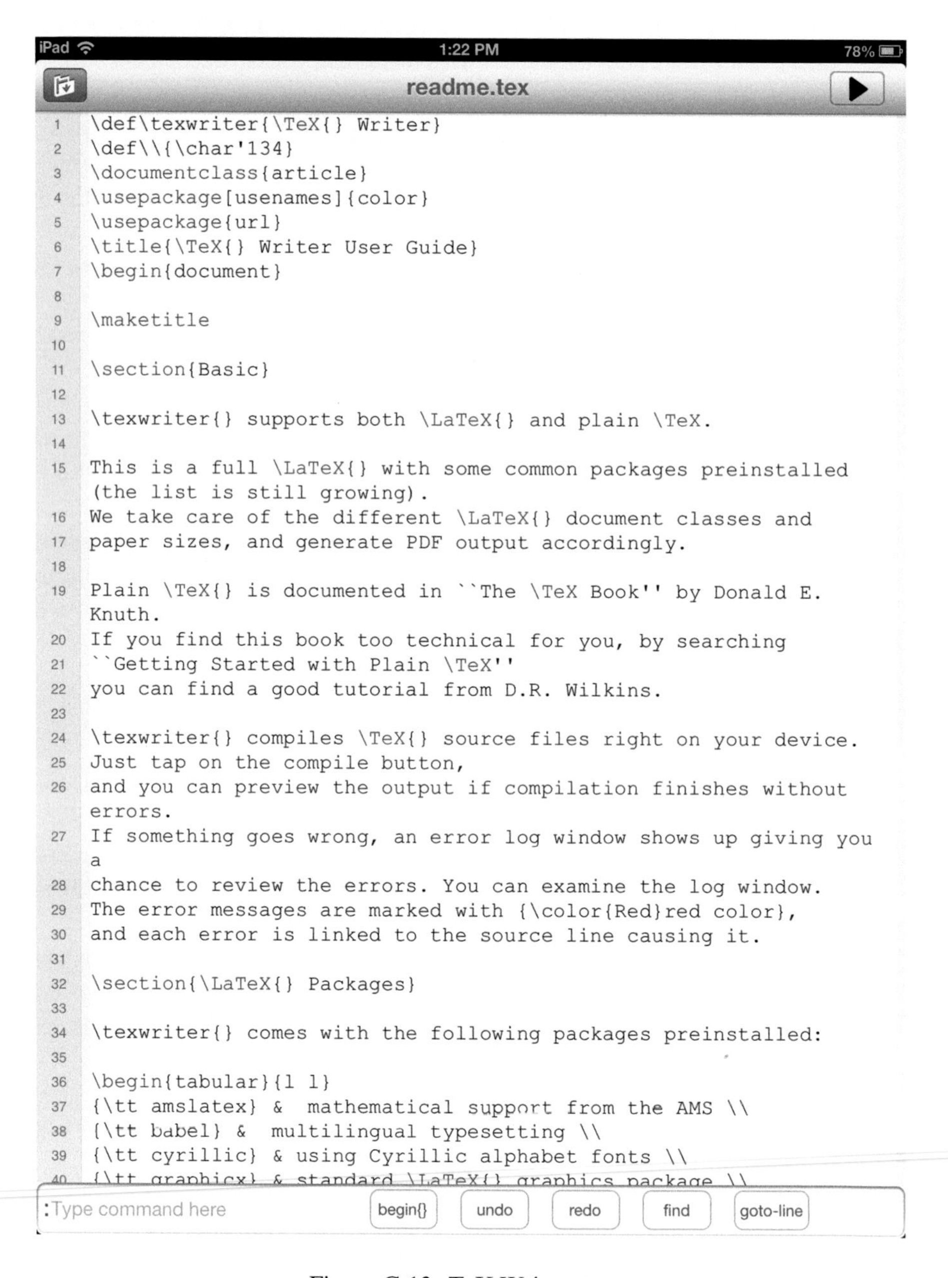

Figure G.13: TeX Writer startup

APPENDIX

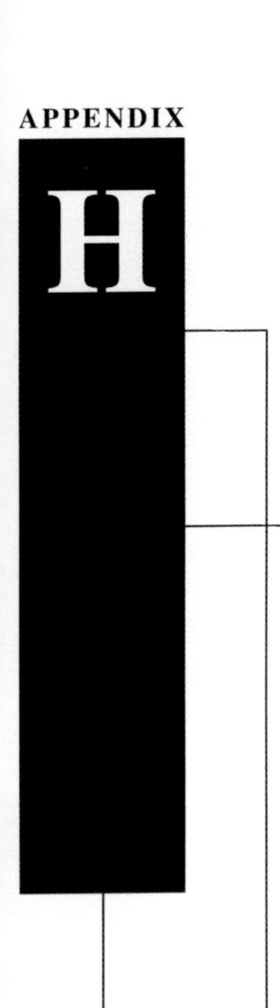

Final thoughts

In this final appendix, I will outline some of the material I did not discuss and suggest some additional reading to learn more about LaTeX, typesetting, and writing. We conclude by looking at some projects that may come to fruition soon.

H.1 What was left out?

The mission statement in the introduction stated that my goal for this book was to provide you with a good foundation in LaTeX including the AMS packages, and that we would not cover programming or visual design. As a result, I have omitted a great deal of material.

H.1.1 LaTeX *omissions*

LaTeX has some additional features that I have not discussed in this book:

1. The `picture` environment allows you to draw simple pictures with lines and circles.

2. The `array`, `tabular`, and `tabbing` environments have a number of additional features.

G. Grätzer, *More Math Into* LaTeX, DOI 10.1007/978-3-319-23796-1

3. LaTeX makes the style parameters of a document and of most LaTeX constructs available to the user for modification. Very few of these parameters have been mentioned in this book.

4. Low-level NFSS commands provide finer control over fonts.

The following are some pointers to additional information on these topics:

1. Drawing with the `picture` environment has the advantage of portability. This environment is described in Leslie Lamport's *LaTeX: A Document Preparation System,* 2nd edition [53]. A very advanced internal drawing system is TikZ by Till Tantau, see

   `http://sourceforge.net/projects/pgf/`

   However, I believe that the best approach is to use a drawing application that can save your illustrations in EPS or PDF format so that you can include them in your document using the `graphicx` package (see Section 8.4.3).

2. The `tabbing`, `tabular`, and `array` environments—and their extensions—are described in detail in Leslie Lamport's *LaTeX: A Document Preparation System,* 2nd edition [53] and Chapter 5 of *The LaTeX Companion,* 2nd edition [56].

3. The style parameters for LaTeX are set by the document class. When a publisher changes the document class loaded by your document, the style parameters are set to its specifications. If you explicitly change style parameters in your document, a publisher will have trouble getting your source file to conform to their publishing style. If you must change any basic style parameters, be sure to explain what you did with comments.

4. There are two types of commands defined in the NFSS, high-level and low-level commands. The latter are, by and large, meant for style designers and package writers. Nevertheless, anyone who wants to use fonts other than Computer Modern (the default) would do well to read Chapter 7 of *The LaTeX Companion,* 2nd edition [56].

   Low-level NFSS commands are briefly mentioned in Section 3.6.9 and are used in Appendix E.

H.1.2 TeX *omissions*

Almost all discussions of Plain TeX were omitted from this book. TeX is a powerful programming language, allowing you to design any page layout or formula. Remember, however, that to change any design feature, you should be knowledgeable not only about TeX, but also about document design. Also keep in mind that making such changes may make it difficult or impossible for a publisher to make your document conform to its own specifications.

H.2 Further reading

Much documentation is included with the LaTeX and the AMS distributions and many third-party packages are also well documented. You will also find a great deal of documentation on CTAN.

As you have no doubt noticed, there are many references to *The LaTeX Companion,* 2nd edition [56] in this book. While it is not a beginner's book, it is indispensable for advanced LaTeX users with special needs. It is also the best overview of more than a hundred important packages. For package writers and students of NFSS, it is *the* basic textbook. For graphics work, read *The LaTeX Graphics Companion* [17], and on Web publishing *The LaTeX Web Companion* [18].

Learning TeX is a bit more complicated than learning LaTeX. You may want to start with Wynter Snow's *TeX for the Beginner* [67]. It introduces many of TeX's basic concepts in a very relaxed style with many examples. The notes on LaTeX make the book especially useful, and the author gives many examples of writing macros. The use of TeX as a programming language is not discussed.

Raymond Seroul and Silvio Levy's *A Beginner's Book of* TeX [66] is another good introduction. This book also includes a chapter on TeX programming. Donald E. Knuth's *The TeXbook* [48] provides a nice introduction to TeX.

Paul W. Abrahams, Karl Berry, and Kathryn A. Hargreaves' TeX *for the Impatient* [1] explains many TeX commands, grouped by topic. This book has a very useful, nonsequential approach. Finally, Victor Eijkhout's TeX *by Topic: A TeXnician's Reference* [14] is an excellent reference book on TeX, mainly for experts. For many tutorial examples, see the articles and columns in *TUGboat* (see Section D.2).

For advice to authors of mathematical articles and books, see *Mathematics into Type* [68] by Ellen Swanson (updated by Arlene Ann O'Sean and Antoinette Tingley Schleyer). You may find it interesting to see how many of the rules in Swanson's book have been incorporated into LaTeX. The definitive book on style (in North America) is *The Chicago Manual of Style,* 16th edition [11]. Two other views on copy editing are presented in Judith Butcher's *Copy Editing: The Cambridge Handbook* [9] and *Hart's Rules for Compositors and Readers at the University Press, Oxford* by Horace Hart [45], updated in R. M. Ritter's *New Hart's Rules: The Handbook of Style for Writers and Editors* [64]. The special problems of writing about math and computer science are admirably dissected in Lyn Dupré's *BUGS in Writing: A Guide to Debugging Your Prose,* 2nd edition [13].

Most people who write math have little or no background in typography, the art of printing with type. But when you become a typesetter, it can be useful to learn a little bit about typography. I would highly recommend Robert Bringhurst's *The Elements of Typographic Style* [8]. See also Ruari McLean's *The Thames and Hudson Manual of Typography* [54] and Alison Black's *Typefaces for Desktop Publishing: A User Guide* [6].

Harley Hahn's *A Student's Guide to Unix* [44] provides an excellent introduction to UNIX.

Bibliography

[1] Paul W. Abrahams, Karl Berry, and Kathryn A. Hargreaves, TEX *for the Impatient.* Addison-Wesley, Reading, MA, 1990.

[2] Adobe Systems, *PDF Reference, Version 1.7.* 1st edition. Adobe Press, 2009.

[3] American Mathematical Society, *AMSFonts, Version* 2.2 *User's Guide.* Providence, RI, 1997.

[4] ———, *User's Guide for the* `amsmath` *Package* (*version* 2.0). Providence, RI, 1999. (Revised 2002.)

[5] ———, *Using the* `amsthm` *package* (*version* 2.20). Providence, RI, 2004.

[6] Alison Black, *Typefaces for Desktop Publishing: A User Guide.* Architecture Design and Technology Press, London, 1990.

[7] Johannes Braams, *Babel, a multilingual package for use with* LATEX*'s standard document classes.* 2005, on CTAN.

[8] Robert Bringhurst, *The Elements of Typographic Style.* Hartley & Marks Publishers, 2004.

[9] Judith Butcher, Caroline Drake, Maureen Leach, *Butcher's Copy-editing: The Cambridge Handbook for Editors, Copy-editors and Proofreaders.* 4th edition. Cambridge University Press, London, 2006.

[10] Pehong Chen and Michael A. Harrison, *Index preparation and processing.* Software Practice and Experience **19** (9) (1988), 897–915.

[11] *The Chicago Manual of Style.* 16th edition. University of Chicago Press, Chicago, 2010.

[12] Michael Doob, TEX *Starting from* 1. Springer-Verlag, New York, 1993.

[13] Lyn Dupré, *BUGS in Writing. A Guide to Debugging Your Prose.* 2nd edition. Addison-Wesley Professional, Reading, MA, 1998.

G. Grätzer, *More Math Into* LATEX, DOI 10.1007/978-3-319-23796-1

[14] Victor Eijkhout, TEX *by Topic: A TEXnician's Reference.* Addison-Wesley, Reading, MA, 1991. Free download at `http://www.eijkhout.net/tbt/`

[15] Michel Goossens, Frank Mittelbach, and Alexander Samarin, *The* LATEX *Companion.* Addison-Wesley, Reading, MA, 1994. Second edition 2004.

[16] Enrico Gregorio, *Horrors in* LATEX*: How to misuse* LATEX *and make a copy editor unhappy, TUGboat* **26** (2005), 273–279.

[17] Michel Goossens, Sebastian Rahtz, and Frank Mittelbach, *The* LATEX *Graphics Companion.* Addison-Wesley, Reading, MA, 1997.

[18] Michel Goossens and Sebastian Rahtz (with Eitan Gurari, Ross Moore, and Robert Sutor), *The* LATEX *Web Companion: Integrating* TEX*, HTML and XML.* Addison-Wesley, Reading, MA, 1999.

[19] George Grätzer, *Math into* TEX*: A Simple Introduction to* AMS-LATEX . Birkhäuser Boston, 1993.

[20] ———, AMS-LATEX . Notices Amer. Math. Soc. **40** (1993), 148–150.

[21] George Grätzer, *Advances in* TEX *implementations. I. PostScript fonts.* Notices Amer. Math. Soc. **40** (1993), 834–838.

[22] ———, *Advances in* TEX *implementations. II. Integrated environments.* Notices Amer. Math. Soc. **41** (1994), 106–111.

[23] ———, *Advances in* TEX *implementations. III. A new version of* LATEX*, finally.* Notices Amer. Math. Soc. **41** (1994), 611–615.

[24] ———, *Advances in* TEX*. IV. Header and footer control in* LATEX. Notices Amer. Math. Soc. **41** (1994), 772–777.

[25] ———, *Advances in* TEX*. V. Using text fonts in the new standard* LATEX. Notices Amer. Math. Soc. **41** (1994), 927–929.

[26] ———, *Advances in* TEX*. VI. Using math fonts in the new standard* LATEX. Notices Amer. Math. Soc. **41** (1994), 1164–1165.

[27] George Grätzer, *Math into* LATEX*: An Introduction to* LATEX *and* AMS-LATEX . Birkhäuser Boston, 1996. 2nd printing, 1998.

[28] George Grätzer, *General Lattice Theory.* 2nd edition. Birkhäuser Verlag, Basel, 1998. xix+663 pp.

[29] George Grätzer, *First Steps in* LATEX. Birkhäuser Boston, Springer-Verlag, New York, 1999.

[30] George Grätzer, Pervije Sagi v LaTeX'e. Mir Publisher, Moscow, 2000. (Russian)

[31] George Grätzer, *Math into* LATEX. 3rd edition. Birkhäuser Verlag, Boston, Springer-Verlag, New York, 2000. xl+584 pp. ISBN: 0-8176-4131-9; 3-7643-4131-9

[32] George Grätzer, *Turbulent transition, TUGboat* **21** (2001), 111–113.

[33] ———, *Publishing legacy document on the Web, TUGboat* **22** (2001), 74–77.

[34] G. Grätzer, *More* Math into LaTeX. 4th edition. Springer-Verlag, New York, 2007. xxxiv+619 pp.
ISBN-13: 978-0-387-32289-6, e-ISBN: 978-0-387-68852-7.
Kindle Edition 2007, ASIN: B001C3ABDA

[35] G. Grätzer, *A gentle learning curve for* LaTeX. PracTeX Journal Number 3, 2008.

[36] ———, *What Is New in LaTeX? I. Breaking Free.* Notices AMS **56** (2009), 52–54.

[37] ———, *What Is New in LaTeX? II. TeX implementations, Evolution or Revolution.* Notices AMS **56** (2009), 627–629.

[38] ———, *What Is New in LaTeX? III. Formatting references.* Notices AMS **56** (2009), 954–956.

[39] ———, *What Is New in LaTeX? IV. WYSIWYG* LaTeX. Notices AMS **58** (2011), 828–830.

[40] ———, *What Is New in LaTeX? V.* LaTeX *on an iPad. Foundation.* Notices AMS **60** (2013), 332–334.

[41] ———, *What Is New in LaTeX? VI.* LaTeX *on an iPad. Empire.* Notices AMS **60** (2013), 434–439.

[42] George Grätzer, *Practical* LaTeX. Springer-Verlag, New York, 2014.
ISBN: 978-3-319-06424-6

[43] George Grätzer, *What Is New in LaTeX? VII. The STIX math symbols.* Notices AMS **62** (2015).

[44] Harley Hahn, *Harley Hahn's Student's Guide to Unix.* 2nd edition. McGraw-Hill, New York, 1993.

[45] Horace Hart, *Hart's Rules For Compositors and Readers at the University Press, Oxford.* Oxford University Press, Oxford, 1991.

[46] Uwe Kern, *Extending* LaTeX*'s color facilities: the* `xcolor` *package.* December 21, 2005.
`http:\\www.ukern.de\tex\xcolor.html`

[47] Donald E. Knuth, *The Art of Computer Programming.* Volumes 1–3. Addison-Wesley, Reading, MA, 1968–1998.

[48] Donald E. Knuth, *The* TeX*book.* Computers and Typesetting. Vol. A. Addison-Wesley, Reading, MA, 1984, 1990.

[49] Donald E. Knuth, TeX*: The Program.* Computers and Typesetting. Vol. B. Addison-Wesley, Reading, MA, 1986.

[50] Donald E. Knuth, *The METAFONTbook.* Computers and Typesetting. Vol. C. Addison-Wesley, Reading, MA, 1986.

[51] Donald E. Knuth, *METAFONT: The Program.* Computers and Typesetting. Vol. D. Addison-Wesley, Reading, MA, 1986.

[52] Donald E. Knuth, *Computer Modern Typefaces.* Computers and Typesetting. Vol. E. Addison-Wesley, Reading, MA, 1987.

[53] Leslie Lamport, LaTeX*: A Document Preparation System.* 2nd edition. Addison-Wesley, Reading, MA, 1994.

[54] Ruari McLean, *The Thames and Hudson Manual of Typography.* Thames and Hudson, London, 1980.

[55] Frank Mittelbach, *An extension of the* LaTeX *theorem environment. TUGboat* **10** (1989), 416–426.

[56] Frank Mittelbach and Michel Goosens (with Johannes Braams, David Carlisle, and Chris Rowley), *The* LaTeX *Companion.* 2nd edition. Addison-Wesley, Reading, MA, 2004.

[57] Frank Mittelbach and Chris Rowley, LaTeX *2.09* → LaTeX *3. TUGboat* **13** (1) (1992), 96–101.

[58] ———, LaTeX—*A new version of* LaTeX. TeX and TUG NEWS **2** (4) (1993), 10–11.

[59] ———, *The* LaTeX *3 project.* Euromath Bulletin **1** (1994), 117–125.

[60] ———, LaTeX *3 in '93.* TeX and TUG NEWS **3** (1) (1994), 7–11.

[61] Frank Mittelbach and Rainer Schöpf, *The new font family selection—user interface to standard* LaTeX. *TUGboat* **11** (1990), 297–305.

[62] Oren Patashnik, *BIBTeXing.* Document in the BibTeX distribution.

[63] ———, *BIBTeX 1.0. TUGboat* **15** (1994), 269–273.

[64] R. M. Ritter, *New Hart's Rules: The Handbook of Style for Writers and Editors, Oxford.* Oxford University Press, Oxford, 2005.

[65] Rainer Schöpf, *A new implementation of the* LaTeX `verbatim` *and* `verbatim*` *environments. TUGboat* **11** (1990), 284–296.

[66] Raymond Seroul and Silvio Levy, *A Beginner's Book of* TeX. Springer-Verlag New York, 1995.

[67] Wynter Snow, TeX *for the Beginner.* Addison-Wesley, Reading, MA, 1992.

[68] Ellen Swanson, *Mathematics into Type.* Updated edition. Updated by Arlene Ann O'Sean and Antoinette Tingley Schleyer. American Mathematical Society, Providence, RI, 1999.

[69] Till Tantau, *User's Guide to the Beamer Class.* 2005.
`http://latex-beamer.sourceforge.net`

[70] Gérard Tisseau and Jacques Duma, *TikZ pur l'impatient*, 2015.
`http://math.et.info.free.fr/TikZ/index.html`

Index

Italic numbers indicate figures or tables, **bold** numbers indicate definitions.

G. Grätzer, *More Math Into LATEX*, DOI 10.1007/978-3-319-23796-1

Printed in the United States
By Bookmasters